Information and Communications Technologies in Agriculture

Information and Communications Technologies in Agriculture

Editor: Michelle Jacobs

STATES
ACADEMIC PRESS
www.statesacademicpress.com

States Academic Press,
109 South 5th Street,
Brooklyn, NY 11249, USA

Visit us on the World Wide Web at:
www.statesacademicpress.com

ISBN: 978-1-63989-287-7 (Hardback)

Cataloging-in-Publication Data

Information and communications technologies in agriculture / edited by Michelle Jacobs.
p. cm.
Includes bibliographical references and index.
ISBN 978-1-63989-287-7
1. Agricultural information networks. 2. Communication in agriculture--Technological innovations.
3. Agriculture--Data processing. 4. Agriculture--Computer simulation.
5. Agricultural innovations. I. Jacobs, Michelle.
S494.5.I47 I54 2022
630.7--dc23

Table of Contents

Preface

The use of information and communication processes for rural and agriculture development is the primary objective of information and communications technologies in agriculture. It is also referred to as e-agriculture. ICT in agriculture aims to improve livelihood by enhancing agricultural productivity and income. It includes designing, development, application and evaluation of innovative ways that are used in the rural domain while focusing primarily on agriculture. Standards, methodologies, norms, policy support and development of institutional and individual capacities are the primary components of this field. ICT in agriculture uses various tools such as networks, devices, mobiles, applications and services ranging from new and innovative technologies to pre-existing instruments like televisions, satellites, fixed telephones and radios. The book strives to provide a fair idea about this discipline and helps develop a better understanding of the latest advances within this field. The various studies that are constantly contributing towards advancing technologies and evolution of information and communications technology in agriculture are examined in detail. Those with an interest in this field would find this book very helpful.

This book is a result of research of several months to collate the most relevant data in the field.

When I was approached with the idea of this book and the proposal to edit it, I was overwhelmed. It gave me an opportunity to reach out to all those who share a common interest with me in this field. I had 3 main parameters for editing this text:

1. Accuracy – The data and information provided in this book should be up-to-date and valuable to the readers.

2. Structure – The data must be presented in a structured format for easy understanding and better grasping of the readers.

3. Universal Approach – This book not only targets students but also experts and innovators in the field, thus my aim was to present topics which are of use to all.

Thus, it took me a couple of months to finish the editing of this book.

I would like to make a special mention of my publisher who considered me worthy of this opportunity and also supported me throughout the editing process. I would also like to thank the editing team at the back-end who extended their help whenever required.

Editor

1

Assessment of Use of Selected Information Communication Technologies (ICTs) for Extension Service Delivery: Implication for Agricultural Development in Nigeria

Adetumbi, Saheed.Ige, Olaniyi, Olumuyiwa., Akin and Adewale, Jacob., Gbemiga*

Abstract

The study was carried out to assess the implication of the use of selected Information and Communication Technologies (ICTs) for extension service delivery in Nigeria. The simple random technique was used to select fifty percent of the total extension agents which equivalent to forty-two respondents as sample size for the study. Frequency counts, percentages, mean and standard deviation were used as descriptive statistic. Also, Chi-Square test and Spearman's rho correlation were employed as inferential statistic to test for the hypotheses. Findings indicated that access to various ICTs tools especially Radio and Mobile phone and were found to be relevant to farming operations. The result of Chi square test revealed that significant relationship exist between sex, age, level of education, years of working experience of the extension agents and level of use of ICTs. Furthermore, the result of Spearman rho Correlation showed that there was no significant relationship between ICT training of extension workers and the level of use of ICTs. The study recommended among others that, there should be a periodic review of the use of current ICTs in extension service delivery to facilitate effectiveness in the use of ICTs for extension service and adequate funding of extension service should also be ensured to enhance the maintenance of ICTs made available to extension personnel.

Keywords:
Extension service, Information, , Technology, Nigeria

Department of Agricultural Extension and Rural Development, Ladoke Akintola University of Technology, P.M.B. 4000, Ogbomoso, Nigeria.
** Corresponding author's email: oaolaniyi@lautech.edu.ng*

INTRODUCTION

Most African countries depend on agriculture for their economic growth. International Food and Agriculture Programme (IFAP, 2004) asserted that about 80 percent of the population who live in rural areas of developing nations depend directly on agriculture for their livelihood. In Nigeria, the Federal Ministry of Agriculture and Rural Development (FMARD) (2002) stated that agriculture remains the largest contributor to the Nigerian economy, accounting for over 38 percent of non-oil foreign earnings and employing 58 percent of the active labour-force of the population. In the recent years, a number of Sub-Sahara African countries experienced slow agricultural development. The decline in the agricultural development could be attribute to a number of constraint which include: inappropriate agricultural policies, inadequate information provision, low adoption of agricultural technologies and institutional frameworks (Kiplang and Wallance, 2003).

In addressing the factors that limit agricultural production, Low External Input and Sustainable Agriculture (LEISA) now Farming Matters (2005) emphasized that the underlying factor is related to information and knowledge. Access to information is one of the most valuable resources in agricultural development. Recently, the demand for agricultural information is of great importance than ever before, thereby making information a prominent factor in agricultural production.

In view of this, this makes the extension service to be relevant most especially in making information available to the clientele. Communication is the essence of agricultural extension which seeks to provide knowledge and information for rural dwellers in order to modify their behaviour in a way to provide sustainable benefits to them and the society at large (Alex, *et al.,* 2004). Knowledge and communication are central to this process, indeed without effective communication there can be no meaningful development. The extension service system play an important role in enabling rural communities to identify and articulates their needs and to access relevant information for the creation of knowledge.

Communication of information and knowledge is an important aspect of extension service and extension service providers must be able to access a continuous flow of new information and innovation if they are to be of continuous benefit to their clients. The Subject Matter Specialists (SMS) traditionally links this system being a specialist within the extension service who maintains contact with recent innovation and latest technologies and in turn feed these to village extension personnel (Alex *et al.*, 2004). This linkage and information flow between different actors in agricultural production can be enhanced through the use of Information Communication Technologies (ICTs). Agricultural extension is evidently an area of agricultural production where dissemination of information through the use of Information Communication Technologies (ICTs) cannot be overruled.

The term Information and Communication Technologies (ICTs) has been described as including a wide range of media used to denote the use of computer and communication systems within computers. Communication Technology Association (CTA) (2002), defined ICTs as the combination of hardware, software and all other means of production that enable the exchange, processing and management of information and knowledge. ICTs thus connote technologies and methods involved in storing, managing creating, processing and disseminating information, example of this include, Computers, Books, E-mails, Radio, Telephones, Television, Pager, Personal Digital Assistants (PDAs) and the "web".

Agricultural Extension, depends largely on information exchange between and among farmers and a broad range of other actors, is an area in which ICT can have significant impact. Research Scientists can relate directly with the farmers through ICTs. Frontline extension workers, who are the direct link between farmers and other actors in the agricultural knowledge and information system, are well positioned to make use of ICT to access expert knowledge or other types of information that could be beneficial to the farmers (Salau and Saingbe, 2008). The potential applications of ICTs in agricultural extension include the following:

• Capacity to reach a large audience, e.g. the use of radio, TV and Internet

• Can be effectively used for training and demonstrations e.g T.V., Video, VCD, and CD-ROM.

• Can be used to make the extension systems and structures more efficient through better management of information and scarce resources e.g the use of Data bases for MIS and Networking softwares .

• For the search and packaging of information on demand and for exploring of alternative production options and technologies e.g the use of search engines, the web and data bases.

• ICT may be used for normal weather fo recasts and as a warning system for disease/pests outbreaks and other disasters before they occur and also for the provision of timely and sensitive market information e.g. with the use of Radio, TV, and SMS.

• ICTs are important for networking among and between the key stakeholders in the Research-Extension-Farmers-Inputs-Linkage System (REFILS) e.g. with the use of Telephone, Video, SMS, and;

• ICTs can also be effectively used for community mobilization, learning and action e.g Radio, TV, public address systems and the Web (Arokoyo, 2005).

In spite of the constraints to the use of ICT, ICTs have enhance efficient and effective dissemination and access to scientific, technical and marketing information which are previously difficult and often expensive to come by (Aina, 2002). ICTs offers tremendous capabilities to the agricultural sector in terms of vast information storage, fast, and in-expensive communication channel, links between different media, easy and enjoyable use at comparatively low and declining cost (Kiplang and Wallance, 2003). ICTs resource centres meant for extension service system could be owned by government or its agencies, Educational Institutions, Community organization and Non-government organizations. These resource centres may be located in Churches, Mosques, Community centres, Local Government Headquarters, Post Office, Farmers co-operative houses and Radio and/or Television Stations. In Oyo State for example, ICT resource centres are established especially in Oyo State Agricultural Development Programme (OYSADEP) resource centre located at Saki, International Institute of Tropical Agriculture (IITA) controlled resource centre in Ago-are, Atisbo Local Government Area. However, lack of adequate and timely information is one of the factors that could be responsible for low agricultural production in the rural areas of developing nations (Arokoyo, 2004), this could be sequel to the fact that many extension service providers are handicapped by administrative bottlenecks, declining public funds, farmers' conservativeness, lack of accessibility and network connectivity, high rate of illiteracy of the farmers, outrageous extension agents/ farmers ratio, coupled with extension agents' disposition to change which could affect their use of ICTs.

Objectives of the Study

The main purpose of this study was to determine the usefulness of selected Information and Communication Technologies (ICTs) for extension service delivery in Oyo State. The specific objectives are to

i. identify the socio economic characteristics of the respondents

ii. identify the various ICTs available for extension service delivery in the study area

iii. ascertain the frequency of use of available ICTs facilities

iv. examine the usefulness of ICTs for extension service delivery

v. ascertain the constraints faced by the extension workers in the use of ICTs.

Hypotheses of the study

Based on the objectives of the study, the following hypotheses are stated in null form:

1- There is no significant relationship between selected socio economic characteristics of the respondents and level of use ICTs

2- There is no significant relationship between level of ICTs training of the extension agents and level of use of ICTs.

MATERIALS AND METHODS

Site of study

This study was carried out in Oyo State, Nigeria. The State with the appellation 'The Pace Setter State' was created on February 1976 consequent upon the break-down of the old Western Nigeria. Geographically, Oyo State is located in the South West region of Nigeria, latitude 8 degree North and longitude 4 degree East bisects the State into four equal parts. The State covers a total of 27,249 square kilometer landmass and it is bounded partly by Ogun State in the south, Kwara State in the North, to the West. It is bounded partly by Ogun State and partly by the Republic of Benin while in the East it is bounded by Osun State. The topography of the study area is one of gentle lowland in the south, rising to a plateau 40 metres and above in the north. The vegetation pattern is that of rain forest in the south and Guinea Savannah in to the north, the climate is equatorial, notably with wet and dry season with relatively high humidity, the dry season starts in September while the wet season starts in April and terminate in September / October, average daily temperature ranges between 29°C and 35°C almost throughout the year. Consequent upon the prevailing climate and soil characteristics the following crops are cultivated; maize, yam, cassava, cowpea, sorghum, vegetable and tree crops like cocoa, oil palm, kola nut, walnut and citrus. Other agricultural activities in the state include fishery and livestock production. Animals are commonly reared on free range and these animals include poultry, goat, sheep, cattle and pigs amongst others.

Sampling procedure and sample size

The population for the study comprised of extension service providers in Oyo state. As at the time of this study, there were eighty-two agricultural extension personnel in the study area and fifty percent (50%) of these extension workers were selected from each of the agricultural zones in the study area, thereby giving a total of forty-two (n=42) extension agents (Table 1).

Table 1: Sampling of extension agent in Oyo State for the study

Agricultural Zones	Total Number of Extension Agents	Number of Extension agents sampled (50%)
Oyo	21	11
Saki	22	11
Ogbomoso	18	9
Ibadan/Ibarapa	21	11
Total	82	42

Source: Field Survey (2009)

Method of data collection

Primary data for this study were collected using structured questionnaire. The questionnaire contained both open and close ended questions.

Measurement of variables

There were two types of variables measured in this study, they were independent and the dependent variables. The independent variables include the socioeconomic characteristics of respondents such as age, sex, marital status, religion, academic qualification, ICT training while the dependent variable was the level of use of ICTs by the respondents. This was measured on a 4 point rating scale of Never=0, rarely=1, Sometimes=2 and always= 3. The respondents were categorized into three namely: Low, average and High Users based on mean and standard deviation of utilization scores. The dispersion statistics of the index are mean (46.9), median (47.0) and mode (47.0) were found to be close hence a normal distribution is assumed. The standard deviation was found to be 1.8. Descriptive statistical tools such as frequency counts, percentage were used to describe the data. Chi-square and Spearman rho Correlation were used to test the hypothesis.

RESULTS AND DISCUSSION

Personal characteristics

The analysis of the data collected indicated that 82.9% of the extension agents were male while 17.1% was female. The mean age of the extension agents was 44 year while 15.2% of them were of Islamic faith, 48.8% were Christians, 92.7% were married and 7.3% were either single of divorced, the result also revealed that 92.7% have a BSc or its equivalent 7.3% have a master

Table 2: Distribution of extension agents according to personal characteristics

Variables	Frequency	Percentage	Statistic
Sex			
Male	34	82.9	
Female	7	17.1	
Age (years)			Mean = 44.0
34-39	8	19.5	SD = 4.41
40-45	14	34.1	
46-50	19	46.3	
Religion			
Christianity	20	48.8	
Islam	21	51.2	
Marital Status			
Single	2	4.9	
Married	38	92.7	
Divorced	1	2.4	
Educational background			
HND	7	17.1	
B.Sc	31	75.6	
M.sc	3	7.3	
Work Experience (years)			
1-10	10	24.4	
11-20	29	70.7	Mean=14.2
21 and above	2	4.9	
Level of training on ICTs			
Certificate	20	48.8	
Learning on the job	9	21.9	
Learning on my own	12	29.3	
Access to ICT tools			
Yes	42	100.0	
No	0	0.00	

Source: Field Survey 2009

degree and the mean of their year of working experience is 14.2 years. The result further revealed that 48.8% of the extension agents had a certificate in the use of ICTs while 21.9% of them are learning on the job and 29.3% of the respondents are learning on their own. All the respondents indicated that they all have access to the use of ICT tools

Distribution of extension agents according to the frequency of use of various ICTs

Table 3 shows the frequency of use of each of the identified ICTs tools in the discharge of their duties. This was achieved by assessing how frequently the itemized ICTs are employed in the discharge of extension service on a five scale of weekly, fortnightly, monthly, Quarterly and never. It was revealed that Radio (M= 4.0) was the most used ICT as it ranked first while television closely followed with a Mean score of 3.9,ranked second and mobile phone ranked third because virtually all the respondents could at least afford to procure and maintain a transistor radio in case there is no electricity or reception to access television signal, next is mobile phone (2.8) which is fact gaining popularity among the respondents in information dissemination sequel to its time

Table 3: Distribution of respondents by frequency of use of various ICTs tools for extension delivery.

ICTs	Mean of frequency of use	Rank
Handbill & Fliers	0.90	5th
Newspaper	1.85	4th
Internet	0.73	6th
Mobile phone	2.85	3rd
Radio	4.00	1st
Television	3.93	2nd
Cinema	0.22	7th
Multimedia Projector	0.22	7th
CD ROM Technology	0.00	9th

Source: Field Survey (2009)

saving and cost effectiveness, in the fourth position is newspapers (1.8) which could only be used because of their literacy level and next is handbills and fliers which are rarely in circulation. The use of the internet with mean score (0.73) ranked sixth and the use of this have been found to be constrained by such factors as, lack of infrastructure facilities, poor connectivity, low technical know- how etc. and sharing the seventh used ICTs are cinema and multimedia projector with mean score (0.2) respectively, which are rarely used for information dissemination in the study area.

However, all the respondents indicated that they never employed the use of CD-ROM technology neither for information sourcing or dissemination of extension information to their clientele.

Categorisation of respondents on level of use of ICT

Table 4 shows high percentage (80.5%) of the respondents were categorized as high and average ICT users respectively and others (19.5%) were low users of ICT tools for extension delivery in the study area.

Usefulness of ICTs to in dissemination of agricultural information by extension agents Relative to the satisfaction of farmers' information needs

In the context of the study, farmers information needs are categorized into Pre-planting operations, Planting operations and Livestock issues and the study revealed that mobile phone and radio ranked first as the most useful means of information dissemination to farmers on post planting operation (M= 2.73), livestock issues (M=2.40) and pre-planting operations (M=2.39) respectively. Radio and mobile phone ranked second most useful ICT for dissemination of information on post –planting operation (M=2.63), pre-planting operation (M=2.37) and livestock issues (M=2.39) respectively.

Television ranked third as the most useful ICT tool used by extension agents in dissemination of information to farmers on pre –planting operations (M=2.15), livestock issues (M=2.15) and post planting operation (M=1.93). The internet ranked fourth for all operations and multimedia projector was the least used ICT for disseminating information to farmers on livestock issues (M=0.49), post-planting operations

Table 4: Distribution of respondents by categorisation on level of use of ICTs.

Level of use of ICTs by Extension agents	Computation	Frequency	Percentage
High level of use (X +1SD)	48.7 to 51.0	9	21.9
Average (X -1SD to X +1SDqi)	45.2 to 48.69	24	58.6
Low level of use (X -1SD)	44.0 to 45.1	8	19.5

Table 5: Distribution of respondents according to usefulness of ICTs to in dissemination of agricultural information by extension agents relative to the satisfaction of farmers' information needs

ICT Tools	Pre-planting Operation		Post-Planting Operation		Livestock Issues	
	Mean score	Rank	Mean score	Rank	Mean score	Rank
Newspapers	0.73	5th	1.05	6th	0.85	5th
Handbill/ Fliers (Under ICT tools)	0.73	5th	0.66	7th	0.78	6th
Journals	0.73	5th	1.22	5th	0.78	6th
Mobile Phone	2.37	2nd	2.73	1st	2.40	1st
Radio	2.39	1st	2.63	2nd	2.39	2nd
Television	2.15	3rd	1.93	3rd	2.15	3rd
Cinema	0.66	6th	0.44	8th	0.66	7th
Multimedia Projector	0.18	7th	0.32	9th	0.49	8th
Internet	0.86	4th	1.80	4th	1.70	4th

Source: Field Survey, 2009.

(M=0.32) and pre-planting operations (M=0.18). The finding of this study shows that mobile phone, radio and television are the most useful ICTs used for information dissemination to farmers on pre-planting operations, post-planting operations and livestock issues. Although, mobile phone is the only modern ICT used while others (radio and television) are analogue ICTs that are still found very relevant in information dissemination, perhaps because of their ease of access and cost of procurement and maintenance (Table 5).

Constraints to the use of ICTs for extension service delivery

Constraints to the use of ICTs for extension service delivery were discussed under various headings viz: high illiteracy level among farmers, high cost of procurement of ICTs, cost of maintenance of ICTs, negligence on the part of the farmers and inadequate technological know-how. It was however discovered that the use of internet was most constrained by illiteracy (M = 2.88), followed closely by newspapers (M = 1.80), handbills and journals (M = 1.73), then mobile phone, cinema and television in that order with the use of radio least affected by illiteracy. Also, cost of procurement was a very serious constraint to the use of internet (M = 2.76), cinema (M = 2.00) and mobile phone (M = 1.93) while radio, journals and handbills were least affected by cost of procurement, perhaps for the fact that handbills are given free of charge. Also, internet, multimedia projector, and cinema were most constrained by cost of maintenance while newspapers, handbills and journals were least affected. Negligence of use of ICTs by the farmer was another constraint that most affect the use of internet, this could be furtherance to the fact that illiteracy and cost of procurement and maintenance have deprived the respondents the use of the internet, the use of journals, newspapers and handbills were also neglected by the farmers perhaps the print media are rarely and hardly accessed by them. The least neglected ICTs been radio, mobile-phone and television. Inadequate technological know-how is another constraint to the use of the internet, multimedia projector and cinema whereas this does not limit the use of radio, television and mobile phone (Table 6).

Ranking of severity of constraint against various ICTs by extension agents

The constraints to use of ICT tools in order of severity as experienced by extension agents were identified as reported in table 7. It was found that the internet was the most constrained ICT, followed by the use of CD-ROM Technology, and Multimedia Projector. Other ICT in order of descending of severity in terms of con-

Table 6: Distribution of respondents according to constraints to the use of ICTs for extension service delivery

Constraints	News-papers	Hand-bills	Jour-nals	Mobile phone	Radio	Television	Cinema	Multimedia Projector	internet
High Illiteracy Level	(2) 1.80	(3) 1.73	(3) 1.73	(4) 0.27	(7) 0.09	(5) 0.24	(6) 0.17	(6) 0.17	(1) 2.88
High cost of Procurement of ICTs	(6) 1.32	(8) 0.44	(7) 0.96	(3) 1.93	(9) 0.24	(5) 1.51	(2) 2.00	(4) 1.83	(1) 2.76
High cost of maintenance of ICTs	(6) 1.56	(7) 1.36	(6) 1.56	(4) 2.44	(5) 2.42	(5) 2.42	(3) 2.46	(2) 2.61	(1) 2.87
Negligence of use of ICTs on the part of farmers	(3) 1.27	(3) 1.27	(2) 1.37	(7) 0.12	(8) 0.09	(6) 0.15	(5) 0.19	(4) 0.44	(1) 1.76
Inadequate technological know-how	(4) 0.44	(5) 0.39	(5) 0.39	(7) 0.15	(7) 0.15	(6) 0.19	(3) 1.22	(2) 1.95	(1) 2.65

Source: Field survey, 2009 Figure in parenthesis is ranks

Table 7: Ranking of severity of constraint against various ICTs used by extension agents

ICTs	Internet	CD-ROM Tech	Multimedia Projector	Cinema	Journals	Mobile Phone	Newspaper	Handbill and fliers	TV	Radio
Level of severity of constraint	6.844	4.515	3.498	2.124	1.814	1.749	1.560	1.496	1.497	1.389
Rank Order	1st	2nd	3rd	4th	5th	6th	7th	8th	9th	10th

straints include: Cinema, Journal and Bulletin, Mobile Phone, Newspaper Handbill and Fliers. Television and Radio was rated the least constrained ICT. This corroborates the finding by Munyna, (2007), in which she stated that lack of adequate infrastructure militates against the use of sophisticated ICTs by extension personnel in the developing nations (Table 7).

Hypothesis 1

Relationship between social economic characteristics of extension agents and level of use of ICTs

The results of the chi – square analysis revealed that there significant relationship exists between selected personal characteristics variables of the respondents such as Sex ($X^2 = 17.78, P<0.05$), Age ($X^2 = 10.68, P<0.05$), level of Education ($X^2 = 33.56, P<0.05$), years of working experience ($X^2 = 14.66, P<0.05$) and level of use of ICTs by extension agents and This implies that the null hypothesis is hereby rejected for aforementioned variables. It is also implies that Sex, Age, level of education and years of working experience significantly influenced level of use of ICTs by the sampled extension agents in the study area.

Conversely, other variables such as marital status and religion were found not significant to the level of use of ICTs by extension agents. The implication of this is that variables such as marital status and religion were found not to significantly influence the level of use of ICTs by extension agents (Table 8).

Hypothesis 2

The hypothesis stated that there is no significant relationship between the level of ICT training of extension agents and their level of use of ICT devices

Spearman's rho correlation analysis was performed to test the relationship between the level of ICTs training of the extension agents and their level of use of ICTs in their extension service delivery. The result of the analysis shows that there is no significant relationship between the level of ICT training of extension agents ($r= 0.110, P< 0.05$) and their level of use of ICT devices.

CONCLUSION AND RECOMMENDATIONS

The study assessed the use of ICTs in disseminating agricultural related information among extension agents in the Oyo State, Nigeria. Based on the findings of the study, it could be concluded that majority of the extension workers in the study area are average users of ICT tools for extension information delivery especially, Radio, Television and Mobile phones are the most relevant ICT for extension delivery. These tools are useful in dissemination of Pre planting, post planting operations as well as livestock issues information. Age, sex, education and years of working experience are important

Table 8: Summary of Chi – square analysis establishing relationship between personal characteristics of extension agents and level of use of ICTs

Personal characteristics	Df	Chi – square X^2	P-value	Remark	Decision
Sex	1	17.780	0.000	Significant	H0 Rejected
Age	12	10.683	0.002	Significant	H0 Rejected
Religion	1	0.024	0.876	Not-significant	H0 Accepted
Marital status	2	65.024	0.329	Not-significant	H0 Accepted
Education	2	33.561	0.000	Significant	H0 Rejected
Work Experience	13	14.659	0.000	Significant	H0 Rejected

Source: Field survey, 2009 Figure in parenthesis is ranks

factors that significantly influenced the level of use of ICT for extension delivery among extension agents in the study area.

Sequel to the findings of the study, it is recommended that

1- Extension institutions should intensify efforts towards disseminating agricultural information meant for the farmers in the study area through the use of radio, television and Mobile phone which are the ICTs that are most accessed by all concerned in the study area.

2- There should also be a periodic review of the use of ICTs in the extension service delivery so as to identify which particular, ICT is most suitable for a particular purpose at a given point in time.

3- Also the extension service providers should be made to undergo periodic on- the- job training on the use of ICT so as to keep them abreast of latest information that could facilitate an efficient extension service delivery.

REFERENCES

1- Aina, L.O. (2004). Decision Makers and Agricultural Information in Africa the Potential Role of the internet. IAALD Quarterly Bulletin, XLIV 96-99.

2- Alex, G. Zijp, W. & Byerlee, K. (2004). Rural Extension and Advisory Services: New Directions RURAL Development Strategy Background Paper No 9 World Bank, Washington D.C.URL.http://www.worldbank.org.

3- Arokoyo, T. (2003). Response to the e-Discussion: ICT Application in Rural Extension and information Exchange. 6th consultative Expert Meeting, April.

4- Arokoyo, T. (2004). ICTs in the Transformation of Agriculture in Nigeria. URL http://www.gisdevelopment.net

5- Arokoyo, T. (2005). ICTs Application in Agricultural Extension Service delivery. In: Adedoyin, F.S (ed) Agricultural Extension in Nigeria. AESON, Ilorin.

6- Federal Ministry of Agricultural and Rural Development, (2002). Federal Ministry of Agricultural and Rural development opportunities in Nigeria's Agricultural Sector (Review, 4th NW).URL.http://www.nigeriabusinessinfo.com

7- International food and Agriculture Programme, (2004). Reducing Hunger in Africa (MDGs) view from farmers Organisations in Africa. Members of IFAP. Contribution to Blair Commission for Africa. ITU (2004) ICT in Africa Status Report available URL http:// www.itu.net.

8- Kiplang, S. & Wallance, T. (2003). An Analysis of the Opportunity for Information Telecommunication in improving access, transfer and use of agricultural information in the Rural Areal note Present at ICT Observatory 2003, ICTs Transforming Agricultural Extension. Wakening 23-35 September, 200.

9- Low External input and sustainable Agriculture (2005). Changing Information Flows LEISA magazine July, URL. Ilia.org.

10- Munyua, H. (2007). Application of ICTs on Africa's Agricultural Sector. the International Development Research Centre.

11- Salau, E.S. & Saingbe, N.D. (2008). Access and Utilization of Information and Communication Technologies (ICTs) Among Agricultural Researchers and Extension Workers in Selected Institutions in Nasarawa State of Nigeria. PAT 2008; 4 (2): 1-11. Accessed on July 22nd 2012. Retrieved from: http://www.patnsukjournal.com/currentissue.

2

Encouraging and Discouraging Factors in Application of Information and Communication Technologies in Agricultural Extension in Isfahan Province of Iran

Gholamreza Pezeshki Rad [1], Zahra Haji Hashemi [2*] and Mohammad Chizari [3]

Abstract

The benefits of Information and Communication Technologies (ICTs) for agricultural extension purposes and activities are obvious. The main objective of this article is to identify encouraging and discouraging factors in application of ICTs in agricultural extension in Isfahan province of Iran. This study investigated areas of ICTs application by Extension Workers (EWs), too. The research design was a descriptive method. The population includes 320 Isfahan province EWs that 200 person were choose by census and stratified sampling methods. In our contribution we focused on finding. Face and content validity of the questionnaire was established using a panel of experts consisting of faculty members in the Departments of Agricultural Education and Extension, Tarbiat Modares University and extension officers in the Isfahan Province Agricultural Jihad Organization. Cronbach's Alpha to estimate the reliability of the instrument was among 82.6 - 87.8. On the ranking of principal areas of applying ICTs by EWs in extension "typing letters/report", and "prepare and protect file on process basic agricultural and rural information", and "access to the latest agricultural news and information", had most usage. The main result of the study revealed that the EWs in the study approximately use ICT quite a bit. For EWs ICT Encouraging Items Principal Factor Analysis extracted four factors which accounted for 76.24% of the total variance. And Factor analysis for EWs ICT discouraging Items extracted six factors which accounted for 62.84% of the total variance.

Keywords:
Agricultural Extension, Information and Communication Technologies (ICTs), Encouraging and Discouraging Factors, Extension Workers (EWs), ICT Application, Iran

[1] *Associate Professor of Agricultural Extension and Education Department of Tarbiat Modares University, Peykanshahr, Tehran, Iran.*

[2] *Graduated Student of Tarbiat Modares University.*

[3] *Professor of Agricultural Extension and Education Department of Tarbiat Modares University, Peykanshahr, Tehran, Iran.*

* *Corresponding author's email: haji.hashemii@yahoo.com*

INTRODUCTION

In today's world, Information and Communication Technology (ICT) is significantly important in an ever increasing speed and precision in different economic and social tasks. Particularly, in many organizations having several branches scattered in different physical locations, or organizations responsible for doing multiple and versatile tasks. ICT systems are able to solve the problems of such organizations. Experience from organizations across the world show that ICT easily solves many problems related to information orientated organizations, and it also provides powerful tools in doing so (madadi *et al.*, 2012). The benefits of ICT for agricultural extension purposes and activities are obvious. ICT can be used to increase the effectiveness and efficiency of extension work (Akpabio *et al.*, 2007). ICT use enables extension officers to play complimentary roles of accessing and transferring relevant information to farmers (Meera *et al.*, 2004) and also helps farmers to utilize such information to solve pressing problems (Klink in Leeuwis, 2004).

Gelb and Offer (2006) reported that adoption of ICT in rural areas for agricultural purposes is a unique challenge and has long been of specific public concern with regional, national and international strategic significance. Studies in recent years have shown more and more that the ownership and use of a computer as well as access to the Internet have become rather the rule than the exception (Doluschitz and Pape, 2002; and Nass, 2001). ICT adoption in agriculture has increased considerably over the last several years. (Rosskopf and Wagner, 2003), but In spite of an obviously increasing usage of hardware, usage of ICT still faces many obstacles. From a scientific views point the reasons for this discrepancy have started to be examined (Gelb *et al.*, 2001). In Iran, it becomes necessary to investigate factors affecting ICT application by EWs.

Therefore, some studies did to recognize main areas, factors affecting of applying ICT in agriculture and agricultural extension system (Falaki, 2005; Fallah, 2006; Hedjazi *et al.*, 2006; Annor- Frempong *et al.,* 2006; Meera *et al.*, 2004; Bheenick and Brizmohan, 2003; Maru and, 2003; and Chieochan *et al.*, 2000).

Shabanali Fami *et al.*, (2009) in their study investigated areas for potential application of ICTs in the agricultural extension system of Iran (2009). The findings revealed that six factors/components containing 32 variables determined about 68.91% of the variations in the areas of application. The six factors were Management of documentation and preparation of extension information (14.88% of variance), Planning and implementation of extension education programs (13.6%), Professional empowerment of extension agents (10.8%), Expansion of intra-organizational relationships (10.27%), Facilitation of organizational processes (9.7%), and improving communication through e-mail services (9.64%).

Saheed *et al.*, (2013) in assessment of use of selected ICTs for extension service delivery, indicated constraints of ICTs application for extension service delivery such as: high illiteracy level among farmers, high cost of procurement of ICTs, cost of maintenance of ICTs, negligence on the part of the farmers and inadequate technological know-how. It was found that the Internet was the most constrained ICT, followed by the use of CD-ROM technology, and multimedia projector. Finally, they ranked severity of constraint against various ICTs by extension agent.

Faghiharam *et al.*, (2012) in their study for assessment of socio- psycho factors on use of ICT showed nine variables that affect the application of ICT. These variables clarify 84 percent of the dependent variable's variance totally. Computer competence, computer self–efficiency, management support and perceived ease of use of computer were recognized as strong predictors of ICT application.

In another study, Patil *et al.*, (2008) at investigate for adoption of ICT for Indian agriculture, asked the factors limiting the use of ICT by farmers, extension working with farmers, and research working with farmers. An evaluation of the Indian scene suggests that market information and weather updates are of prime interest; illiteracy, cost and lack of awareness are the major adoption constraints. Human capital enhancement was understood to be the main remedial factor to change the low rate of ICT adoption and its effectiveness.

Akpabio *et al.*, (2007) in their study focused

on constraints affecting the utilization of Information and Communication Technologies (ICTs) for agricultural extension activities by Agricultural Extension Officers in Nigeria's Niger Delta Region. Findings revealed important specific constraints to include: poor ICT infrastructure development, high cost of broadcast equipment, high charges for radio/television presentations, high cost of access/interconnectivity and electricity power problems. The use of Factor Analysis aided to crystallize identified constraints into three factors of 'poor enabling environment', 'lack of accesses and 'dissemination of unrelated information'. Against the background of several national government initiatives to develop the infrastructure base of the country, recommendations are proffered to ameliorate ICT use constraints.

Another survey was carried out to identify immediate applications and constraints to the application of IT in extension in Mauritius. All extension officers recognize the use of computers in their work as a tool to store, process and retrieve information, with a range of objectives more pertinent to the field of extension and the main issues raised by farmers. The crop extension officers in this study would presently use computers to access information on new technology and research findings, especially on pest and disease control. Sugarcane and livestock extension officers see the immediate applications of computers in their work as accessing a database on the individual farmers and as a means of speeding up the flow of ICT. Also extension officers ranked the constraints that might hinder the application of IT in extension. Training of extension officers constitutes the most important constraint to be overcome before IT can be applied in extension. Other constraints include lack of familiarity of extension officers with computer and lack of infrastructure. Finally this study concluded that, the institutions will have to devise Information Technology strategies to overcome current obstacles dealing with issues such as training, infrastructure, appropriate human resources, and the adoption of a policy on information sharing (Bheenick and Brizmohan, 2003).

Several studies have endorsed the some barriers of ICT application in education. BECTA (British Educational Communications and Technology Agency) in its report (2004), categorized barriers to ICT use into two groups: external barriers and internal barriers. External barriers included: lack of access to resources; lack of time; lack of effective training; technical problems; whereas internal barriers included: lack of confidence; resistance to change and negative attitudes; no perception of benefits. Findings showed that although students and teachers in their study had good ICT skills in terms of their own personal use, they were unable to transfer these skills to using ICT in the classroom.

A recent study in Singapore (Teo, 2006) based on the observations of ICT-mediated lessons and face-to-face interviews with teachers, ICT heads-of-department and school principals, identified six major barriers to teacher ICT-integration: (a) inadequate appointment of technical support staff, (b) inadequate appointment and training of student ICT helpers, (c) lack of sufficient time for teachers to prepare for ICT-mediated lessons, (d) insufficient collaboration among teachers in preparing ICT-mediated lessons, (e) lack of support provided by school leaders in addressing teachers' ICT concerns, and (f) insufficient training, demonstrations or advice for teachers on how to incorporate ICT into classroom instruction.

Another studies indicated barriers to use ICT among farmers. In one study all respondents (farmers and non-farmers) were asked to identify factors limiting the use of ICT by farmers. While the scientists saw the cost of technology and the lack of user friendliness as the main problems, the participants of this study thought that lack of training and failure to understand the possible benefits were the greatest impediments. Both farmers and non-farmers from agriculture-related areas agreed on that point (Rosskopf and Wagner, 2003). Mc Carthy (2003) in a sample survey of 35 Teargas beef farmers also found that lack of computer skills is still a major problem for farmers. 37% of respondents identified lack of computer skills as the most significant problem with using a computer on the farm. 23% felt the major problem was that too much work was involved to get a benefit, followed by 20% who identified security fears as being the major problem. Another research is: Gelb and Parker, 2006.

Morris (2002) in a sample survey of 162 ICMSA members found that the lack of appropriate training (28%) was perceived to be the most common barrier inhibiting present/future Internet use. Second was the cost of the Internet connection (28%), third was the cost of the computer (14%), fourth was finding time to use (13%), fifth was privacy (12%), sixth was security (4%) and last was other (1%).

Maru (2003) in his paper "ICT Use in Agricultural Extension in India" indicated the key issues affecting ICT Initiatives in Agriculture and particularly in dissemination of marketing information are:

1. The lack of reliable connectivity in rural areas
2. The sheer lack of purchasing power in the rural segment, which makes it necessary to find sponsors (govt, co-operative, private sector companies) who benefit by way of advertising, goodwill creation or fulfillment of certain objectives.
3. The time element in establishing e-governance services in districts, which has a cascading impact on investment cost.
4. The need to have high usage of graphics and voice-overs (less confidence with computers, low literacy levels) in content.

This study was therefore conceived to identify perceived encouraging and discouraging factors in application of ICT in agricultural extension and principle areas of application in Isfahan Province of Iran.

Purpose and objectives

The overall purpose of this study was to investigate and analyze the factors encouraging and discouraging in application of information and communication technologies (ICT) in agricultural extension in Isfahan Province of Iran as perceived by extension workers (EWs). The specific objectives of the study were to:

1. Describe job and demographic characteristics of EWs;
2. Determine application of ICT in agricultural extension by EWs;
3. Identify factors encouraging and discouraging to the use of ICT by EWs;
4. Determine relationship between encouraging (four factors) and discouraging (seven factors) and application of ICT by EWs.

MATERIALS AND METHODS

This study used a descriptive-co relational design. The population for the study consisted of all extension workers (N = 320) at the extension organization, Isfahan Province, Iran among whom 42 individuals were located in the agricultural organization of Isfahan province and the branch of the county management (N=70), were studied by census. For the selection of the required sample of the centers of providing agricultural services, the appropriate stratified sampling method was used that 200 answered a self-completion questionnaire were returned, totally. Factors encouraging and discouraging the application of ICT by EWs in agricultural extension were assessed using an instrument. The instrument was divided into four parts; part one was designed to assess ICT application among EWS in their job, using 21 statements. The 21 statements were measured on a four-point, Likert-type scale, that ranged from; 1 = "Not at All" 2 = "Slightly" 3 = "Quite a Bit" and 4 = "Very Much."

Parts two and three were measured factors encouraging and discouraging ICT adoption in agricultural extension by extension worker. The 31 and 30 statements were measured on a four-point type scale respectively. The scale ranged from; 1 = "Not at All" 2 = "Somewhat" 3 = "Slightly" and 4 = "Very Much." Finally, part four was designed to identify demographic and job's characteristics. The following were demographic and job's characteristics assessed: gender, age, highest education completed major area of study, work experience, current work location, and employment status. The instrument was assessed for content and face validity by a panel of experts consisting of faculty members in the Departments of Agricultural Education and Extension, Tarbiat Modares University and extension officers in the Isfahan Province Agricultural Jihad Organization. The instrument was pilot tested using extension worker (N = 30) at Isfahan Province who were not included in the study. Minor changes were made to improve the clarity and readability of the instrument. Cronbach's alpha, an internal consistency measure, was used to estimate the reliability. The reliability coefficient for part one, two and three of the instrument were among 82.6 - 87.8. respectively. Each of these is considered to be very acceptable level of reliability for research purposes. Data were analyzed by SPSS software 16 and using frequencies, percentages, means, and standard deviation. Descriptive statistics and Pearson Moment Correlation tests were computed to describe the relationships. The scale developed by Davis (1971) was used to describe the magnitude of the relationships.

Factor analysis was computed to group variables for encouraging and discouraging factors for ease of presentation and discussion purposes. The details of factor analysis for encouraging and discouraging factors are shown in Tables (3 and 4).

RESULTS

Objective 1: Demographic Profile of Respondents

Majorities (87%) of the respondents were male (see Table 1). Age of respondents was distributing evenly among the three groups (23.5% under 30 years; 36.0% between 31-40 years and 35.0 % between 41-50 years). Three of five respondents (59%) had completed Bachelor of

Table1: Frequency and Percentage for Demographic and Job's Characteristics.

Demographics Profile		f	%
Gender			
Male		154	87.0
Female		23	13.0
	Total	177	100.0
Age*			
30 years and less		42	23.5
31 – 40 years		64	35.8
41 – 50 years		63	35.2
51 years and more		10	5.6
	Total	179	100.0
Highest education completed			
High school diploma		18	10.2
Bachelors of art		33	18.6
Bachelors of science		104	58.8
Masters of science		22	12.4
	Total	177	100.0
Major area of study			
Extension and education		17	9.5
Agronomy		50	27.9
Animal science		21	11.7
Other agricultural majors		64	35.8
Not related to agricultural major		27	15.1
	Total	179	100.0
Work experience in agricultural extension**			
5 years and less		75	41.2
6 – 15 years		68	37.4
16 years and more		39	21.4
	Total	182	100.0
Work location			
Provincial level		36	19.7
District (Shahrestan) level		51	27.9
County (Dehestan) level		96	52.5
	Total	183	100.0
Employment			
Tenured		87	48.9
Tenured track		91	51.1
	Total	178	100.0

*Mean of age = 37.83, SD = 8.16 and range between 23 to 55 years old.

** Mean of work experience = 9.42, SD = 7.57 and range between 1 to 29 years of experience.

Table 2: ICT Application by EWs in their Job

ICT Application	n	Mean*	S.D	Rank
Typing letters/reports	183	3.24	0.92	1
Prepare and protect file on process basic agricultural and rural information	181	2.89	0.87	2
Access to the latest agricultural news and information	183	3.00	0.97	3
Upgrade EWs' professional competencies	183	2.70	0.92	4
Prepare and present published brochure needed at the local level	183	2.83	0.98	5
Improve the administrative information activities at regional level	183	2.67	0.93	6
Record and exchange of information in regard for extension centers	182	2.84	0.99	7
Document activities and extension programs delivered	183	2.69	0.95	8
Improve accessibility of EWs to agricultural specialized resources	181	2.80	0.99	9
Summarize information using figures and charts	183	2.66	0.95	10
Prepare slides through multimedia software	183	2.61	1.04	11
Increase the speed of EWs' information acquisition from research	183	2.70	1.08	12
Prepare short video films through multimedia Software	183	2.62	1.05	13
Improve EWs' skills	182	2.35	0.95	14
Learn about presentation skills for conferences/seminars	183	2.62	1.06	15
Improve discussion among EWs through computer network	183	2.41	0.99	16
Facilitate accessibility to legislative information	182	2.51	1.04	17
Updates administration rules	182	2.51	1.04	18
Improve linkages between EWs and researcher	181	2.37	1.03	19
Present group training to farmers through video conference	182	2.33	1.00	20
Present advisory services to different farmer groups	182	2.24	1.02	21
Summated score	**183**	**55.45**	**15.89**	

*Mean of age = 37.83, SD = 8.16 and range between 23 to 55 years old.
** Mean of work experience = 9.42, SD = 7.57 and range between 1 to 29 years of experience.

Science degree. A little over one-third of the respondents (35.8%) indicate their major area of study in other agricultural majors such as farm machinery, plant pathology, food science, irrigation, and horticulture. Nearly 28% reported agronomy as a major followed by a major other than agriculture (15.1%). On average, respondents had 9.42 years of work experience. A little over one-half of the respondents (52.5%) worked in a county level. Employment status of respondents was more or less evenly divided (49.0%) among tenured and (51%) non tenure personnel.

Objective 2: Application of ICT in agricultural extension by EWs;

This part of the survey concerned the principal areas of applying ICTs by EWs in extension and identified from previously reported research findings. EWs responded to 21 items using the following type scale: 1 = not at all, 2 = slightly, 3 = quite a bit and 4 = very much.

A summery is shown in Table 2. It shows means and standard deviations for each of the 21 ICT applications. The statements which the respondents used most were: "typing letters/report" (M=3.24, SD=0.92), and "Prepare and protect file on process basic agricultural and rural information"(M=2.89, SD=0.87), and "access to the latest agricultural news and information" (M=3.00, SD=0.97). Overall, the result of the Mean (55/45), and Standard Deviation (15/89), indicate that the EWs in the study approximately use ICT quite a bit.

Objective 3: Factors encouraging and discouraging to the use of ICT by EWs;

At first, Table 3 presents means and standard deviations for each of the 25 encouraging factors to the use of ICT by EWs. The most effective items to use of ICT which the respondents statement were: "Provide a competitive environment for presenting ICT services"(M=3.40, SD=0.80), and "Identify appropriate applications for effect use of ICT"(M=3.39, SD=0.81), and "Develop step by step policies for logical hieratical application of ICT"(M=3.39, SD=0.82), and "Develop implementation plans for ICT in extension"(M=3.34, SD=0.81).

As is shown in Table 4 "EWs low level of knowledge and skills in regard to information" (M=3.64, SD=0.64), and "The cost of establish-

Table 3: Means and Standard Deviations Factors encouraging to the use of ICT by EWs; (N = 183).

Items	n	Mean*	S.D	Rank
Provide a competitive environment for presenting ICT services	183	3.40	0.80	1
Identify appropriate applications for effect use of ICT	181	3.39	0.81	2
Develop step by step policies for logical hieratical application of ICT	181	3.39	0.82	3
Develop implementation plans for ICT in extension	182	3.34	0.81	4
Develop program guidelines for ICT activities	182	3.33	0.81	5
Reduces taxes and communication expenses for ICT	180	3.42	0.84	6
Provide secure appropriate hardware and software for extension	180	3.40	0.84	7
Develop appropriate websites in Persian languages	183	3.51	0.87	8
Establish continuous flow of information from research to key stakeholders	183	3.41	0.85	9
Use government's free policies on ICT services	183	3.25	0.82	10
Establish a central portal for coordinates of organizations	183	3.35	0.87	11
Provide access to rural social centers as an ICT center	180	3.24	0.85	12
Appoint responsible and specialist people for ICT units	183	3.26	0.86	13
Encourage collaboration of EWs with computer specialists	183	3.37	0.89	14
Create maintenance services and desire support for ICT users	181	3.25	0.86	15
Establish ICT unit in organization structure of extension	183	3.25	0.88	16
Involve Clients in all the phases of ICT development applied	179	3.21	0.87	17
Encourage private sectors to invest in Extension-ICT activities	179	3.36	0.92	18
Maintain equal possible accessibility to ICT for EWs	183	3.30	0.92	19
Educate EWs for the appropriate use of ICT	179	3.23	0.92	20
Basic structure for distance education at different levels	183	3.26	0.93	21
Increase capacity of information exchange and communication	181	3.29	0.94	22
Revise social and cultural procedures in educational and social structure of extension	180	3.22	0.96	23
Develop local networks for ICT-Extension	183	3.22	1.01	24
Increase the level of English proficiencies among EWs	181	3.22	1.04	25
Summated score	**183**	**145.98**	**22.9**	

* Mean computed on a scale 1 = not at all, 2 = somewhat, 3 = slightly, and 4 = very much

ing and operating data centers specialized in agricultural extension" (M=3.93, SD=0.72), and "Limited accessibility to technical and supportive person" (M=3.57, SD=0.68) were the most effective items to ICT usage that statement with respondent. Finally, "Lack of up-to-date information in extension" (M=3.51, SD=0.68), and "Lack of holistic information in extension" (M=3.51, SD=0.69), and "Lack of precise information in extension" (M=3.50, SD=0.69) allocated 4th to 6thranks in priority.

In continuous, principal factor analysis was performed to identify factors that underlie encouraging and discouraging to the use of ICT by EWs. The main objective of this technique is to classify a large number of variables into a small number of factors based on relationships among variables. For reporting factor analysis data, the procedures suggested by Warm rod (2000) were used. These included: sample size, list and scale of measurement used for the observed variables, means and standard deviations for the observed variables, factor loadings, final communalities, Eigen values for each factor extracted, proportion of common variance explained for each factor, total variance, and correlation matrix. Factors with Eigen values greater than 1 were retained and rotated to a varimax solution for interpretation. In addition, Cronbach's alpha for each extracted factor was reported.

The variables which loaded on each of the four encouraging factors and six discouraging factors to the use of ICT by EWs were given to a panel of experts to "name" the factors. The panel consisted of faculty members, and graduate students in the department of Agricultural and Extension Education. Making sense of the panel's judgment was very difficult. In addition, to give a specific name for each factor was even more difficult. The names of the two categories and Their factors are given below.

1- Encouraging Factors:

Factor 1 - Providing easy access to the ICT Unit

Factor 2 - Develop strategic plan for developing ICT units

Table 4: Means and Standard Deviations Factors discouraging to the use of ICT by EWs; (N = 183)

Items	n	Mean*	S.D	Rank
EWs low level of knowledge and skills in regard to information	183	3.64	0.64	1
The cost of establishing and operating data centers specialized in agricultural extension	181	3.93	0.72	2
Limited accessibility to technical and supportive person	179	3.57	0.68	3
Lack of up-to-date information in extension	181	3.51	0.68	4
Lack of holistic information in extension	181	3.51	0.69	5
Lack of precise information in extension	181	3.50	0.69	6
Lack of appropriate policy for collaborating between government and private sectors regarding ICT	183	3.50	0.71	7
Lack of communications equipment in rural areas and extension	183	3.55	0.74	8
Lack of subject matter software in Persian language	181	3.55	0.75	9
High maintenance cost relative to linkage and up-to-date appropriate content	183	3.43	0.73	10
Lack of appropriate Persian websites	179	3.52	0.76	11
Limited in-service training in the area of ICT for EWs	180	3.44	0.77	12
Lack of commitment of some stakeholders in country's agricultural knowledge and information system	176	3.29	0.76	13
Lack of interest to the application of ICT among EWs	183	3.40	0.79	14
Lack of interest to the application of ICT among policy makers	183	3.40	0.80	15
Weak structure in extension organization at different levels	183	3.47	0.82	16
Poor network security	183	3.21	0.79	17
High cost of buying and maintenance of ICT hardware and software	183	3.33	0.83	18
Existence of social and cultural barriers at different level	183	3.10	0.84	19
Social and cultural Weak to look at (viewed) ICTs as a goal not as a tool in agricultural extension	182	3.07	0.84	20
Anxiety around risks as a result of misuse of ICT	182	3.12	0.88	21
Lack of EWs self-esteem in utilizing ICT	181	3.08	0.88	22
ICT is viewed as "fancy" in extension	182	2.99	0.87	23
ICT is not a priority for EWs	183	3.09	0.91	24
Summated score	**183**	**81.20**	**19.3**	

* Mean computed on a scale 1 = not at all, 2 = somewhat, 3 = slightly, and 4 = very much

Factor 3 - Provide the requirements and tools

Factor 4- Provide government's free competitive environment for service delivery

2- Discouraging Factors

Factor 1 - Poor infrastructure

Factor 2 – Extension Weakness in Providing Information

Factor 3- Extension Weakness in providing requirements.

Factor 4- Social and cultural barriers

Factor 5- Lack of interest in the use of ICT

Factor 6- The high cost of installation of ICT

For EWs ICT Encouraging Items factor analysis extracted four factors which accounted for 76.24% of the total variance. The final statistics (Eigen values, percent of variance explained, Reliability (alpha), and factor loadings for each of the four factors are shown in Table 5. Factor 1 explained 64.56% of the total variance. Thirteen variables loaded on this factor. The variable, maintain equal possible accessibility to ICT for EWs, was most representative of Factor 1 (r = 0.767). Factor 2 explained 4.31% of the total variance. Five variables loaded on this factor. The variable, develop program guidelines for ICT activities, was most representative of Factor 2 (r = 0.847). Factor 3 explained 4.08% of the total variance and five variables loaded on this factor. The variable, increase the level of English proficiencies among EWs, was the most representative of Factor 3 (r = 0.796). Factor 4 explained 3.29% of the total variance with two variables loaded on this factor. The statement, use government's free policies on ICT services, was most representative of Factor 4 (r = 0.696).

Upon examination of the factor loadings and the items which loaded on each of the four factors, internal consistency estimates were computed for the four factors. The reliability estimates for the four factors ranged from a low

Table 5: Factors, Reliabilities, Variance Explained, Eigen Values, and Factor Loadings for EWs ICT Encouraging Items (N = 183).

Factor, Reliability Variance explained Eigen Value	Items	Factor Loading
	Establish ICT unit in organization structure of extension	.600
Factor 1 (13)	Appoint responsible and specialist people for ICT units	.594
0.97	Maintain equal possible accessibility to ICT for EWs	.767
64.56%	Identify appropriate applications for effect use of ICT	.594
9.016	Reduces taxes and communication expenses for ICT	.702
	Encourage private sectors to invest in Extension-ICT activities	.597
	Establish continuous flow of information from research to key stakeholders	.628
	Develop local networks for ICT-Extension	.696
	Revise social and cultural procedures in educational and social structure of extension	.608
	Develop appropriate websites in Persian languages	.751
	Establish a central portal for coordinates of organizations	.628
	Provide access to rural social centers as an ICT center	.511
	Encourage collaboration of EWs with computer specialists	.688
Factor 2 (5)	Involve clients in all the phases of ICT development applied	.525
0.93	Create maintenance services and desire support for ICT users	.623
4.31%	Develop program guidelines for ICT activities	.847
5.735	Develop implementation plans for ICT in extension	.843
	Develop step by step policies for logical hieratical application of ICT	.608
Factor 3 (5)	Basic structure for distance education at different levels	.642
0.91	Increase capacity of information exchange and communication	.677
4.08%	Provide secure appropriate hardware and software for extension	.582
5.187	Increase the level of English proficiencies among EWs	.796
	Educate EWs for the appropriate use of ICT	.578
Factor 4 (2)	Use government's free policies on ICT services	.696
0.79	Provide a competitive environment for presenting ICT services	.628
3.29%		
3.698		

Note. Figures in parentheses in the left column indicate total number of items in each factor.

0.79 (Factor 4) to a high of 0.97 (Factor 1). (See Table 5).

Factor analysis for EWs ICT discouraging Items extracted six factors which accounted for 62.84% of the total variance. Eigen values, percent of variance explained, alpha, and factor loadings for each of the six factors are shown in Table 6. Factor 1 explained 31.79% of the total variance. Six variables loaded on this factor. The variable, lack of subject matter software in Persian language, was most representative of Factor 1 (r = 0.812). Factor 2 explained 9.24% of the total variance. Three variables loaded on this factor. Three variables, were equal (r = 0.951). Factor 3 explained 7.97% of the total variance and four variables loaded on this factor. The variable, weak structure in extension organization at different levels, was the most representative of Factor 3 (r = 0.820). Factor 4 explained 5.69% of the total variance with four variables loaded on this factor. The statement, ICT is viewed as "fancy" in extension, was most representative of Factor 4 (r = 0.822). Factor 5 explained 4.16% of the total variance with four variables loaded on this factor. The statement, Low education level among EWs, was the most representative of Factor 5 (r = 0.768). Factor 6 explained 3.99% of the total variance. Three variables loaded on this factor. The variable, limited in-service training in the area of ICT for

Table 6: Factors, Reliabilities, Variance Explained, Eigen Values, and Factor Loadings for EWs ICT Discouraging Items (N = 183).

Factor, Reliability Variance explained Eigen Value	Items	Factor Loading
Factor 1 (6)	Poor network security	0.565
0.82	Limited accessibility to technical and supportive person	0.569
31.79%	Lack of commitment of some stakeholders in country's agricultural knowledge	0.476
3.656	and information system	0.463
	Anxiety around risks as a result of misuse of ICT	0.771
	Lack of appropriate Persian websites	0.812
	Lack of subject matter software in Persian language	
		0.951
Factor 2 (3)	Lack of holistic information in extension	0.951
0.99	Lack of precise information in extension	0.951
9.24%	Lack of up-to-date information in extension	
3.387		
Factor 3 (4)	Lack of communications equipment in rural areas and extension	0.775
0.78	Weak structure in extension organization at different levels	0.820
7.97%	EWs low level of knowledge and skills in regard to information	0.696
3.170	Lack of appropriate policy for collaborating between government and private sectors regarding ICT	0.527
Factor 4 (4)	Lack of EWs self-esteem in utilizing ICT	0.462
0.76	Existence of social and cultural barriers at different level	0.718
5.69%	ICT is viewed as "fancy" in extension	0.822
3.040	Social and cultural Weak in Considering ICTs as a goal not as a tool	0.764
Factor 5 (4)	Lack of interest to the application of ICT among policy makers	0.573
0.82	Lack of interest to the application of ICT among EWs	0.573
4.16%	ICT is not a priority for EWs	0.715
2.389	Limited in-service training in the area of ICT for EWs	0.768
Factor 6 (3)	The cost of establishing and operating data centers specialized in agricultural extension	0.488
0.58	High cost of buying and maintenance of ICT hardware and software	0.835
3.99%	High maintenance cost relative to linkage and up-to-date appropriate content	0.507
1.851		

Note. Figures in parentheses in the left column indicate total number of items in each factor.

EWs, was most representative of Factor 6 (r = 0.835).

Internal consistency estimates were computed for the six factors to upon examination of the factor loadings and the items which loaded on each of the six factors. The reliability estimates for them are shown in Table 6. The reliability estimates ranged from a low 0.58 (Factor 6) to a high of 0.99 (Factor 2).

Objective 4: Relationships between encouraging (four factors) and discouraging (six factors) and application of ICT by EWs.

Pearson Correlation Coefficients were calculated to describe the relationships between encouraging, discouraging ICT application factors and application of ICT by EWs. In spite of Factor 2 (r = 0.165), in encouraging ICT factors none of the four factors of encouraging items was found to be related to ICT application (Table 7). However, some of the discouraging factors such as factor 3 (r = -0.152), and factor 5(r =-0.169), were significant at the .05 level. Multiple regression analyses using the ENTER procedure (simultaneous entry of all variables) was conducted to determine the importance of variables in explaining variance in the ICT application.

Table 7: Summary of ICT Scores and Pearson Correlation Values between Encouraging and Discouraging ICT Adoption by ICT Application Scores (n = 183)

Factors	Summated Mean Score	Summated SD	r	p
ICT Application (21 Items)	**55.45a**	**15.89**	-	-
Encouraging ICT Adoption (25 Items)	81.20b	19.30	0.155	0.037
Factor 1 (13- 68)d	55.81	11.98	0.130	0.078
Factor 2 (5 - 24)d	19.81	4.23	0.165	0.026
Factor 3 (5 - 20)d	16.33	3.99	0.133	0.073
Factor 4 (2 - 12)d	9.82	2.23	0.137	0.065
Discouraging ICT Adoption (24 Items)	**145.98c**	**22.90**	**- 0.134**	**0.070**
Factor 1 (6 - 28)d	23.44	3.77	-0.072	0.330
Factor 2 (3 - 12)d	10.53	2.06	-0.023	0.759
Factor 3 (4 - 16)d	14.17	2.30	-0.152	0.040
Factor 4 (4 - 16)d	12.50	2.50	-0.060	0.421
Factor 5 (4 - 16)d	13.26	2.65	-0.169	0.022
Factor 6 (3 - 12)d	10.11	1.72	-0.067	0.370

a) Mean computed on a scale 1 = Not at All to 4 = Very Much and could range from a low of 21 to high of 84 with theoretical mid-point of 52.5

b) Mean computed on a scale 1 = Not at All to 4 = Very Much and could range from a low of 31 to high of 124 with theoretical mid-point of 77.5

c) Mean computed on a scale 1 = Not at All to 4 = Very Much and could range from a low of 29 to high of 116 with theoretical mid-point of 72.5

d) Numbers in parentheses indicates the range of scores for each of the factors.

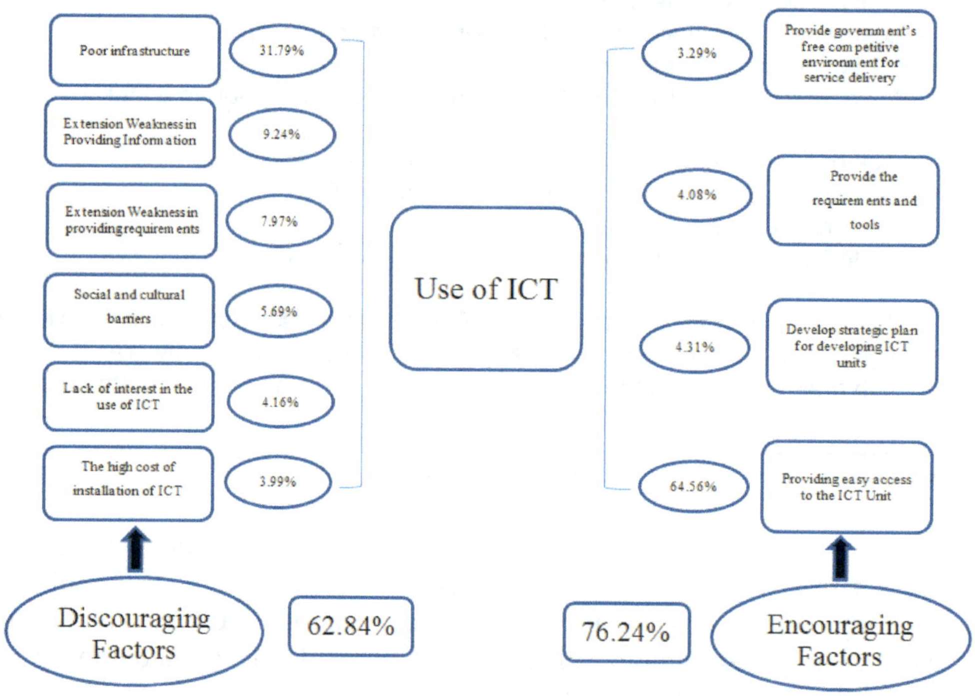

Figure 1: The Encouraging and Discouraging Factors to the Use of ICT by EWs.

CONCLUSIONS AND RECOMMENDATIONS

Extension increasingly acknowledges the importance of new ICTs. In fact, technologies are now being considered as a basic component of any extension program. ICTs modify the practice of extension services and open interesting avenues for the collection and dissemination of local knowledge and its integration with modern and global knowledge, requiring a redefinition of knowledge management.

Overall, findings indicated that majorities (87%) of the respondents were male. Age of respondents was distributing evenly among the three groups (23.5% under 30 years; 36.0% between 31-40 years and 35.0 % between 41-50 years). Three of five respondents (59%) had completed Bachelor of science degree. A little over one-third of the respondents (35.8%) indicate their major area of study in other agricultural majors such as farm machinery, plant pathology, food science, irrigation, and horticulture. Nearly 28% reported agronomy as a major followed by a major other than agriculture (15.1%). On average, respondents had 9.42 years of work experience. A little over one-half of the respondents (52.5%) worked in a county level. Employment status of respondents was more or less evenly divided (49.0%) among tenured and (51%) non tenure personnel. Overall, the results of the study indicate that the EWs in the study approximately use ICT quite a bit.

Our study identified the typing letters/ report, access to the latest agricultural news and information, prepare and protect file on process basic agricultural and rural information as a main area for ICT application by EWs in extension activities.

Based on our findings, we also recommend the present advisory services to different farmer groups, present group training to farmers through video conference and Improve linkages between EWs and researcher as well as for typing letters/report, access to the latest agricultural news and information and other priorities. This ICT applications requires a substantial investment in ICTs. Information and communication technologies allow extension workers to enter information concerning events and upload it through the Internet to a central database. In addition, we recommend the providing easy access to the ICT units.

To achieve the goal of ICT application by EWs, there are some main encouraging (namely providing easy access to the ICT Unit, develop strategic plan for developing ICT units, provide the requirements and tools, and provide government's free competitive environment for service delivery) and discouraging (namely poor infrastructure, extension weakness in providing information, extension weakness in providing requirements, social and cultural barriers, lack of interest in the use of ICT, high cost of installation of ICT) factors. The agricultural extension in Isfahan Province should overcome and transform the challenges to the opportunities and straight encouraging factors, too. In an environment, agricultural extension organization familiar with the challenges could transform them to the opportunities, and lack of familiarity with the nature of discouraging factors would transform them to threats. Future research is needed to deepen the analysis of these factors and in order to pursue the major research question - on what is of particular importance in settings with low experience and exposure to ICT. Despite many limitations, there are prospects to incorporate ICTs into the extension system in Iran. The requisite investment should be recommended in agricultural development plans and a certain percentage of the total budget of state-sponsored projects should be allotted for ICTs. For extension to realize the full potential of ICTs in agricultural development, the current discouraging factors must be addressed. Extension needs appropriate infrastructure, information, requirement, increase interest and financial support from the government.

In order to incorporate ICT into the extension system, EWs should be provided government's free competitive environment for ICT application. Every extension center should be equipped with an established minimum standard for ICT capability in order to conduct ICT based courses and improve communication skills of extension staff. Lastly, agricultural extension planners and policy makers should pay attention to the 10 encouraging and discouraging factors when integrating ICTs with extension processes.

The study recommended among others that, there should be a periodic review of the use of current ICTs in among EWs to facilitate effectiveness in the use of ICTs for extension service and adequate funding of extension service should also be ensured to enhance the maintenance of ICTs made available to extension personnel.

Finally, in this study Pearson Correlation Coefficients were calculated to describe the relationships between encouraging, discouraging ICT application factors and application of ICT by EWs. None of the four factors of encouraging items was found to be related to ICT application in spite of factor 2. In related to discouraging factors, some of them such as factor 3, and factor 5, were significant at the .05 level.

REFERENCES

1- Akpabio, I. A. (2007). Agricultural extension officials and utilization of information and communication technologies in the Niger Delta, Nigeria [Abstract]. Article 3. Retrieved January 25, 2008, *from http://jesonline.org/current.htm*

2- Annor, F., Kwarteng, J., Agunga, R., & Zinnah, M. M. (2006). *Challenges and Prospects of Infusing Information Communication Technologies (ICT) in Extension for Agricultural and Rural Development in Ghana.* Paper presented at the Proceedings of the 22th Annual Conference of the Association for International Agricultural and Extension Education (AIAEE), Clearwater Beach, Florida.

3- British Educational Communications and Technology Agency (BECTA). (2004). *A Review of the Research Literature on Barriers to the Uptake of ICT by Teachers,* Retrieved August 13, 2008. From http://www.becta.org.uk.

4- Bheenick, K.J., & Brizmohum, R. (1998). *The scope of information technology applications in agricultural extension in Mauritius.* Paper presented at the Third Annual Meeting of Agricultural Scientists, Leduil, Mauritius. Retrieved February 12, 2008, from http://www.uom.ac.mu/faculties/fao/AIS/AMAs98/Index.htm

5- Chieochan, O., Lindley, D., & Dunn, T. (2000), May. Factors affecting the use of information technology in Thai agricultural cooperatives: A work in progress, 2, Article 1.Retrieved February 12, 2008, from http://www.ejisdc.org/ ojs2/ index. php/ ejisdc/issue/views/62

6- Doluschitz, R., & Pape, J. (2002): E-Business in der Agrarwirtschaft zwischen Euphoria und Resignation, in: Zeitschrift fuer Agrarinformatik, 10: 13-20.

7- Faghiharam, B. & Zargham, M. & Woan Yi, L. & Norani, M. S. (2012). Assessment of Socio- Psycho Factors on Use of ICT (Case Study: Educational Faculty Members). *International Conference on Education and Educational Psychology (ICEEPSY 2012)*. Procedia - Social and Behavioral Sciences 69 (2012) 763 – 769. [on line], Available on: www.sciencedirect.com

8- Falaki, M. (2005). *A study to recognize main areas, driving forces and obstacles of applying information technology (IT) in Agricultural extension system, MSc thesis*. University of Tehran, Iran.

9- Fallah Haghighi, N. (2006). *Analysis of factors affecting the extent of IT application in agricultural extension in Gilan Province of Iran, MSc thesis.* University of Tehran, Iran.

10- Gelb, E.M., Parker, C., Wagner, P., Rosskopf, K. (2001). Why is the IT adoption rate by farmers still so slow? (Summary of third EFITA Montpellier conference - prepublication).

11- Gelb, K. Offer, A. (2006). Constraints Affecting ICT Utilization by Agricultural Extension Officers in the Niger Delta, Nigeria. *The Journal of Agricultural Education and Extension.* 13(4), 263-272.

12- Gelb, E., Parker, C., (2006). Is ICT Adoption for Agriculture Still an Important Issue? Proceedings, 5the AFITA Conference, Bangloru, India Gell, E., Schiefer, G., Parker, C., & Rosskopf, K. (1999). Why Is the Adoption Rate by Farmers so Slow? *Second Conference of the European Federation for Information Technology in Agriculture, Food and the Environment (EFITA)*.Bonn, September. [on line], Available on: *http://www.efita.dk/papers/ep4/Efita-Paper-4.asp*

13- Hedjazi, Y., Rezaee, R. & Zamani, N. (2006). Factors affecting the use of ICTs by Iranian Agriculture Extension Specialists. *Journal of Extension Systems*, 22 (1), 1- 15.

14- Leeuwis, C. (2004). Communication for Rural Innovation: Rethinking Agricultural Extension, 3rd end. Oxford: Blackwell.

15- Maru, A. (2003). Information and Communication Technology Use in Agricultural Extension in India. [on line], Available on: *http://www.cta.int/observatory2003/ppt_presentation/country_case_india .pdf*

16- Madadi, Y. & Iravani, H. & Nazari Nooghabi, S. (2011). Factors effective on Familiarity and Usage of Information and Communication Technology (ICT) University College of Agriculture and Natural Resources, University of Tehran, Iran. Procedia Social and Behavioral Sciences 15 (2011) 3625–3632. [on line], Available on: www.sciencedirect.com

17- Maru, A., & Ehrle. (2003)."Building a Framework for ICT Use in Agricultural Research and Development. Is the North Different from the South?" *EFFITA 2003 conference*, July 5-9. Debrecen, Hungary.

18- Mc Carthy, P. (2003). *A Supply Chain Framework to Secure the Future of the Irish Beef Farming Sector,* Masters in Information Technology Thesis, National University of Ireland, Galway.

19- Meera, N., Jhamtani, A., & Rao, D. (2004). Information and Communication Technology in Agricultural Development: A Comparative Analysis of Three Projects from India. *Agricultural Research & Extension Network, ODI Network Paper NO.135*
20- Morris, E. (2002). The Uptake of Information Technologies in Irish Agriculture, *ISITA conference*, Portlaoise, Ireland. [on line], Available on: *http://www.iol.ie/~harkin/2002paplizzy.htm*
21- NASS (2001) Farm Computer Usage and Ownership. National Agricultural Statistics Service Agricultural Statistics Board, USDA. Source: http://usda.mannlib.cornell.edu/usda/reports/nassr/other/computer/
22- Patil, V.C., Gelb, E., & Maru, A., Yadaraju, N.T., Moni, M. & Harekrishna, M. (2008). Adoption of Information and Communication Technology (ICT) for Agriculture: An Indian case stud. *IAALD AFITA WCCA 2008. World Conference on Agricultural Information and IT.*
23- Rosskopf, K. & Wagner, P (2003). Requirements for Agricultural Software and Reasons for Adoption Constraints – Results of racial Studies- EFFITA 2003 conference, July 5-9. Debrecen, Hungary.
24- Sadik, A. (2005). Factors influencing teachers' attitudes towards personal use and schools use of computers: News evidence from a developing nation. Evaluation.
25- Saheed, A. Olaniyi, I. Akin, O., & Gbemiga, A. J. dew sale, Jacob. (2013). Assessment of Use of Selected Information Communication Technologies (ICTs) for Extension Service Delivery: Implication for Agricultural Development in Nigeria. *International Journal of Agricultural Management & Development (IJAMAD)*. 3(2),131-139. Available on: www.ijamad.com
26- Shabanali Fami, H., Falaki, M., Iravani, H., & Mohammadi, H. M. (2009). Principal Areas of Applying Information and Communication Technologies in the Agricultural Extension System of Iran. *Journal of Agricultural & Food Information. 10: 247- 258.*
27- Teo, T. (2006). Attitudes towards computers: A study of post-secondary students' in Singapore. Interactive Learning Environments. 14(1), 17-24.

3

The Preparation of Updated Vegetation Maps by Processing Satellite Images: A Way in Sustainable Management of Agriculture

Ali Mohammadi Torkashvand * and Shahryar Sobhe Zahedi

Abstract

An important factor in sustainable agriculture and economic management is to calculate areas under different crops that the inputs of agriculture connect to this topic. Planning of agricultural mechanization, fertilizer and pesticide requirements, pests and diseases control, estimates of agricultural production, income and tax and financial planning, all linked to the cultivated areas and estimation of agricultural products. One of the problems in the agricultural section of Iran is the lack of accurate statistics of cultivated crops areas that this is much higher for horticultural products. Over time, it varies the area of land under cultivated crop, and orchards and bare lands; consequently the estimation of yield is not done as well due to these changes caused some problems in planning and management. Land Surveying is time-consuming and expensive, while mapping farms and orchards lands through classified satellite images is a high speed and low cost way. Nowadays, the satellite image processing techniques have developed for the estimation of crops, pest control, agricultural macro planning and preparing updated maps. A principal problem is the interference of plants spectral reflections that different methods have been proposed by researchers to differentiate vegetation on satellite images. At this paper, remote sensing imagery in mapping vegetation or various plants are investigated.

Keywords:
Agriculture management, Satellite images, Sustainable agriculture, Vegetation

[1] *Department of Horticulture, Rasht Branch, Islamic Azad University, Rasht, Iran.*
[2] *Research Center of Agriculture and Natural Resources, Guilan, Iran.*
* *Corresponding author's email: m.torkashvand54@yahoo.com*

INTRODUCTION

In a general concept, planning is the allocation of resources in order to change of present situation into a desirable condition that its ultimate goal is the development and construction. The first and the most important step in a planning system are designing operations in achievement to a goal or goals. Planning should be done for sustainable development of a country or region. In the process of sustainable development, five subset resources, environment, population, economy and society are involved. Because of the limited resources, they have remarkable importance. Environment is also an important part of system that should be conserved in the process of sustainable development (Alavipanah, 2004). One of the goals of the first, second, third and fourth programs in development of Iran was the increase in agricultural production. All policies and arrangements in various themes such as soil and water, mechanization, research, education and promotion etc. are programmed in developing agricultural products.

Various factors influence on production process that among them, the climatic factors have the special importance particularly in arid and semi-arid parts of Iran, since drought were caused some problems in achieving the goals of development plan in Iran's agriculture. The population growth has increased the land exploitation, extensively It is not used the cultivated land in the agricultural part in an appropriate manner. Degradation of soil and water resources, degradation of pastures, vegetation and forest is due to improper exploitation of these resources as resulting immediate and short-term benefits. Assessing and monitoring the state of the earth surface is a key requirement for global change research (Committee on Global Change Research, National Research Council, 1999; Jung *et al.*, 2006; Lambin *et al.*, 2001). Land use planning is a plan of agricultural development with special emphasis on the accurate recognition of soil and water resources, the current use of lands, recognition priorities and the allocation of land to sustainable use.

Nowadays the application of advanced technologies in the land evaluation and natural resources management and decision making has increased and improved. Therefore, special attention to science and new technologies is very important especially for developing countries. Remote sensing and spectral reflections recorded by satellite sensors are new technologies. Maps of vegetation in earth's surface are the biophysical phenomenon that the date and accuracy of these maps can be a significant way in environmental management and planning (Naghibi *et al.*, 2010). Recognition of vegetation covers characteristics and relationships between plant species and environmental factors have been considered by ecologists. This is due to the importance of habitat of vegetation covers, energy and other important characteristics of plants on Earth (Ahmadpour *et al.*, 2011). Vegetation cover is the ratio of surface area covered by plant to total area (Zhang *et al.*, 2003). Fortunately, many researchers have attempted to mapping and zoning vegetation and land use (Naghibi *et al.*, 2010).

The information contained in digital imagery, acquired by Remote Sensing technology, can be used for mapping, monitoring and assessing the properties of the environmental and terrestrial features elements. Vegetation mapping also presents valuable information for understanding the natural and man-made environments through quantifying vegetation cover from local to global scales at a given time point or over a continuous period. It is critical to obtain current states of vegetation cover in order to initiate vegetation protection and restoration programs (Egbert *et al.*, 2002; He *et al.*, 2005).

Land use and land cover is found as significant information in an area that may assist managers and decision-makers to take drastic measures. One of the most important land uses is agriculture lands as well as orchards, which may play an important role in providing man's food. Land-use changes in farming areas that derive from declining of agricultural practices or even from frequently unreasonable over exploitation of water resources have led many semi arid regions of the world to boundary conditions demanding urgent strategies of water management towards sustainable development (Brandt and Thornes, 1996; Hugget, 1993). As increase in human demands, the sustainability of land use is questionable. Better land management involves identifying land-use changes, understanding current land-use patterns or features, and assessing economic and ecological benefits and costs that arise from land-use practices, as well as finding the best alternatives for each area (Wu *et al.*, 2001). Determination of plant cover percent by traditional methods don't present full view of plant cover and also it consumes time and cost. Thus, human errors in determining vegetation by traditional methods can be very large. Remotely sensed data frequently are used to map land surface cover for use in a variety of resource assessment, land management, and modeling applications. Mapping from coarse spatial resolution images and with multispectral instruments necessarily has focused on land cover and broad vegetation types (Loveland, 2000) rather than discrimination of vegetation at a species level.

Crop yield forecast in an area usually requires crop area estimation, which is mainly concerned by some relating organizations. Satellite data along with remote sensing technique may be employed as a useful and effective tool to estimate crop area. Recent developments of image processing technique as well as availability of high resolution satellite imageries avail to use this technique as a quick and low cost method in compare to conventional methods for crop area estimation. This paper presents different methods or models in estimating vegetation and various crops.

DISCUSSION

One of the common methods in calculating the vegetation percentage by satellite images is the use of vegetation indices. So far several vegetation indices have been proposed which depending on the region followed different results. Many indices have been investigated in various studies that the most important indices presented by researchers (Baret and Guyot, 1991; Chen, 1996; Clevers, 1989; Crripen, 1990; Huete, 1988; Huete, 1997; Jordan, 1969; Kaufman and Tanre, 1992; Major *et al.*, 1990; Pinty and Verstraete, 1992; Richardson and Wieg, 1977; Rondeaux *et al.*, 1996; Rouse *et al.*, 1974; Tucker, 1979; Qi *et al.*, 1994) as following:

NDVI (Normalized difference vegetation index) $= (b_4-b_3)/(b_4+b_3)$

b_1, b_3, b_4, b_5 and b_7 are five bands of octet bands related to ETM+ sensor

DVI (Difference vegetation index) $= b_4-b_3$

RVI (Ratio vegetation index) $= b_4/b_3$

IPVI (Infrared percentage vegetation index) $= b_4/(b_4+b_3)$

MIR (Leaf water content, Mid-IR) $= b_5/b_7$

MSI (Moisture stress index) $= MIR/b_5$

RAI (Reflectance absorption index) $= b_4/(b_3 + MIR)$

IR_2 (Infrared Index) $= (b_4-b_7)/(b_4+b_7)$

NRVI (Normalized Ratio vegetation index) $= (RVI-1)/(RVI+1)$

TVI (Transformed vegetation index) $= NDVI+0.5$

ARVI (Atmospherically resistant vegetation index) $= (b_4-R_{b1})/(b_4+R_{b1})$

That $R_{b1} = -\lambda(b_1-b_3)$

GEMI (Global environment monitoring index) $= [\mu(1-0.25\mu)-(b_3-0.125)]/(1-b_3)$

That $\mu = [(b_4^2-b_3^2)+1.5b_4+0.5b_3]/(b_4+b_3+0.5)$

MSAVI (Modified soil adjusted vegetation index 1) $= [b_4-b_3(1+L)]/(b_4+b_3+L)$

That L=1-2

$MSAVI_2$ (Modified soil adjusted vegetation index 2) $= 0.5[2b_4+1-[(2b_4+1)2-8(b_4-b_3)]0.5]$

SAVI (Soil adjusted vegetation index) $= [b_4-b_3(1+L)]/(b_4+b_3+L)$

SARVI (Soil & atmospherically resistant vegetation index) $= [b_4- R_{b1}(1+L)]/(b_4+ R_{b1}+L)$

WDVI (Weighted difference vegetation index) $= b_4-\alpha b_3$

TSAVI $= [\alpha(b_4-\alpha b_3-\beta)]/[\alpha b_4 + b_3 + \alpha\beta + 0.08(1+\alpha^2)]$

$SAVI_2$ (Soil adjusted vegetation index 2) $= b_4/[b_3+(\alpha/\beta)]$

OSAVI (optimized soil adjusted vegetation index) $= 1.16[(b_4-b_3)/(b_4+b_3+0.16)]$

α and β are the slope of soil line and soil line intercept, respectively.

MSR (Modified simple ratio) $= (RVI-1)/(RVI^{0.5}+1)$

Ebrahimi (2010) on research in the arid regions of central Iran concluded that among 17 common indicators, MSAVI1 index with highest precision can be used to estimate the percentage of vegetation. Another research by Baugh and Groeneveld (2009) in an arid region of Colorado, USA, using TM images showed that plant cover percent can be estimated by NDVI indices with an accuracy R^2=0.77. Darvishzadeh *et al.* (2008) introduced the indices related to soil reflections in vegetation mapping as better indexes. The vegetation percent of Sistan region, southeast Iran, estimated by Shafiei and Hosseini (2012), they found that WDVI index had the highest precision (R^2=0.89) in estimating vegetation between 83 indices. Darvishzadeh *et al.* (2012) evaluated 17 vegetation indexes to compute vegetation percentage. They concluded that indexes having soil line indices (PVI, TSAVI, $SAVI_2$, WDVI) had the more accurate in estimation of vegetation percentage in arid region (R^2>0.63).

A study by Mohammadi *et al.* (2012) showed that GNDVI and DVI indexes have the highest correlation with total cover, and GNDVI and GI with millet cover. The studies of Baugh and Groeneveld (2009) and Lawrence and Ripple (1998) introduced NDVI index in estimating vegetation with acceptable accuracy.

Another way in providing vegetation map is the classification of spectral reflections (digital numbers) on satellite images. Of course, the supervised classification of images by training points of surface covers presents more accurate results than unsupervised classification. Unsupervised classification methods are purely relying on spectrally pixel-based statistics and incorporate no prior knowledge of the characteristics of the themes being studied. The benefit of applying unsupervised classification methods is to automatically convert raw image data into useful information so long as higher classification accuracy is achieved (Tso and Olsen, 2005). Algorithms of unsupervised classification were investigated and compared with regard to their abilities to reproduce ground data in a complex area by Duda and Canty (2002). Despite its easy application, one disadvantage of the unsupervised classification is that the classification process has to be repeated again if new data (samples) are added. By contrast, a supervised classification method is learning an established classification from a training dataset, which contains the predictor variables measured in each sampling unit and assigns prior classes to the sampling units (Lenka and Milan, 2005). The supervised classification is to assign new sampling units to the priori classes. Thus, the addition of new data has no impact on the established standards of classification once the classifier has been set up (Sohn and Rebello, 2002; Xu *et al.*, 2005).

Extensive field knowledge and auxiliary data may help improve classification accuracy. Studies have shown that classification accuracy can be greatly improved after applying expert knowledge (empirical rules) and ancillary data to extract thematic features (e.g. vegetation groups) (Gad and Kusky, 2006; Shrestha and Zinck, 2001). In a regional scale vegetation classification conducted in the Amanos Mountains region of southern central Turkey using Landsat images, Domacx and Suzen (2006) incorporated vegetation-related environmental variables and considerably improved classification accuracy when compared with the traditional MLC method.

Muschen *et al.* (2001) tried to separate agricultural area from non-agricultural area using controlled classification of integrated images of TM5 with IRS IC PAN land sat and ERS_2 radar by maximum likelihood method. In addition to this separation, they tried to separated wheat, maize farm and rangelands. Rembold *et al.* (2000) investigated land cover changes in a 22 years period at Lakes region in south of Ethiopia by aerial photographs (1972) and classifying TM land sat images (1994). The analyses indicate that cultivated surface had been increased and more erosion had been occurred in new cultivated lands.

Mohammdi Torkashvand and Eslami (2012) focused on identification and mapping olive in the part of Roodbar region, Guilan, Iran, based on supervised classification of IRS images. The indicated that there was an interference of spectral reflections between olive orchards, paddies and other orchard caused to low accuracy in providing olive orchards map. Mohammdi Torkashvand (2011) investigated preparation of paddy lands and computed accuracy 73% by supervised classification of IRS Satellite images.

Using AIF (adaptive image fusion) index,

Fletcher (2005) used high resolution Quick Bird satellite images to recognize citrus with black mold (Capnodiumcitri) in Texas region of America and identified it as a suitable method. Das *et al.* (2009) tried to prepare map for regions with reducing citrus production capacity in Meghalaya region of India using IRS satellite images. The map of regions where citrus production capacity had been reduced was prepared using soil erosion information, vegetation condition and humidity tension. Due to some changes which are created above time in paddy surface preparing updated map of paddy is one of the most important requirements in the management and region agricultural planning. With regard to this that land surveying required high cost and time and also preparing the map through aerial photographs is required to prepare aerial photograph which still along with high cost, use of satellite data along with remote sensing technique may be employed as a useful and effective tool to estimate crop area.

Olive area estimation has been carried out in a few olive-growing countries as Turkey, Spain and Portugal. Unal *et al.* (2004) used Land sat 7 and IRS data to survey olive, pistachio and vineyard in Gazin-Tab area Turkey. He used image supervised classification technique to estimate olive plantation area. Teresa Barata and Granado (2000) has compared land sat images with aerial photo to map olive gardens in Portugal. Ramos *et al.* (2007) quantified the eventual land movement and the subsequent displacement of olive trees produced by continuous tillage erosion. They analyzed these movements on a property of olive orchards located on variable sloping land. Berni *et al.* (2009) applied models based on canopy temperature estimated from high resolution airborne imagery to calculate tree canopy conductance (Gc) and the crop water stress index (CWSI) of heterogeneous olive orchards. Darvishsefat and Zareh (1998) used remotely sensed data in providing plant cover map in Qaen, Iran. Senay and Elliott (2002) evaluated ability AVHRR sensor data to identify combinations of grassland vegetation and found this sensor is able to detect different species of rangeland with different types of trees and bushes.

There are other methods in providing vegetation map that can refer to neural network, fuzzy logic and mixed pixels. Artificial Neural Network (ANN) and fuzzy logic approaches are also seen in literature for vegetation classifications. ANN is appropriate for the analysis of nearly any kind of data irrespective of their statistical properties. ANN is very useful in extracting vegetation-type information in complex vegetation mapping problems (Filippi and Jensen, 2006), though it is at the expense of the interpretability of the results since ANN deploys a black-box approach that hides the underlying prediction process (Cerna and Chytry, 2005). Berberoglu *et al.* (2000) combined ANN and texture analysis on a per-field basis to classify land cover and found the accuracy could be 15% greater than the accuracy achieved using a standard per-pixel ML classification. Jung *et al.* (2006) in preparing clouds cover and Carpenter (1999) in vegetation mapping used neural network method. Matkan *et al.* (2011) evaluated efficiency using neural network and remote sensing in estimation of vegetation and calculated accuracy 0.74.

A fuzzy classification approach is usually useful in mixed-class areas and was investigated for the classification of suburban land cover from remote sensing imagery (Zhang and Foody, 1998). Fuzzy classification is a kind of probability-based classification rather than crisp classification. Unlike implementing per-pixel-based classifier to produce crisp or hard classification, Xu *et al.* (2005) employed a Decision Tree (DT) derived from the regression approach to determine class proportions within a pixel so as to produce a soft classification. Theoretically, probability-based or soft classification is more reasonable for composite units since those units cannot be simply classified to one type but to a probability for that type. While soft classification techniques are inherently appealing for mapping vegetation transition, there is an unresolved issue of how best to present the output. Rather than imposing subjective boundaries on the end-member communities, transition zones of intermediate vegetation classes between the end member communities were adopted to better represent the softened classification result (Hill *et al.*, 2007).

CONCLUSION

Although there are various methods for image preprocessing, but they cannot prepare vegetation map with a full accuracy and precision. The interferences spectral reflections are caused to reduce the accuracy of map; particularly it increases in preparing map of a special plant. Thus, ancillary data, including field samples, topographical features, environmental characteristics and other digital (geographic information system) data layers, have been proved very helpful to get a more satisfactory result or increase classification accuracy.

ACKNOWLEDGMENT

This study was supported by the grant of Islamic Azad University, Rasht Branch, Rasht, Iran, and it is related to a research design entitled: the possibility preparing tea orchards map by satellite images and GIS.

REFERENCES

1-Ahmadpour, A., Solaimani, K., Shokri, M., & Ghorbani, J. (2011). Comparison of three common methods in supervised classification of satellite data for vegetation studies. *Journal of Applied RS and GIS Techniques in Natural Resource Science*, 2(2), 69-81.

2- Alavipanah, S.K. (2004). *Application of remote sensing in the earth sciences (soil)*. University of Tehran Inc., Tehran, Iran.

3- Baret, F., & Guyot, G. (1991). Potentials and limits of vegetation indices for LAI & APAR assesment. *Remote Sensing of Environment*, 35, 161-173.

4- Berni, J.A.J., Zarco-Tejada, P.J., Sepulcre-Cantó, G. Fereres, E., & Villalobos, F. (2009). Mapping canopy conductance and CWSI in olive orchards using high resolution thermal remote sensing imagery. *Remote Sensing of Environment*, 113, 2380-2388.

5- Baugh, W.M. & Groeneveld, D.P (2009). Broadband vegetation index performance evaluated for a low-cover environment. *International Journal of Remote Sensing,* 27(3), 4715-4730.

6- Berberoglu, S., Lloyd, C.D., Atkinson, P.M., & Curran, P.J. (2000). The integration of spectral and textural information using neural networks for land cover mapping in the Mediterranean. *Comput. Geoscience,* 26, 385–396.

7- Brandt, C.J., & Thornes, J.B. (1996). *Mediterranean desertification and land Use*. John Wiley Inc., London, UK.

8- Cˇerna′, L., & Chytry′, M. (2005). Supervised classification of plant communities with artificial neural networks. *Journal of Vegetation Science*, 16, 407-14.

9- Chen, J. M. (1996). Evaluation of vegetation indices and a modified simple ratio for boreal applications. *Remote Sensing of Environment*, 22, 229-242.

10- Clevers, J.G.P. (1989). The application of a weighted infrared-red vegetation index for estimating leaf area index by correcting soil moisture. *Remote Sensing of Environment,* 29, 25–37.

11- Crippen, R.E. (1990). Calculating the vegetation index faster. *Remote Sensing of Environment*, 34, 71–73.

12- Committee on Global Change Research, National Research Council (1999). Global environmental change: *Research pathways for the next decade.* National Academy Press, Washington, DC.

13- Darvishsefat, A.A., & Zareh A. (1998). Investigating potential of satellite data in the preparation of vegetation in arid and semiarid regions (case study, Qaen region). *Journal of Natural Resources,* 51(2), 39-45.

14- Darvishzadeh, R., Matkan, A.A., Hosseiniasl, A., & Ebrahimi Khusefi, M. (2012). Estimation of vegetation fraction in the Central arid region of Iran using satellite images (Case study: Sheitoor basin, Bafgh). *Arid Biome Scientific and Research Journal,* 2(1), 25-38.

15- Darvishzadeh, R., Skidmore, A., Atzberger, C., & Wieren, S. (2008). Estimation of vegetation LAI from hyperspectral reflectance data: Effects of soil type and plant architecture. I*nternational Journal of Applied Earth Observation and Geoinformation,* 10, 358-373.

16- Das, P.T., Tajo, L. & Goswami, J. (2009). Assessment of citrus crop condition in umling block of Ri-bhoi district using RS and GIS technique. *Journal of the Indian Society of Remote Sensing.* 37(2), 317-324

17- Domacx A., & Su¨zen, M.L. (2006). Integration of environmental variables with satellite images in regional scale vegetation classification. I*nternational Journal of Remote Sensing,* 27, 1329-50.

18- Duda, T., & Canty, M. (2002). Unsupervised classification of satellite imagery: choosing a good algorithm. *International Journal of Remote Sensing,* 23, 2193-2212.

19- Ebrahimi, M. (2010). *Estimation of vegetation fraction using vegetation indices and artificial neural network in order to improve modified perpendicular index*. RS & GIS department, Shahid Beheshti University, Tehran, Iran.

20- Egbert, S.L., Park, S., & Price, K.P. (2002). Using conservation reserve program maps derived from satellite imagery to characterize landscape structure. *Computer and Electronic in Agriculture*, 37, 141-56.

21- Filippi, A.M., & Jensen, J.R. (2006). Fuzzy learning vector quantization for hyperspectral coastal vegetation classification. *Remote Sensing Environment*, 100:512–30.

22- Fletcher, R.S. (2005). Evaluating high spatial resolution imagery for detecting citrus orchards affected by sooty mold. *International Journal of Remote Sensing,* 26(3), 495-502.

23- Gad, S., & Kusky, T. (2006). Lithological mapping in the Eastern Desert of Egypt, the Barramiya area, using Landsat thematic mapper (TM). *Journal of African Earth Science,* 44, 196-202.

24- He, C., Zhang, Q., & Li, Y. (2005). Zoning grassland protection area using remote sensing and cellular automata modeling: a case study in Xilingol steppe grassland in northern China. *Journal of Arid Environment*, 63, 814-26.

25- Hill, R.A., Grancia, K., Smith, G.M., & Schardt, M. (2007). Representation of an alpine treeline ecotone in SPOT 5 HRG data. *Remote Sensing of Environment,* 110, 458-467.

26- Huete, H. (1988). A Soil-Adjusted Vegetation Index (SAVI). *Remote Sensing of Environment*, 25, 295–309.

27- Huete, H., Liu, Q., Batchily, K., & Van Leeuwen, W. (1997). A comparison of vegetation indices over a global set of TM images for EOS-MODIS. *Remote Sensing of Environment,* 59, 440-451.

28- Jordan, C. F. (1969). Derivation of leaf area index from quality of light on the forest floor. *Ecology,* 50, 663-666.

29- Hugget, R.J. (1993). *Modelling the human impact on nature.* Oxford University Inc. Oxford, UK.

30- Kaufman, Y.J., & Tanre, D. (1992). Atmospherically Resistant Vegetation Index (ARVI) for EOS-MODIS. *IEEE Transactions on Geoscience & Remote Sensing,* 30, 261-270.

31- Jung, M., Churkina, G., & Henkel, K. (2006). Exploiting synergies of global land cover products for carbon cycle modeling. *Remote Sensing Environment*, 101, 534-53.

32- Lambin, E.F., Turner, B.L., Helmut, J. (2001). The causes of land-use and land-cover change: moving beyond the myths. *Global Environment Change,* 11, 261-9.

33- Lawrence, R.L., & Ripple, W.J (1998). Comparisons among Vegetation Indices andBandwise Regression in a Highly Disturbed، Heterogeneous Landscape: Mount St. Helens Washington. *Remote Sensing of Environment*, 64, 91-102

34- Lenka, C., & Milan, C. (2005). Supervised classification of plant communities with artificial neural networks. *Journal of Vegetation Science*, 16, 407-414.

35- Loveland, T.R. (2000). *The patterns and characteristics of global land cover.* In: M.J. Hill and R.J. Aspinall, Editors, Spatial Information and Land Use Management, Gordon and Breach, Reading 13–24.

36- Major, D.J., Baret, F., & Guyot, G. (1990). A ratio vegetation index adjusted for soil brightness. *International Journal of Remote Sensing*, 11, 727-740.

37- Mohammadi, M.,Ebrahimi, A., & Haqzade, A. (2012). Capability of IRS satellite on vegetation cover estimation (Case study: Chaharmah-va-Bakhtiari). *Renewable Natural Resources Research,* 3(1), 41-53.

38- Matkan, A.A., Darvishzadeh, R., Hosseini, A.A. & Ebrahimi, M. (2011). Capability using satellite images and neural network in estimating vegetation percentage in arid region. *Journal of Environmental Erosion*, 1, 7-27.

39- Mohammadi Torkashvand, A. (2011). The Preparation of Paddy Map by Digital Numbers of IRS images and GIS. *Journal of American Science,* 7(1), 382-385.

40- Mohammadi Torkashvand, A., & Eslami, A. (2012). Proposing a methodology in preparation of olive orchards map as an important medicinal plant in Iran by remote sensing (RS) and geographical information system (GIS). *Journal of Medicinal Plants Research,* 6(5), 680-688

41- Muschen, B., Flugel, W.A., Hochschild, V., Steinnocher, K. & Quiel, F., (2001). Spectral and spatial classification methods in the ARSGISIP project. *Physics and Chemistry of the Earth,* 26 (7-8), 613-616.

42- Naghibi, S.J., Habibian,S.H., & Habibian, S.M.R.(2010). Determination of optimum vegetation indices for vegetation cover percentage modelling, using satellite spectral reflectance. *Journal of Ecophysiology,* 1 (3), 63-74.

43- Pinty, B., &Verstraete, M. (1992). GEMI: A nonlinear index to monitor global vegetation from satellites. *Vegetation*, 10, 15-20.

44- Qi, J., Chehbouni, A., Huete, A. R., & Kerr, Y. H., Sorooshian, S. (1994). A Modified Soil Adjusted Vegetation Index (MSAVI). *Remote Sensing of Environment,* 48, 119-126.

45- Ramos, M.I., Gil, A.J., Feito, F.R., & Garcia-Ferrer, A. (2007). Using GPS and GIS tools to monitor olive tree movements. *Computer and Electronics in Agriculture,* 57, 135-148.

46- Rembold, F., Carnicelli, S., Nori, M., & Gioranni, A.F. (2000). Use of aerial photographs, Landsat TM imagery and multidisciplinary field survey for land-cover change analysis in the lakes region (Ethiopia). *International Journal of Applied Earth Observation and Geoinformation*, 2(3-4), 181-189.

47- Richardson, A.J., & Wieg, C.L. (1977). Distinguishing vegetation from soil background information. *Photogrammetric Engineering & Remote Sensing*, 43, 1541-1552.

48- Rondeaux, G., Steven, M., & Baret, F. (1996). Optimisation of soil-adjusted vegetation indices. *Remote Sensing of Environment*, 55, 95–107.

49- Rouse, J.W., Haas, R.H., Schell, J.A., Deering, D.W., & Harlan, J.C. (1974). *Monitoring the vernal advancement of retrogradation of natural vegetation.* NASA/GSFC, Type III, final report, Greenbelt, MD.

50- Senay, G.B., & Elliott, R.L. (2002). Capability of AVHRR data in discriminating rangeland cover mixtures. *International Journal of Applied Earth Observation and Geoinformation*, 23(2), 299-312.

51- Shafei, H., & Hosseini, S.M. (2012). A study of vegetation in Sistan region through satellite data. *Journal of Plant Ecophysiology,* 3, 91-105.

52- Shrestha, D.P., & Zinck, J.A. (2001). Land use classification in mountainous areas: Integration of image processing, digital elevation data and field knowledge (application to Nepal). *International Journal of Applied Earth Observation and Geoinformation,* 3, 78–85.

53- Sohn, Y., & Rebello, N.S. (2002) Supervised and unsupervised spectral angle classifiers. *PhotogrammEng Remote Sens*, 68, 1271–80.

54- Teresa Barata, P., & Granado, I. (2000). Classification of forest cover types in ortho and satellite images using a methodology based on mathematical morphology. 11th Portuguese Conference on Pattern Recognition, Lisbon, Portugal.

55- Tso, B., & Olsen, R. (2005). Combining spectral and spatial information into hidden Markov models for unsupervised image classification. *International Journal of Remote Sensing,* 26, 2113–33.

56- Tucker, C.J. (1979). Red and photographic infrared linear combinations for monitoring vegetation. *Remote Sensing of Environment,* 8, 127–150.

57- Unal, E., Mermer, A., & Mete Dogan, H. (2004). Determining major orchard (pistachio, olive, and vineyard) areas in Gaziantep province using remote sensing techniques. The International Archives of the Photogrammetry, Remote sensing and Spatial Information Sciences. 34, part XXX.

58- Wu, J., Ransom, M.D., Kluitenberg, G.J., Nellis, M.D., & Seyler, H.L. (2001). Land-use management using a soil survey geographic database for Finney County, Kansas. *Soil Science Society American Journal,* 65, 169–177.

59- Xu, M., Watanachaturaporn, P., & Varshney, P.K. (2005) Decision treeregression for soft classification of remote sensing data. *Remote SensingEnvironment*, 97, 322-36.

60- Zhang, J., & Foody, G.M. (1998).A fuzzy classification of sub-urban land cover from remotely sensed imagery. *International Journal of Remote Sensing,* 19, 2721–38.

61- Zhang, Y.X., Li, X.B., & Chen, Y.H. (2003). Overview of field and multi-scale remote sensing measurement approaches to grassland vegetation coverage. *Advanced Earth science*, 18, 85-93.

4

Evaluation of Land Cover Changes Using Remote Sensing Technique (Case study: Hableh Rood Subwatershed of Shahrabad Basin)

Khadijeh Abolfathi [1], Marzieh Alikhah-Asl [2], Mohammad Rezvani [2] and Mohammad Namdar [3]*

Abstract

The growing population and increasing socio-economic necessities creates a pressure on land use/land cover. Nowadays, land use change detection using remote sensing data provides quantitative and timely information for management and evaluation of natural resources. This study investigates the land use changes in part of Hableh Rood Watershed of Iran using Landsat 7 and 8 (Sensor ETM+ and OLI) images between 2001 and 2013. Supervised classification was used for classification of Landsat images. Four land use classes were delineated including rangeland, irrigated farming and plantations land, and dry farming lands,urban. Visual interpretation, expert knowledge of the study area and ground truth information accumulated with field works to assess the accuracy of the classification results. Overall accuracy of 2001 and 2013 image classification was 81.48 (Kappa coefficient: 0.7340) and 87.04 (Kappa coefficient: 0.7841), respectively. The results showed considerable land cover changes for the given study area. Land cover change detection showed that in a period of 12 years, 277.57 hectares of dry farming lands and 340 hectares of dense range have been lost. But, 341 hectares for low dense range, 280 hectares for semi dense range and 1.4 hectares for urban areas, have been added in area.

Keywords:
Land cover, Land use change, Supervised classification, Remote sensing

[1] *M.Sc. Student in Assessment and Land Use Planning, Natural Resources and Environmental Engineering Department, Faculty of Agricultural Science, Payame Noor University (PNU), Tehran, Iran.*
[2] *Assistant Professor, Natural Resources and Environmental Engineering, Faculty of Agricultural Science, Payame Noor University (PNU), Tehran, Iran.*
[3] *Forest, Rangeland and Watershed Management Organization, Tehran, Iran.*
* *Corresponding author's email: abolfathi15@yahoo.com*

INTRODUCTION

Biophysical materials and anthropogenic features are often subject to rapid changes (Roy *et al.*, 1996; Tahir and Hussain, 2008; Podobnikar *et al.*, 2009). The land cover changes occur naturally in a progressive and gradual way; however, sometimes it may be rapid and sudden due to anthropogenic activities (Butenuth *et al.*, 2007). Thus, analyzing the land cover changes and understanding the subsequent trends of change contribute to present complex dynamics of land cover and are important for policy making, planning and implementation of natural resource management (Ioannis and Meliadis, 2011; Knorr *et al.*, 2011; Reddy and Gebreselassie, 2011). In recent years, land use/cover changes in Iran, have occurred increasingly and in some respects unsatisfactorily so that the rate of land degradation has been intensified. So, it is important to know land cover changes across time, while there is not a recent report about land use changes in this area. Nowadays, integration of Geographic Information System (GIS) and remote sensing has provided accurate information about land use changes (Alqurashi and Kumar, 2013; Imam, 2011). Considering the importance of remote sensing and GIS in evaluating the changes in land cover, this technique is used in the present study. There are various classification algorithms for the detection of land use changes by satellite imagery (Aplin and Atkinson, 2004; Lu and 2004; Singh and Khanduri, 2011); however, there is not guarantee to use the best algorithm in all conditions (Srivastava *et al.*, 2012; Yang and Lo, 2002). In this study, post classification change analysis was applied. This method provides "from" to "to" transition rules (White *et al.*, 2013; Yuan *et al.*, 2005). Post classification method is also known to be more appropriate method for change detection (Lillesand *et al.*, 2004). In this study, supervised classification has been selected, which are widely used for the classification of remote sensed images for mapping land use changes (Abd Elcavy *et al.*, 2011; Jackson *et al.*, 2004; Ramadan *et al.*, 2004; Shalaby and Tateishi, 2007; Zhang *et al.*, 2011). To survey the vegetation canopy, Normalized Difference Vegetation Index (NDVI) was used. Past studies have demonstrated the potential of using NDVI to study vegetation dynamics (Townshend and Justice, 1986; Verhoef *et al.*, 1996). The purpose of this study was to detect land cover changes from 2001 to 2013 using Landsat ETM+/OLI images in the Hableh Rood Subwatershed of Shaharabad Basin of Tehran located in Iran. Maximum Likelihood and Parallelepiped algorithms were used to provide the land use maps in the ERDAS IMAGINE software.

MATERIALS AND METHODS

Study area

The study area with 3128.22 hectares in Hableh Rood Subwatershed of Shaharabad basin of

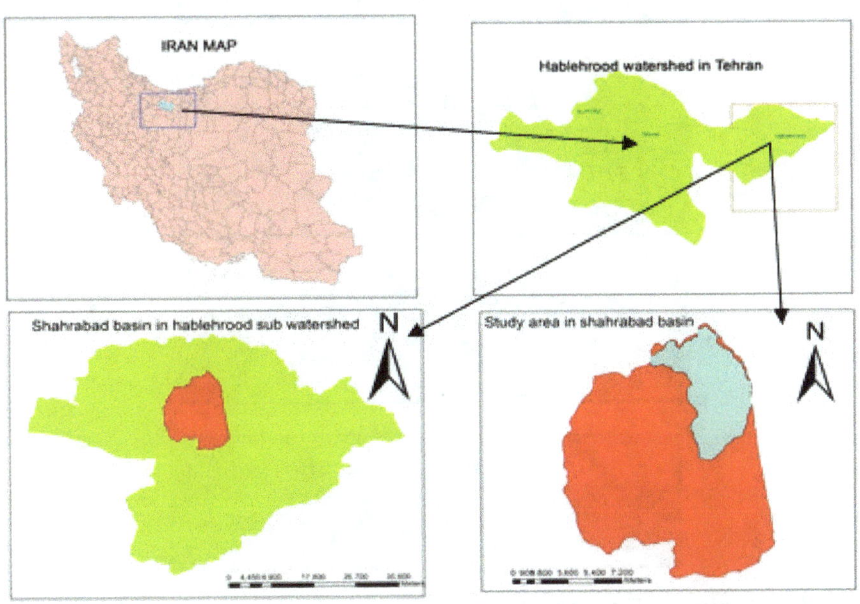

Figure 1: Location of the study area

Tehran located in Iran. The selected subwatershed lies between 52 ° 44' 24" to 52 °35' 10" E longitudes and 35 °43' 06" to 35 °52' 53 " N latitudes (Figure1). It has an average altitude of 2130 meters above sea level and the amount of rainfall is variable in Shahrabad basin, on average, about 354.4 mm. Given the fact that the studied case includes range land, agricultural and rural area, we tried to survey the impacts of droughts on land cover changes in the studied area. This research presented an analytical approach to describe land use changes of the environmentally sensitive Hableh Rood Subwatershed of Shahrabad basin.

Satellite data

The Landsat ETM+ and OLI images used in this study were acquired from www.usgs.gov. The dataset were on 2001 and 2013, with nearly zero percent cloud cover over the region. We tried to select all images almost during the growing seasons for reducing the effect of season on land use change results. The received Landsat images were already geo-referenced with 20 well-distributed ground points using first order (Affine) transformation and nearest neighbor resampling technique. So, the Root Mean Square Error (RMSE) of 0.8 and 0.7 Pixel was estimated for the image 2001 and 2013, respectively.

Methodology

In this study, Erdas Imagine 9.2 and ArcGIS 9.3 were the selected software. Initially, all images were rectified to the UTM coordinate system. Erdas Imagine was used to classify the land uses

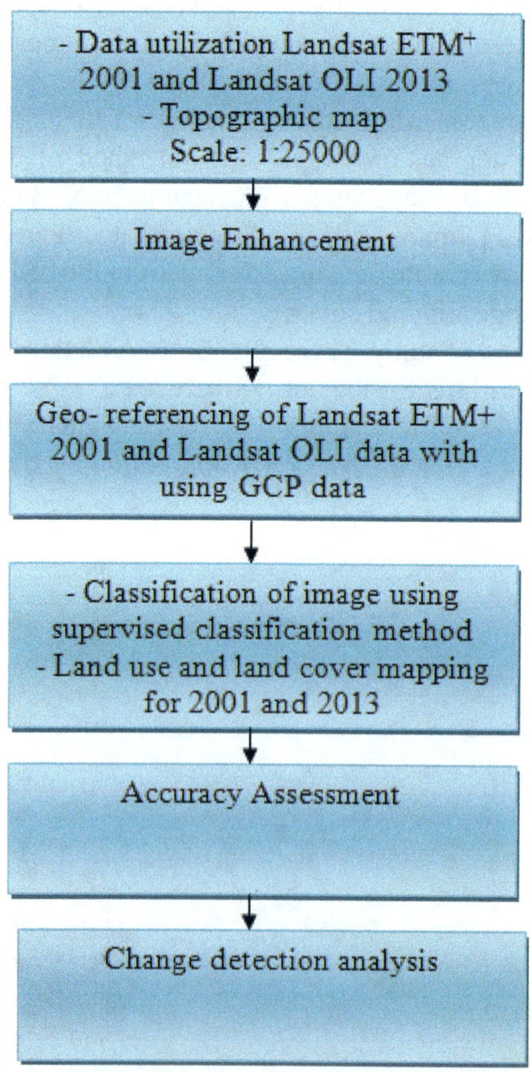

Figure 2: Flow chart showing the major steps research

from the satellite images, after which the raster images were converted to vector maps. The following steps were completed in the ArcGIS environment. An interface was developed in ArcGIS 9.3 environment, which provided the users an efficient and correct area calculation tool. Then, the land use changes were measured and the obtained values were tabulated for all land use changes in the Hableh rood subwatershed of Shahrabad basin. This process was applied for all images. The classification of land cover pattern was one of the prerequisites for analysis. Thus, the initial focus of this study was on classifying the main four types of land use patterns in the Landsat satellite images generated during the time period of 2001-2013. Figure 2 shows the flow chart of this study.

Training site for supervise classification

Selecting training site for supervised classification of remotely sensing images is related to the effective field visit within the local area and collecting exact and useful information (Ahmadizadeh *et al.*, 2014). In order to create training sampling points, stratified random algorithm was utilized. Sample points from various classes in the region with suitable numbers were surveyed by GPS and 1:25000 topographic maps. Among these 86 sampling points used for classification of land class, 43 points were used for classification process and the rests were kept for accuracy assessment.

Supervised classification

The Land Use/Land Cover map of the studied area was prepared through digital analysis of satellite data using a non-parametric classification (Parallelepiped) and supervised maximum likelihood classification technique. In the parallelepiped decision rule, the data file values of the candidate pixel are compared to the a priori set higher and lower limits of every signature in every band (Imagine, 2002). Supervised classification is a procedure for identifying areas on an image by identifying 'training' sites of known targets and then extrapolating those spectral signatures to other areas of unknown targets. Supervised classification relies on the prior knowledge of the location and identity of land cover types that are in the image. Training areas, usually small and discrete compared to the full image, are used to "train" the classification algorithm to recognize land cover classes based on their spectral signatures, as found in the image. The classifier then uses the training statistics to compute a probability value of whether it belongs to a particular land cover category class. This allows for within-class spectral variance. In this algorithm, the image analyst uses prior knowledge to weight the probability function. The MLC usually provides the highest classification accuracies (Lellesand and Kiefer, 1994). The NDVI is a standard "greenness" index that has been used for many applications, and has been shown to work well for detecting significant changes in levels of green vegetation (Coppin *et al.*, 2004). This index is one of the most famous, simplest and functional indexes in vegetation studies (Kassa, 1990). It has a simple calculation process and has the highest dynamic power in comparison with other indexes. This index is more sensitive in connection with vegetation changes and is less sensitive to the effects of the climate and soil, except in cases where the vegetation is low (Kogan,1993). Normalized Difference Vegetation Index (NDVI) was used to prepare a rangeland density map. NDVI is a method of measuring and mapping the density of green vegetation. For its measurement scientists use satellite sensors that observe the distinct wavelengths of visible and near-infrared sunlight which is absorbed and reflected by the plants, then the ratio of visible and near-infrared light reflected back up to the sensor is calculated. The ratio gives a number from minus one (−1) to plus one (+1). An NDVI value of zero means no green vegetation and close to +1 (0.8–0.9) indicates the highest possible density of green leaves. The 'normalized difference vegetation index' is calculated by the formula:

$$NDVI = (\rho NIR - \rho Red)/(\rho NIR + \rho Red)$$

where, ρNIR and ρRed are reflectance values of red and near infrared light received at the sensors. The NDVI was first formulated by Rouse *et al.* (1974) and applied to a wide range of practical remote sensing applications in a series of studies by Tucker *et al.* (1985) in the 1970s and 1980s.

RESULTS

Land cover classification

Land cover classification is one of the most studied topics in remote sensing, and land cover maps provide the basis for many applications like modeling of carbon budgets, management of forests, and estimation of crop yield (Wolter *et al.*, 1995; Lark and Stafford, 1997; Jung *et al.*, 2006; Rogan *et al.*, 2010). While it is relatively easy to generate a land cover map from remotely sensed data, it is not easy to make it accurate. Using multi-temporal images as inputs is reported to help improve classification accuracy (Wolter *et al.,* 1995; Guerschman *et al.*, 2003; Carrao *et al.*, 2008; Zhu and Woodcock, 2012), especially for vegetation because of the unique phonological characteristics of different vegetation types. To achieve higher classification accuracy, most current land cover products are generated using multitemporal images as their inputs (Friedl *et al.*, 2010). In this study, four land cover classes were, in total, established as irrigated and plantations land, dry farming, rangeland and urban. Normalized Difference Vegetation Index (NDVI) was used to prepare a rangeland density map categorized into three canopy density classes: dense range, semi dense range and lowly dense range. The classified images of land use/ land cover in Shahrabad basin for 2001 and 2013 are shown in Figures 3 and 4. The Land use/Land cover classification results are summarized in Table 3 for 2001 and 2013.

Accuracy assessment

Accuracy assessment is critical for a map generated from any remote sensing data. For accuracy assessment, error matrices were used with overall accuracy, user's and producer's accuracies, and the Kappa statistic were then derived from the error matrices. The Kappa statistic represents agreement obtained after removing the proportion of agreement that could be expected to occur by chance (Torahi and Chand Rai, 2011). Table (1) provides the separability results achieved for accuracies by supervised classification of the 2001 and 2013 images. According to Table 2, irrigated and plantations land have minimum producer's, 25%, 73.68% for 2001 and 2013 image respectively, and rangeland with another class has maximum producer's, 100% for 2001 and 2013 image.

Change detection

The changes were detected through the post classification by using the gained maps by the aid of supervised classification method. For this propose, the tabulate area order was used in GIS software as confusion matrix. This provides "from" to "to" transition among defined land use classes. Information on existing land use / land cover types and how they change over time is a prerequisite for sustainable resource development planning. In the study area various categories of land use / Land cover were delineated for 2001 and 2013 (Figures 3 and 4). The study area is classified under four land use /

Table 1: Overall accuracy and Kappa coefficient statistics achieved for the supervised classification method

Accuracy	2001	2013
Overall classification	81.48	87.04
Kappa statics	0.73	0.78

Table 2: Summary of accuracies by supervised classification

		Producer's Accuracy%	User's Accuracy%
Land cover 2001	Irrigated and plantations	25	100
	Dry farming	80	84.21
	Rangeland	100	76
Land cover 2013	Irrigated and plantations	73.68	87.50
	Dry farming	92.86	86.67
	Rangeland	100	100

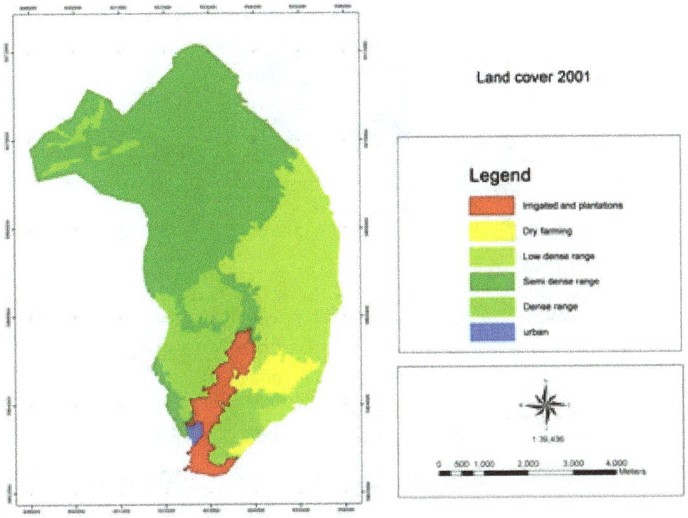

Figure 3: The classification images of land use/ land cover in Shahrabad basin, 2001

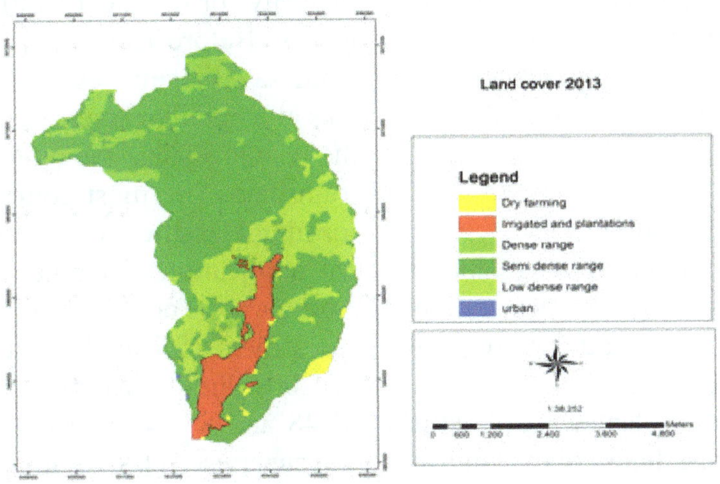

Figure 4: The classification images of land use/ land cover in Shahrabad basin' 2013

Land cover categories. Table 3 shows that 81.05% of the watershed in 2001 and 90.04% in 2013 are covered with rangeland. According to Table 3, the greatest increase was in lowly dense range land about 10.91%, as this class was increased from 203 hectares in the first year (2001) to 544 hectares in final year (2013). Semi dense range class was increased from 1767 hectares in (2001 to 2047 hectares in 2013. Urban class to last 12 years (form 2001 to 2013) has increased about 0.03% (1.04 hectares). Dry farming land was decreased from 412.83 hectares in 2001 to 135.26 hectares in 2013. The largest decrease was observed in dense range land class, nearly 10.87%. The area under irrigated and plantations farming land

Table 3: Summary of Landsat classification area statistics for 2001 and 2013

Land cover class	2001		2013		Relative change	
	Ha	%	Ha	%	Ha	%
Irrigated and plantations	178.03	5.69	174.08	5.56	-3.95	-0.13
Dry farming	412.83	13.19	135.26	4.32	-277.57	-8.82
Dense range	566	18.09	226	7.22	-340	-10.87
Semi dense range	1767	56.48	2047	65.43	280	8.95
Lowly dense range	203	6.48	544	17.39	341	10.91
Urban	1.3	0.04	2.34	0.07	1.04	0.03

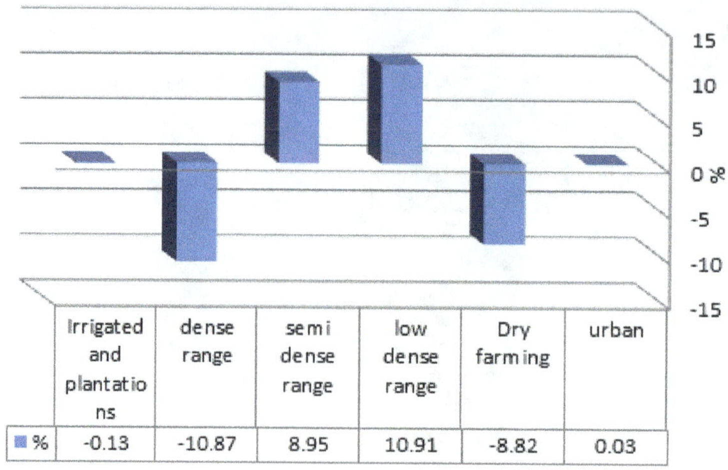

Figure 5: Land use change among 2001 and 2013

was 178.03 hectares in 2001. This class was reduced by about 0.13%; thus, there was little change in the irrigated and plantations farming land from 2001 to 2013. The percentages of land use change between 2001 and 2013 are shown in Figure 5.

DISCUSSION

Regarding the importance of rangelands in the provision of forage for livestock, reliable land use detection is the essential prerequisite and tool for planning these areas.

This study indicated the potential use of remote sensing data in studying land cover/ land use change. GIS techniques integrated in this study proved its undoubted capabilities for spatial analysis. Information from satellite remote sensing can play a useful role in understanding the nature of changes in land cover/use where they are occurring, and projecting possible or likely future changes. Ioannis and Meliadis (2011) used Landsat TM/ETM+ data to monitor the changes in Greece during the period 1999-2008. Nine different land cover/use categories were detected, namely coniferous, broadleaves and mixed forest, agriculture lands, rangelands, grasslands, water bodies, urban areas and others uses. The results showed remote sensing technology in combination with GIS can render reliable information on vegetation cover. The analysis of the spatial extent and temporal change of vegetation cover using remotely sensed data is of critical importance to environmental management.

In this study Landsat images were used satisfactorily for the identification of the four categories. Before using the classification results from satellite images for change detection, it was attempted to assess their validity by testing the results against the reference data or ground truth data. The most common method of performing classification accuracy assessment for any application of remote sensing is creating an error matrix (Foody, 2002). The result of accuracy assessment indicated that the overall accuracy of 2001 and 2013 image classification was 81.48 and 87.04, respectively. It is concluded that supervised classification method can provide acceptable classification accuracy values in this study. Tillmann *et al.* (2012) showed that the supervised classification of multi temporal satellite images is an effective tool to quantify current land use as well as to detect changes in an altering environment. Hence, land use within the test areas could be examined with high accuracy. Shalaby and Tateishi (2007) used supervised classification method to detect land cover changes in the north western coastal zone of Egypt and estimated the change rates. The results showed that coastal zone and cover type were reduced due to agricultural activities and tourist.

Since the drought phenomenon has occurred in the studied area in recent years, dry farming lands have dramatically decreased from 412 hectares to 135 hectares in 2013. As a result, these lands turned into barren agricultural lands. Along with population growth, the urban area has increased from 1.3 hectares to 2.34 hectares in 2013. Increasing population growth in the study area has increased the pressure on range-lands in this area. Ahmadizadeh *et al.* (2014) detected land use changes in Birjand plain and concluded that urban area has increased and dry farming lands has decreased that is similar to the results of this research. Abd El Cavy *et al.* (2011) used supervised classification to detect recent and historical land use/land cover conditions for the western Nile delta. Five land use/land cover categories were identified and mapped. The post-classification comparisons of the classified images indicated that the major change consisted of barren land changing into agricultural land.

A comparative study of the topographic map and satellite imagery of the study area over two time periods revealed that change has been noticed in Hableh Rood subwatershed of Shahrabad basin. Because of the mountainous area and limited irrigated farm land due to lack of slopes between 0-5%, the quantity of this area has not changed dramatically. According to lassifications based on satellite images taken in 2001 and 2013 and according to NDVI index, the quantity of proper rangelands has decreased while rangelands with mediate and weak canopy have increased. According to the taken images and field works, the conditions of canopies in the area clearly reflect overgrazing. Part of the rangeland changes during 2001 and 2013 is due to lower quantity of dry farming lands and their conversion to the low output range. According to the results, dense range class in the area has been decreased from 18.09% in 2001 to 7.22%% in 2013. Semi dense range class has been increased from 56.48% in 2001 to 65.43% in 2013. One reason for the increase is the conversion of barren agricultural lands to rangelands which mostly consist of level 3 rangelands. Lowly dense range has been increased from 6.48% in 2001 to 17.39% in 2013. Because rangelands had been turned into agricultural lands by the villagers in the past, unfortunately, our field works revealed that some of these areas have already been abandoned dry lands. So, the knowledge of the type and percentage of land uses in the watersheds can help the planners in a comprehensive management and sustainable development.

REFERENCE

1- Abd El-Kawy, O.R, Rod, J.K, Ismail, H.A, Suliman, A.S. (2011). Land use and land cover change detection in the western Nile delta of Egypt using remote sensing data. *Applied Geography,* 31(2), 483-494.

2- Ahmadizadeh, S., Yosefi, M., & Saghafi, M. (2014). Land use change detection using remote sensing and artificial neural network-application to birjand, Iran. *Computational Ecology and Software,* 4(4), 276-288.

3- Alqurashi, A.F., & Kumar, L. (2013). Investigating the use of remote sensing and gis techniques to detect land use and land cover change: A review. *Advances in Remote Sensing*, 2 (2), 193-204. DOI:10.4236/ars.2013.22022

4- Aplin, P., & Atkinson, P.M. (2004). Predicting missing field boundaries to increase per-field classification accuracy. *Photogrammetric Engineering and Remote Sensing*, 70(1), 141-149.

5- Butenuth, M., Gosseln, G.V., Tiedge, M., Heipke, C., Lipeck, U., & Sester, M. (2007). Integration of heterogeneous geospatial data in a federated database. *ISPRS Journal of Photogrammetry and Remote Sensing,* 62(5), 328-346.

6- Carrao, H., Gonçalves, P., & Caetano, M. (2008). Contribution of multispectral and multitemporal information from MODIS images to land cover classification. *Remote Sensing of Environment,* 112(3), 986-997.

7- Coppin, P., Jonckheere, I., Nackaerts, K., Muys, B., & Lambin, E. (2004). Review article digital change detection methods in ecosystem monitoring: a review. *International Journal of Remote Sensing,* 25(9), 1565-1596.

8- Foody, G. (2002). Status of land covers classification accuracy assessment. *Remote Sensing of Environment,* 80(1), 185-201.

9- Friedl, M..A., Sulla-Menashe, D., Tan, B., Schneider, A., Ramankutty, N., Sibley, A., & Huang, X. (2010). MODIS Collection 5 global land cover: Algorithm refinements and characterization of new datasets. *Remote Sensing of Environment,* 114(1), 168-182.

10- Guerschman, J.P., Paruelo, J. M., Bella, C.D., Giallorenzi, M.C., & Pacin, F. (2003). Land covers classification in the Argentine Pampas using multi-temporal Landsat TM data. *International Journal of Remote Sensing,* 24(17), 3381-3402.

11- Imagine, E. (2002). ERDAS Field Guide Sixth Edition. Leica Geosystems. Atlanta, Georgia, USA.
12- Imam, E. (2011). Use of geospatial technology in evaluating landscape covers type changes in Chandoli National Park, India. *Computational Ecology and Software,* 1(2), 95-111.
13- Ioannis, M., & Meliadis, M. (2011). Multi-temporal Landsat image classification and change analysis of land cover/use in the Prefecture of Thessaloniki, Greece. *The International Academy of Ecology and Environmental Sciences*, 1(1), 15-25.
14- Jung, M., Henkel, K., Herold, M., & Churkina, G. (2006). Exploiting synergies of global land cover products for carbon cycle modeling. *Remote Sensing of Environment,* 101(4), 534-553.
15- Kassa, A. (1990). *Drought risk monitoring for Sudan using NDVI*, 1982-1993. Unpublished thesis, University College, London.
16- Knorr, W., Pytharoulis, I., Petropoulos, G.P., & Gobron, N. (2011). Combined use of weather forecasting and satellite remote sensing information for fire risk, fire and fire impact monitoring. *Computational Ecology and Software,* 1(2), 112-120.
17- Kogan, F.N. (1993). United States droughts of late 1980's as seen by NOAA polar orbiting satellites In *Geoscience and Remote Sensing Symposium, 1993. IGARSS'93. Better Understanding of Earth Environment. International* (pp. 197-199). IEEE.
18- Lark, R. M., & Stafford, J.V. (1997). Classification as a first step in the interpretation of temporal and spatial variation of crop yield. *Annals of Applied Biology*, 130(1), 111-121.
19- Lillesand, T.M., & Kiefer, R.W. (1994). *Remote sensing and photo interpretation*. John Wiley and Sons: New York, 750.
20- Lillesand, T.M., Kiefer, R.W., & Chipman, J.W. (2004). *Remote Sensing and Image Interpretation*. New York: John Wiley and Sons.
21- Lu, D., & Weng, Q. (2004). Spectral mixture analysis of the urban landscape in Indianapolis with Landsat ETM+ imagery. *Photogrammetric Engineering & Remote Sensing,* 70(9), 1053-1062.
22- Naiman, R.J., Elliott, S.R., Helfield, J.M., & OKeefe, T.C. (1999). Biophysical interactions and the structure and dynamics of riverine ecosystems: the importance of biotic feedbacks. *Hydrobiologia,* 410, 79-86.
23- Podobnikar, T., Schoner, M., Jansa, J., & Pfeifer, N. (2009). Spatial analysis of anthropogenic impact on karst geomorphology (Slovenia). *Environmental Geology*, 58(2), 257-268.
24- Ramadan, E., Feng, X.Z., & Cheng, Z. (2004). Satellite remote sensing for urban growth assessment in Shaoxing City, Zhejiang Province. *Journal of Zhejiang University SCIENCE*, 5(9), 1095-1101.
25- Rana, S.V.S. (2005). Essentials of ecology and environmental sciences. Chaudhary Charan Singh University. PHI Learning Pvt. Ltd.
26- Reddy, T.B., & Gebreselassie, M. A. (2011). Analyses of land cover changes and major driving forces assessment in middle highland Tigray, Ethiopia: the case of areas around Laelay-Koraro. *Journal of Biodiversity and Environmental Sciences*, 1(6), 22-29.
27- Rogan J., Bumbarger, N., Kulakowski, D., Christman, Z.J., Runfola, D.M., & Blanchard, S. D. (2010). Improving forest type discrimination with mixed lifeform classes using fuzzy classification thresholds informed by field observations. *Canadian Journal of Remote Sensing,* 36(6), 699-708.
28- Rouse, Jr., Haas, RH., SchellJ, A., & Deering, D. W. (1974). *Monitoring vegetation systems in the Great Plains with ERTS.* published by NASA, Washington, D.C., p.309.
29- Roy, P.S., Sharma, K.P., & Jain, A. (1996). Stratification of density in dry deciduous forest using satellite remote sensing digital data an approach based on spectral indices. *Journal of Biosciences*, 21(5), 723-734.
30- Shalaby, A., & Tateishi, R. (2007). Remote sensing and GIS for mapping and monitoring land cover and land-use changes in the Northwestern coastal zone of Egypt. *Applied Geography,* 27(1), 28-41.
31- Singh, P., & Khanduri, K. (2011). Land use and land cover change detection through remote sensing & GIS technology: Case study of Pathankot and Dhar Kalan Tehsils, Punjab. *International Journal Geomatics Geosci,* 1(4), 839-846.
32- Srivastava, P.K ., Han, D., Rico-Ramirez, M.A. , Bray, M., & Islam, T. (2012). Selection of classification techniques for land use/land covers change investigation. *Advances in Space Research,* 50(9), 1250-1265.
33- Tahir, M., & Hussain, T. (2008). *Environmental,* Jawahar Publishers and Distributors; New Delhi, India.
34- Tillmann, C., Holzel, N., & Volker, A. (2012). *Supervised Classification and Change Detection of Agricultural Land Use in the Forest Steppe Zone of West Siberia Using Multitemporal Satellite Imagery.* Unpublished thesis in the subject Landscape Ecology. P:44.
35- Torahi, A. A., ChandRai, S.C. (2011). Land cover classification and forest change analysis, using

satellite imagery-a case study in Dehdez area of Zagros Mountain in Iran. *Journal of Geographic Information System,* 3(-1), 1.

36- Townshend, J. R., & Justice, C.O. (1986). Analysis of the dynamics of African vegetation using the normalized difference vegetation index. *International Journal of Remote Sensing,* 7(11), 1435-1445.

37- Tucker, C. J., Townshend, J.R., & Goff, T.E. (1985). African land-cover classification using satellite data. *Science,* 227(4685), 369-375.

38- Verhoef, W., Menenti, M., & Azzali, S. (1996). Cover a colour composite of NOAA-AVHRR-NDVI based on time series analysis (1981-1992). *International Journal of Remote Sensing,* 17(2), 231-235.

39- White, J., Shao, Y., Lisa, M.K., & Campbell, J.B. (2013). Campbell landscape dynamics on the island of La Gonave, Haiti, 1990–2010. *Land,* 2(3), 493-507

40- Wolter, P.T., Mladenoff, D.J., Host, G.E., & Crow, T.R. (1995). Improved forest classification in the Northern Lake States using multi-temporal Landsat imagery. *Photogrammetric Engineering and Remote Sensing*, 61(9), 1129-1144.

41- Yang, X., Lo, C.P. (2002). Using a time series of satellite imagery to detect land use and land cover changes in the Atlanta, Georgia metropolitan area. *International Journal of Remote Sensing,* 23(9), 1775-1798.

42- Yuan, F., Sawaya, K.E., Loeffelholz, B.C., & Bauer, M .E. (2005). Land cover classification and change analysis of the Twin Cities (Minnesota) Metropolitan Area by multitemporal Landsat remote sensing. *Remote sensing of Environment*, 98(2), 317-328.

43-Zhang, W.W., Yao, L., Li, H., Sun, D. F., & Zhou, L. D. (2011). Research on land use change in Beijing Hanshiqiao Wetland Nature Reserve using remote sensing and GIS. *Procedia Environmental Sciences,* 10, 583-588.

44- Zhu, Z., & Woodcock, C. E. (2012). Object-based cloud and cloud shadow detection in Landsat imagery. *Remote Sensing of Environment*, 118, 83-94.

5

Accurate Fruits Fault Detection in Agricultural Products Using an Efficient Algorithm

Hamidreza Saberkari

Abstract

The main purpose of this paper was to introduce an efficient algorithm for fault identification in fruits images. First, input image was de-noised using the combination of Block Matching and 3D filtering (BM3D) and Principle Component Analysis (PCA) model. Afterward, in order to reduce the size of images and increase the execution speed, refined Discrete Cosine Transform (DCT) algorithm was utilized. Finally, for segmentation, fuzzy clustering algorithm with spatial information was applied on the compressed image. Implementation results in MATLAB environment and based on the gathered data showed that the proposed algorithm contains a good capability in de-noising. Also, in the proposed method, identification accuracy of faulty regions in fruit was higher than other methods. The major advantage of the proposed method was its high speed which makes it appropriate for real time applications.

Keywords:
Fault, DCT transform, Noise, Image processing

Department of Electrical Engineering, Rasht Branch, Islamic Azad University, Rasht, Iran.
** Corresponding author's email: saberkari@iaurasht.ac.ir*

INTRODUCTION

In recent decades, image processing systems have been used in order to automatically control many processes and analyze them in different fields such as food, agriculture, and pharmacy and textile industry. Among them, one of the most popular applications of image processing and machine vision is the quality inspection of agricultural products such as fruits and vegetables based on shape, color, and the fault in them (Arumugam *et al.*, 2011). For judging the quality of this type of agricultural products, machine vision systems use digital cameras to take images, and then these images are analyzed through image processing systems (Abbot, 1999).

Among variety of agricultural products, apple suffers from so many faults. This fruit contains so many changes in skin color. Based on the market standard published by the European commission, apple quality depends on the size, color, shape and presence/absence of fault in it (Izadbakhshi and Javadikia, 2014). Traditionally, spoiled fruits are separated from high definition ones by manpower, which is very tedious and time-consuming (Malamas *et al.*, 2003). Thus, mechanizing this process using image processing systems leads to increase the identification speed and in the same time, error cancelation and accuracy growth (Brosnan and Sun, 2002; Graves and Batchelor, 2003). Identifying healthy fruits from faulty ones faces major challenges, which some of them are (Hills, 1995; Halls *et al.*, 1998):

- Lighting conditions when imaging the fruit: Create shadow or lighting spots on the fruit due to glossy surface of fruits,
- Inappropriate background for the fruit when imaging: If the background has different colors or the same color as the fruit, fruit identification and also finding faulty regions becomes difficult,
- Different fruits in one image: In an image containing different fruits with different colors, finding fruits from the background gets difficult. For this reason, this condition is considered that there must be just a single type of fruit, and
- Appearance change of some fruits: This problem causes different distributions of color in different spots of the fruit, which causes problem while extracting feature out of it.

A variety of algorithms is proposed for fruit fault identification in references. (Linker *et al.*, 2012) proposed an algorithm to estimate the number of apples in color images of gardens under natural light which consists of four min steps; identifying pixels that belong to apples with high probability, using color and smoothness, forming and developing core areas which are interconnected sets belonging to apples with high probability, and partitioning these seed areas into disordered parts and arcs, and combining these arcs and comparing the obtained circles with a simple model of an apple. Although the algorithm correctly detects more than 85 percent of visible apples in images, straight light and color saturation leads to so many positive detections. (Payne *et al.,* 2013) proposed a method to estimate mango crop using image analysis. This method is for counting mango from daily images of individual trees for machine vision goals based on estimating the mango tree crops. Images of mango trees are taken in a three-day period, three weeks before the commercial harvest. Fruits of one of each 15 tress are counted manual, and these trees are imaged from four sides. Correlation between number of trees and manual counting is strong (for two sides). Five hundred fifty five trees are imaged from one side. In these images, pixels are partitioned using color partitioning in RGB color space and YCbCr color space. The obtained pallets have been counted to achieve the number of mangos in images. Algorithm efficiency reduces with increasing the number of fruits of tree and when imaging in terms of direct sunlight. (Mizushima and Lu, 2013) presented a segmentation method to arrange apples and classify it using Otsu method and support vector machine (SVM) classifier. This method consists of three sections; color separation using linear SVM, color image transformation into gray-scale and automatic and adjustable segmentation using Otsu algorithm. This method requires the minimum number of training data. Also it prevents challenges that fruit color and light conditions changes can bring. (Ramano *et al.*, 2012) applied laser light and digital images combination to estimate the moist content and color in sweet pepper. Optical devices are increasingly used for more accuracy and faster access when monitoring the quality of agricultural products. The main work done by is evaluation using charge-coupled device (CCD) camera along with light-emitting diode (LEDs) which their radiation wavelength is 532 nm and 632 nm, respectively. So the changes in moist content of green, yellow and red sweet pepper can be analyzed during the drying process. Results show that diffusion area and light intensity have the ability to predict moist content of sweet pepper during drying. In addition, light diffusion changes in accordance with different depths in the surface that the picture is taken. Also the result that a change in tissue structure may make photons scatter over the surface and change distribution levels was achieved. (Moradi *et al.*, 2011) used the statistical histogram based on fuzzy-C means (FCM) to identify fault of apple. First, the input image is transformed from RGB workspace into workspace. After that, shape of the fruit is extracted by active contour model (ACM). Finally, the image is partitioned by SHFCM algorithm. Experimental results displays that FCM and SHFCM have equal number of repeat times to complete the process of partitioning and their results are the same. However, proposed SHFCM algorithm has a higher speed than the standard FCM. The accuracy of the proposed algorithm about resulting images for FCM and SHFCM are 91 percent and 96 percent for intact pixels and defective pixels, respectively. But the problem of this algorithm is its high computation complexity because of using active contour model, and an optimal algorithm is required to minimize energy in active contour model. In the other work done by (Moradi *et al.*, 2012), expectation-maximization (EM) segmentation algorithm is used for fault identification in apple. In this method, the input image is transformed from RGB workspace into workspace. Then, fruit shape is extracted by ACM. Eventually, image is partitioned by SHEM algorithm. Results indicate that EM and SHEM show pretty equal performances with equal repeats. The discussed SHEM algorithm spends less time than standard EM algorithm. The accuracy of the proposed algorithm for EM and SHEM are respectively 91.72 percent and 94.86 percent for intact pixels and defective pixels (Rakun *et al.*, 2011).

In this paper, an optimal algorithm is presented in order to fault identification in fruit images. First, input image is de-noised using black matching and 3D filtering (BM3D) and principle component analysis (PCA) combined model. Then, modified version of discrete cosine transform (DCT) algorithm is used in order to reduce image size and increase speed of the segmentation task. Finally in partitioning stage, we have used the fuzzy clustering algorithm based on the spatial information of neighboring pixels in the compressed image.

The rest of the paper is as follows: In Section 2, materials and methods used in this paper is presented which consist of our proposed algorithm and its different steps. Implementation results consisting of two experimental phases on gathered data by the author is expressed in Section 3. Finally, Section 4 includes the conclusion of the paper.

MATERIALS AND METHODS

Figure 1 shows block diagram of the proposed algorithm. Different steps of the raised algorithm are as follows, which will be explained:

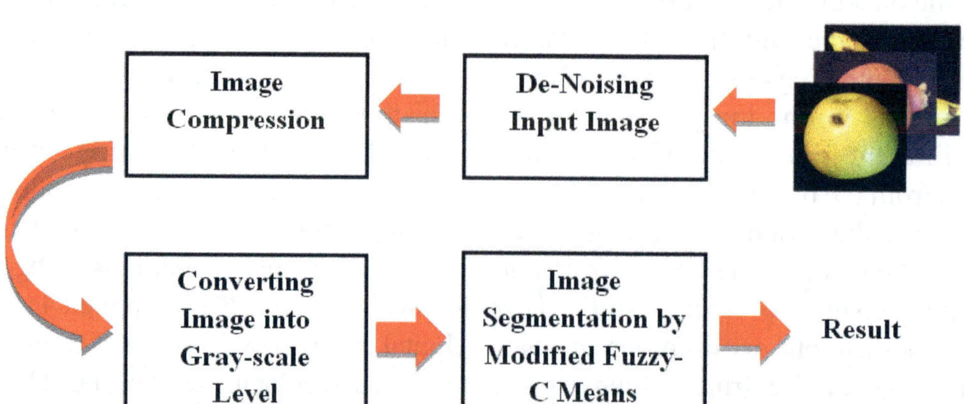

Figure 1: Block diagram of the proposed algorithm

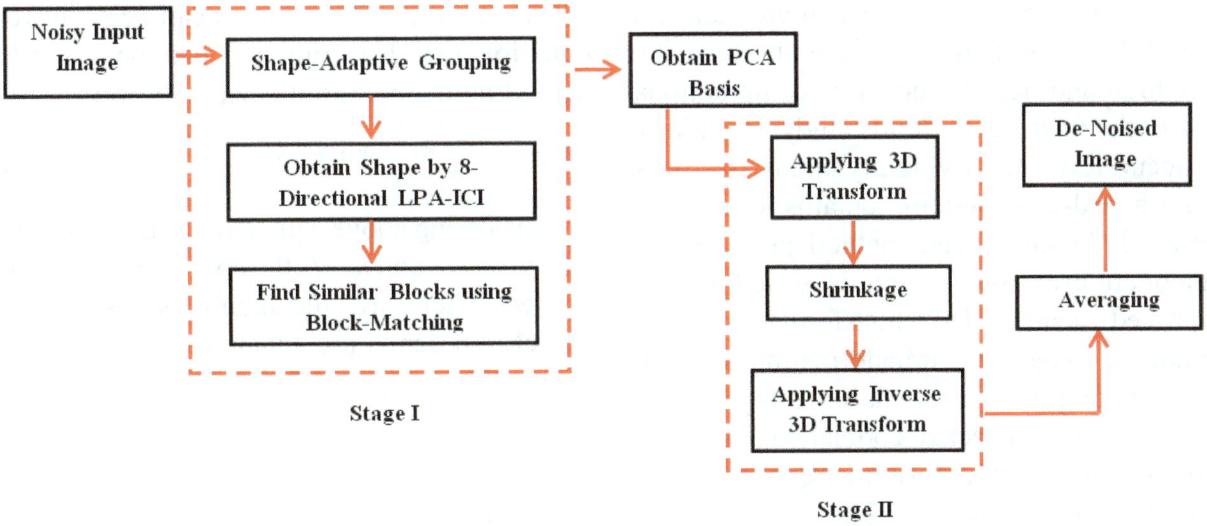

Figure 2: Block diagram of the proposed De-Noising algorithm

- Preprocessing step in order to remove noise using BM3D and PCA combined algorithm,
- Image compression to reduce its size, and
- Fault identification using fuzzy clustering algorithm with spatial data.

Preprocessing stage to De-Noise using BM3D and PCA Combined Algorithm

In the presented algorithm, BM3D and PCA combined model (Dabov *et al.*, 2009) is used to de-noise the image. This model is demonstrated in Figure 2. Assume that the input image is demolished by additive white Gaussian noise with average of zero and variance of σ^2. The input image is scanned array, and for each processed pixel, the following procedure is applied:

- For each pixel being processed, adaptive-shape neighborhood with the centrality of the pixel can be found using eight-fold LPA-ICI model based on (Foi *et al.*, 2007). The neighborhood is located within a square block of a fixed size. This block is called reference block. The number of pixels in the neighborhood is shown by N_{el},
- Blocks similar to the reference block are found using block-matching method and using the method mentioned in the previous step, we extract the adaptive-shape neighborhood from each one of these matching blocks. The number of matching blocks are shown by N_{gr},
- Determining a conversion and applying it on adaptive-shape neighborhoods. For this purpose, we consider a threshold level (τ) and review two states:

1) If the ratio is $N_{gr}/N_{el} \geq \tau$, this means that an acceptable number of neighboring pixels are selected to estimate PCA matrix. Eigenvectors of this matrix form base vectors of PCA. Thus in this step, only eigenvectors (which the eigenvalues corresponding to them are higher than the defined threshold level), are selected.

2) If the ratio is $N_{gr}/N_{el} < \tau$, it means that a sufficient number of similar neighboring pixels are not selected as training data for PCA model. Thus, a certain amount is selected for eigenvector.

- Forming a three-dimensional array by connecting adaptive-shape neighborhoods (min N_{gr}, N_2) with the block reference that has the highest similarity. is a fixed parameter which limits the number of filtered neighborhoods.
- Applying the transformation used in step (3) to each one of adaptive-shape neighborhood groups. In this step, A one-dimensional orthogonal transformation (such as wavelet transformation) is also applied to each one of three-dimensional groups,
- Applying hard-thresholding to three-dimensional spectrum in order to shrinkage the image,
- Applying a three-dimensional reverse transformation from step (5) to find estimations in each of adaptive-shape neighborhood groups, and
- Returning obtained estimations to their original positions using weighed averaging.

Compressing the De-Noised image to reduce its size

The size of the de-noised original image is

3264×1836, which must be compressed. The size of the input image is reduced because it is very high and heavily increases computational cost in the next steps. In this study, in addition to accurately identify faulty areas of fruits, being a real-time system contains a great importance. In this paper, method presented by has been proposed which is based on the improved version of DCT transform to compress de-noised image. DCT transformation is mainly used for compressing signals and images due to energy savings and lack of correlation of samples. More details of this algorithm are explained by (Saberkari and Shamsi, 2012).

Fault identification using fuzzy clustering algorithm with spatial neighboring information

In this paper, fuzzy clustering algorithm (Bezdek, 1987) with spatial information is applied for segmenting the compressed images. Fuzzy clustering algorithm allocates pixels to classes by adapting the fuzzy membership functions. Assume that $X = (x_1, x_2, \ldots x_n)$ is an image with N pixels which belongs to c cluster. Fuzzy clustering algorithm attempts to perform segmentation by minimizing cost function. Cost function in this algorithm is defined as follows:

$$J_m = 2\sum_{i=1}^{c}\sum_{k=1}^{N}(\mu_{ik})^m \|x_k - v_i\|^2 \quad (1)$$

which $V=[v_1, v_2, \ldots, v_c]$ is cluster center vector and N is the number of input data. μ_{ij} Shows membership of x_j pixel in ith cluster. Also, m is constant number which controls fuzziness of partitions. In our proposed algorithm, m changes from 1.2 to 5. Euclidean distance norm ($\|\cdot\|$) which is the distance between pixels and average of clusters is achieved as:

$$D_{ikA}^2 = \|x_k - v_i\|_{A_D}^2 = (x_k - v_i)^T A_D (x_k - v_i) \quad (2)$$

where A_D is a diagonal matrix defined as:

$$A_D = \begin{bmatrix} (1/\sigma^2) & 0 & \cdots & 0 \\ 0 & (1/\sigma^2) & \cdots & 0 \\ \vdots & \vdots & \ddots & \vdots \\ 0 & 0 & \cdots & (1/\sigma^2) \end{bmatrix} \quad (3)$$

The deviation of x_k from v_i can be seen from equation (3). Equation (1) can be rewritten using Lagrange multipliers as follows:

$$J(X;U,V) = \sum_{i=1}^{c}\sum_{k=1}^{N}(\mu_{ij})^m D_{ikA}^2 + \sum_{k=1}^{N}\lambda_k\left(\sum_{i=1}^{c}\mu_{ik} - 1\right) \quad (4)$$

Considering above equation and the conditions $D^2_{ikA} > 0, \forall i, k$ and $m > 1$, the cost function is minimized. Therefore, the memberships function and cluster center are updated as follows:

$$\mu_{ik} = \frac{1}{\sum_{j=1}^{c}\left(\frac{D_{ikA}}{D_{jkA}}\right)^{\frac{2}{m-1}}} \qquad 1 \le i \le c,\ 1 \le k \le N$$

$$v_i = \frac{\sum_{k=1}^{N}(\mu_{ik})^m x_k}{\sum_{k=1}^{N}(\mu_{ik})^m} \qquad 1 \le i \le c \quad (5)$$

The above equation displays as weighted average of data points which belong to a cluster, such as weighs are the membership matrixes.

One of important characteristics of fruit images is high intensity correlation between a pixel and its neighbor pixels. In the other words, these neighboring pixels have similar features; in a way that the probability that a certain pixel belongs to neighboring pixels class is high. Considering the importance of spatial dependence, the need to enter a locative parameter in segmentation algorithms based on intensity appears. Therefore, we enter h_{ij} spatial function into fuzzy clustering algorithm and call it fuzzy clustering algorithm with spatial information (Saberkari *et al.*, 2015).

$$h_{ij} = \sum_{k \in N_{x_j}} \mu_{ik} \quad (6)$$

N_{xj} indicates a square window with centrality of x_j in spatial field, which in this article a 5×5 window is used. If majority of neighborhoods of a pixel belong to the same class, then, spatial function h_{ij} will be maximum for a central pixel. Implementation process of fuzzy clustering algorithm with spatial data is the same as fuzzy clustering algorithm and its membership function changes as below:

$$\mu'_{ij} = \frac{\mu_{ij}^p h_{ij}^q}{\sum_{k=1}^{c}\mu_{kj}^p h_{kj}^q} \quad (7)$$

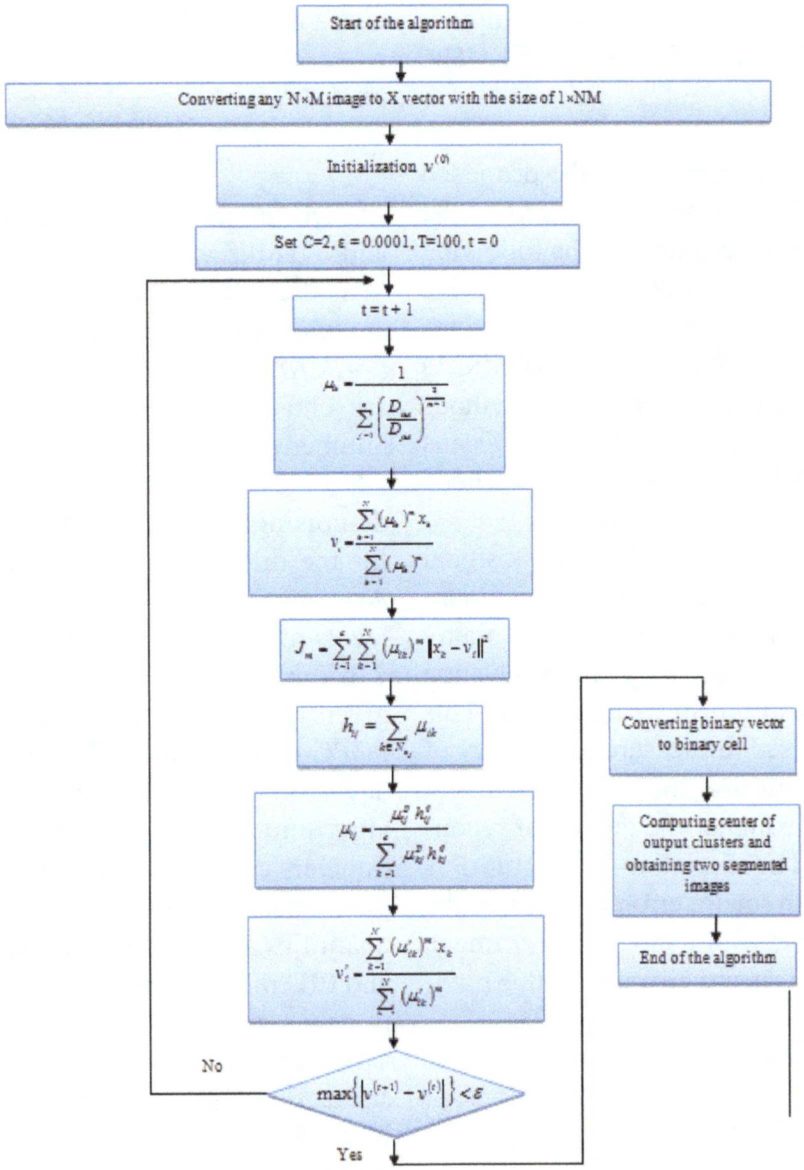

Figure 3: Block diagram of the modified-C means clustering algorithm based on spatial neighboring Information

where p and q are control parameters of both functions. In first step, like fuzzy clustering algorithm, membership functions are computed. In second step, membership function data is mapped into spatial scope and then spatial function is computed out of it. The considerable note is that in homogenous areas spatial function is similar to original membership function. So, clustering result does not change. But for pixels containing noise, the equation above reduces noisy cluster weight by labeling its neighboring pixels. As a result, incorrectly classified pixels are easily corrected from noisy areas. Fuzzy clustering algorithm with spatial data is implemented as the block diagram shown in Figure 3.

RESULTES AND DISCUSSION

Image acquisition device used in this paper is composed of a high resolution (2364×1836 pixels) monochrome digital camera, four interference band pass filters, a frame grabber, a diffusely illuminated tunnel with two different light sources, and a conveyor belt on which fruits are placed. The filters are centered at 450, 500, 750, and 800 nm with the bandwidths of 80, 40, 80, and 50 nm, respectively This device is capable of acquiring only one-view images of fruits. Each of these one-view images was composed of four filter images, which had to be separated by alignment based on pattern matching. All implementation stages of the proposed algorithm are performed on a computer with 3.4 GHz and the RAM memory of 1 GHz.

Experiment I

In the first test, performance of the proposed de-noising algorithm has been compared by the other approaches. For this purpose, the following two criteria have been used for quantitative comparison. Also, some other de-noising algorithms (BM3D, patch based global PCA (PGPCA) and progressive image de-noising (PID) methods proposed by (Charles *et al.*, 2011) and (Knaus and Zwicker, 2014), respectively) have been implemented to compare the results.

Peak signal to noise ratio (PSNR): Peak signal to noise ratio represents the ratio of maximum possible power to noise power. Because many signals have a wide dynamic range, this evaluation criterion arises in the form of logarithm. This criterion is often used to measure the quality of images after reconstruction, the higher the amount of the criterion, the higher the quality of reconstructed image. The signal to noise ratio is defined as the mean square error. If we show the image without noise with I and the noisy image with *K*, we have (Thu and Ghanbari, 2008):

$$MSE = \frac{1}{mn}\sum_{i=0}^{m-1}\sum_{j=0}^{n-1}\left[I(i,j)-K(i,j)\right]^2$$
$$\rightarrow PSNR = 10\log_{10}\left(\frac{MAX_I^2}{MSE}\right) = 20\log_{10}\left(\frac{MAX_I}{\sqrt{MSE}}\right) \qquad (8)$$

where MAX_I is the maximum possible pixel value, and if image pixels are 8-bit, this value will be 255; otherwise, equation (9) is used to compute MAX_I:

$$MAX_I = 2^B - 1 \qquad (9)$$

Structural similarity (SSIM): This criterion is a way to measure the similarity between two images. This criterion is an improved version of previous criterion and the difference is that it considers noises and artifacts in the image as changes in structural information. The structural information is based on the idea that pixels spatially close to each other are greatly dependent on one another and this dependence contains important information. SSIM criterion is defined based on the following equation:

$$SSIM(x,y) = \frac{(2\mu_x\mu_y + c_1)(2\sigma_{xy} + c_2)}{(\mu_x^2 + \mu_y^2 + c_1)(\sigma_x^2 + \sigma_y^2 + c_2)} \qquad (10)$$

where μ_x and μ_x is the average of x and y, σ_x^2 and σ_y^2 is the variance of x and y, respectively. σ_y^x is the covariance between x and y. c_1 and c_2 are two variables which is defined as, $c_1=(k_1l)^2$, $c_2=(k_2l)^2$, respectively. The values of k_1 and k_2 are chosen as 0.01 and 0.03. Also, l is the number of bits of pixel.

In Figures 4, results of applying the proposed de-noising algorithm on input images are given for the different values of standard deviation. As can be seen, noise is greatly removed at all levels of standard deviation.

In Tables 1 and 2, PSNR and SSIM values are given in the proposed de-noising and BM3D-PGPCA and PID algorithms for different standard deviation values. The superiority of the proposed algorithm in improving PSNR and SSIM parameters can be seen in these tables and also in diagrams shown in Figures 5 and 6. For , PSNR quantities are 35.9 and 35.69 in BM3D-PGPCA and PID methods, respectively, while this value is achieved 35.92 in the proposed algorithm. This improvement is also observed for other standard deviations. Similar results were obtained in the SSIM parameter value. By choosing the standard deviation equal to 10, the value of this parameter in the proposed algorithm has improved 5.76 percent and 0.03 percent in comparison to BM3D-PGPCA and PID methods, respectively. There is similar superiority for SSIM parameter in other standard deviation values.

Experiment II

In this test, the performance of our algorithm is compared with other methods in identifying the fruit fault. For this purpose, two following evaluation criteria are utilized. Also two approaches, conventional FCM and Otsu algorithms, are implemented.

Segmentation matching factor (SMF): This parameter measures pixel that have been wrongly segmented and is defined as follows (Ahmed *et al.*, 2002):

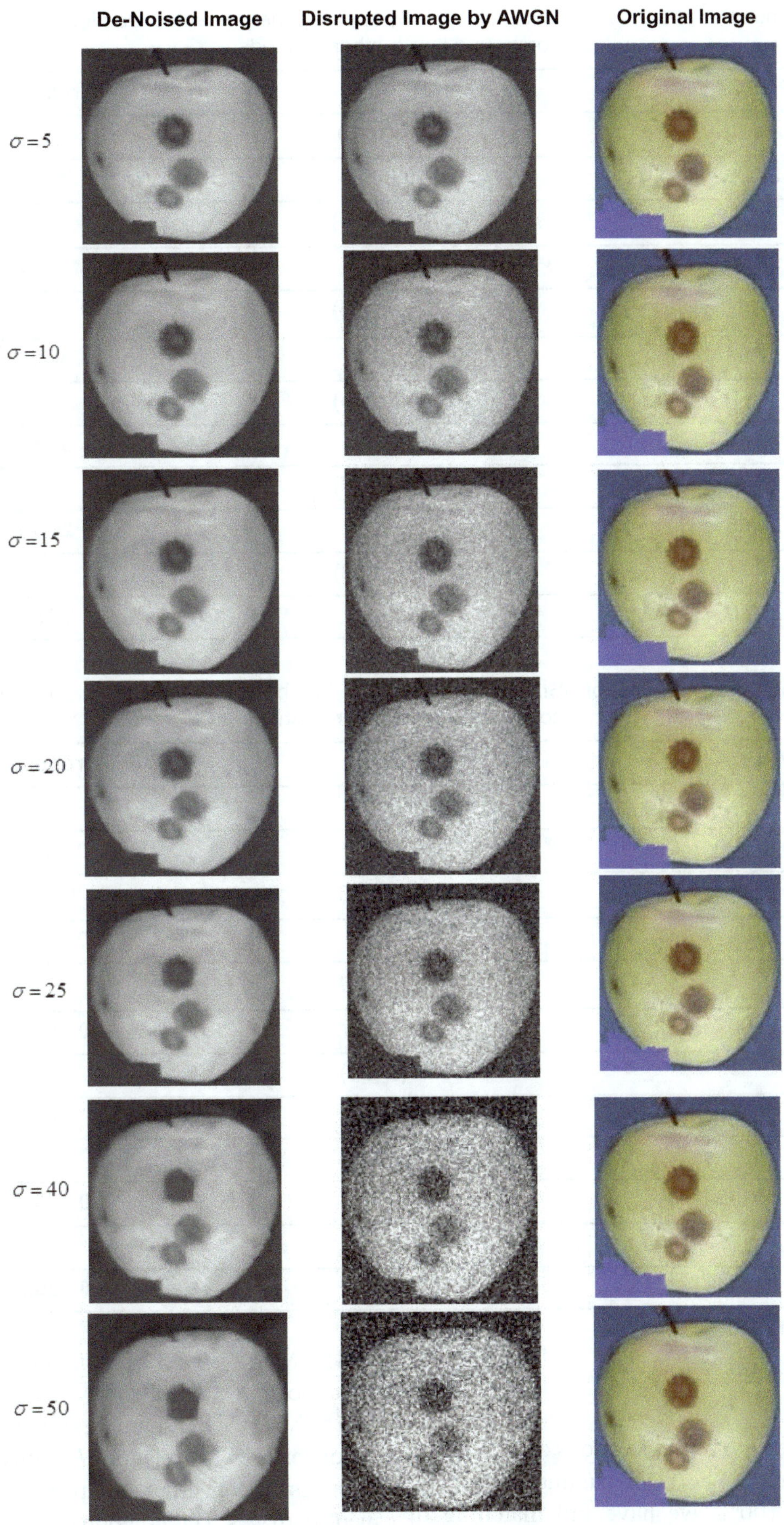

Figure 4: Results of applying the proposed De-Noising algorithm for different levels of standard deviation

Table 1: Quantitative amounts of PSNR achieved by applying the presented De-Noising algorithm and comparing it with other methods

PGPCA	PID	Proposed
	PSNR (σ=5)	
39.09	39.08	**39.39**
	PSNR (σ=10)	
35.09	35.65	**35.92**
	PSNR (σ=15)	
32.90	33.79	**34.02**
	PSNR (σ=25)	
29.67	31.36	**31.56**
	PSNR (σ=40)	
26.65	28.88	**29.16**
	PSNR (σ=50)	
25.40	27.66	**27.99**

Table 2: Quantitative amounts of SSIM achieved by applying the presented De-Noising algorithm and comparing it with other methods

PGPCA	PID	Proposed
	SSIM (σ=5)	
0.94899	0.95686	**0.95772**
	SSIM (σ=10)	
0.8732	0.9232	**0.92352**
	SSIM (σ=15)	
32.90	0.89424	**0.89562**
	SSIM (σ=25)	
0.80685	0.83709	**0.843**
	SSIM (σ=40)	
0.59105	0.7538	**0.77505**
	SSIM (σ=50)	
0.53284	0.70451	**0.73687**

$$SMF = \frac{A_{segment} \cap A_{actual}}{A_{segment} \cup A_{actual}} \tag{11}$$

where and are binary versions of segmented and actual images. For SMF parameter we have:

• If SMF=100%, we have full matching of images,

• If SMF > 50%, segmentation result is acceptable, and

• If SMF < 50%, segmentation result is weak.

Concordance correlation *(P_c)*: this criterion determines the concordance between actual and segmented images. This criterion is applied to evaluate reproducibility of presented segmentation algorithms.

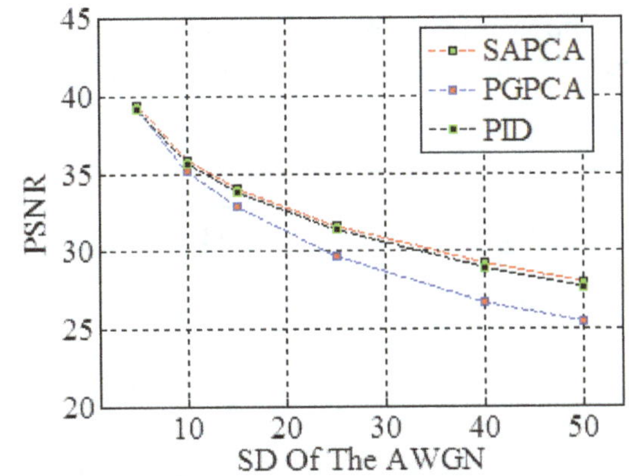

Figure 5: PSNR versus different levels of standard deviation of AWGN

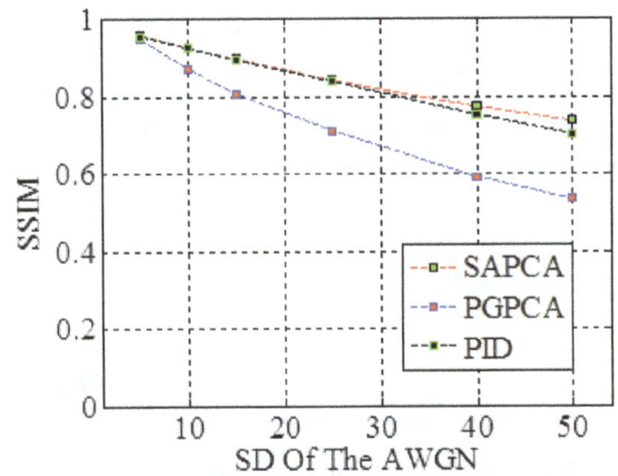

Figure 6: SSIM versus different levels of standard deviation of AWGN

P_c is defined as (Lehmussola *et al.*, 2006):

$$P_c(A,B)=\frac{2S_A\,S_B\,r}{S_A^2+S_B^2+(\bar{A}-\bar{B})^2} \qquad (12)$$

where A and B are two samples, $\bar{A}$ and $\bar{B}$ are the average amounts, S_A and S_A standard deviation of samples. The higher amount of P_c leads to the better the performance of the algorithm.

In figure 7, results of applying our algorithm and also *Otsu* and *FCM* segmentation methods in fault identification of fruit images are compared. To assess the stability of the proposed algorithm in dealing with noise, fruit image have been destroyed with additive white Gaussian noise with the signal to noise ratio of 1, 3, 5, 7 and 9 *(dB)*. In Tables 3 and 4, amounts of *SMF* and P_c are given for the proposed algorithm, *Otsu* segmentation and *FCM* methods. As seen, by increasing signal to noise ratio,

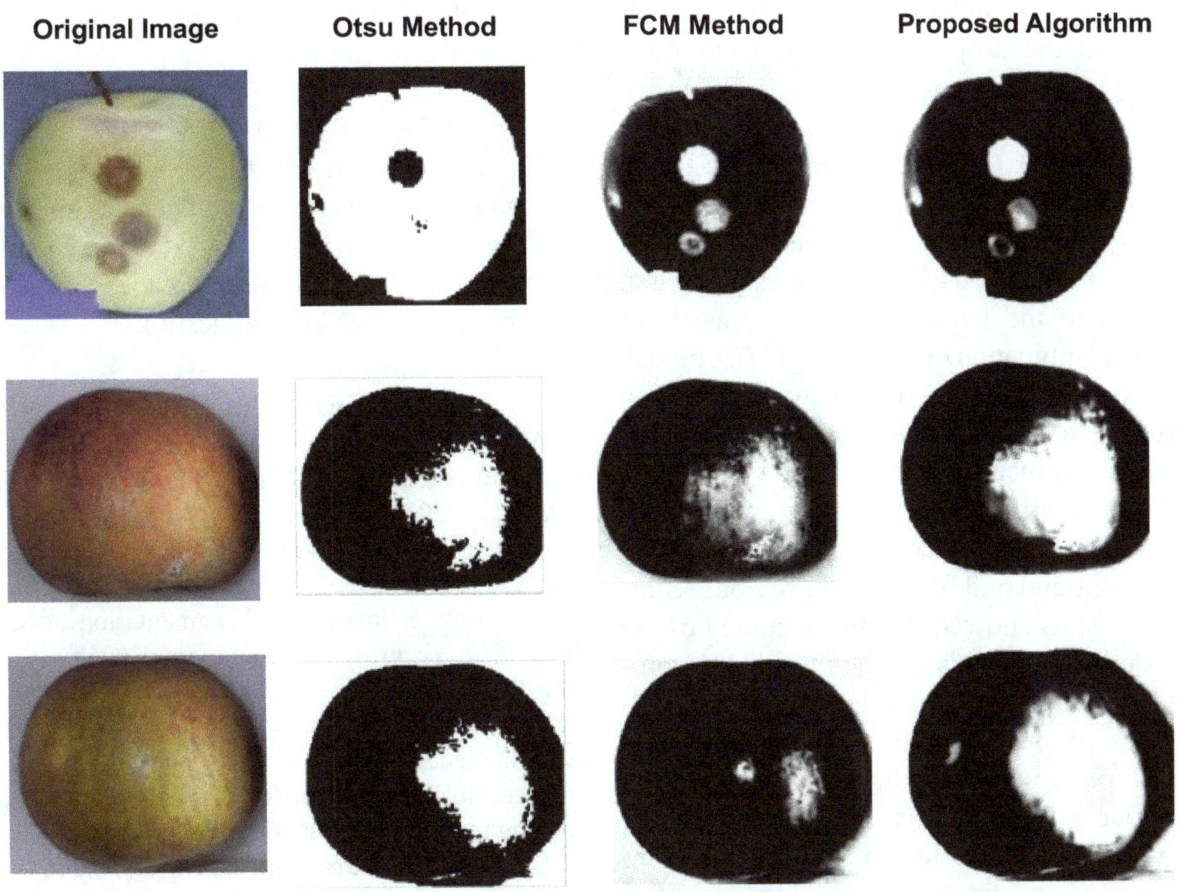

Figure 7: Results of performing the presented algorithm and comparing it with other methods in fruit images

Table 3: Quantitative amounts of SMF achieved by applying the presented algorithm and comparing it with other methods (SD indicates the standard deviation)

SNR	Otsu	FCM	Proposed
1	78.2	81.2	92.6
3	78.9	86.6	93.4
5	82.2	87.1	95.2
7	86.1	89.9	97.4
9	89.6	92.5	98.3
	SD=4.8389	SD=4.2253	SD=2.4641

Table 4: Quantitative amounts of P_c achieved by applying the presented algorithm and comparing it with other methods (SD indicates standard deviation)

SNR	Otsu	FCM	Proposed
1	78.2	81.2	92.6
3	78.9	86.6	93.4
5	82.2	87.1	95.2
7	86.1	89.9	97.4
9	89.6	92.5	98.3
	SD=4.8389	SD=4.2253	SD=2.4641

SMF mount in the presented algorithm increases. For instance, by choosing *SNR* value equal to 7, *SMF* parameter in the presented algorithm improves by factor of 1.13 and 1.08 in comparison to *Otsu* and *FCM*, respectively. This shows that the proposed algorithm has good stability against noise.

CONCLUSION

A new segmentation algorithm has been proposed in this paper to detect faults in the fruits. The algorithm is consisted of three steps. First, to eliminate the background noise, we have used the combination of BM3D and PCA model. Then, we have performed our new compression approach to compress the de-noised image. This leads to increase the execution speed. Finally, the modified version of fuzzy-C means algorithm has been utilized to classify the pixels into foreground and background regions. Simulation results showed that the accuracy of the proposed algorithm is reached to 98.3% compared to other methods. One of the major advantages of our algorithm is its high speed characteristic which results in reduction of the run process. This leads to implementation of the hardware architecture of this algorithm to design an on-line fruit fault embedded identifier system. The other advantage is its stability relative to different levels of noise sources. As we have calculated, our proposed algorithm has the highest amounts of PSNR and SSIM parameters. Segmentation of different kinds of fruits with different skin colors is one of our major works for future.

ACKNOWLEDGEMENT

The author would like to acknowledge M.S. Alireza Farrokhinia for his guidance to improve the language of the manuscript. I also thank the anonymous reviewers of the paper for their valuable comments and suggestions.

REFERENCES

1- Abbot, J.A. (1999). Quality Measurement of fruits and vegetables. *Postharvest Biology and Technology*, 15, 207-225.

2- Ahmed, M.N., Yamany, S.M., Mohamed, N., & Farag, A. (2002). A modified fuzzy c-means algorithm for bias field estimation and segmentation of MRI data. *IEEE Transactions on Medical Imaging,* 21, 193–199.

3- Arumugam, N., Mohamed Arshad, F., Chiew, E., & Mohamed, Z. (2011). Determinates of fresh fruits and vegetables (FFV) farmers participation in contrast farming in Peninsular Malaysia. *International Jornal of Agricultural Management and Development*, 1(2), 65-71.

4- Bezdek, J.C. (1987). Pattern recognition with

fuzzy objective function algorithms. New York/ London, UK, Plenum.

5- Brosnan, T., & Sun, D.W. (2002). Inspection and grading of agriculture and food products by computer vision system-A reviews, *Computer and Electronic in Agriculture*. 36, 193-213.

6- Charles, D., Salmon, J., & Dalalyam, A. (2011). Image denoising with patch based PCA: local versus global. In Jesse Hoey, Stephen McKenna and Emanuele Trucco, *Proceedings of the British Machine Vision Conference,* pages 25.1-25.10. BMVA Press.

7- Dabov, K., Foi, A., Katkovnik, V., & Egiazarian, K. (2009). BM3D image denoising with shape-adaptive principle component analysis. PROC. *workshop on signal processing with adaptive sparse structured presentations* (SPARS'09).

8- Foi, A., Katkovnik, V., & Egiazarian, K. (2007). Pointwise shape-adaptive DCT for high-quality denoising and deblocking of grayscale and color images. *IEEE Transactions on Image Processing*. 16, 1395.1411.

9- Graves, M., & Batchelor, B. (2003). *Machine vision for the inspection of Natural products*. London, Springer.

10- Halls, L., Evans, S., & Nott, K. (1998). Measurement of textural changes of food by MRI relaxometry. *Magnetic Resonance Imaging,* 16, 485-492.

11- Hills, B. (1995). Food processing and MRI perspective. *Trends in Food Science and Technology*. 6, 111-117.

12- Izadbakhshi, M., & Javadikia, H. (2014). Application of hybrid feed-forward neural network (ffnn)-genetic algorithm for predicting evaporation in storage dam reservoirs. *Agricultural Communications*, 2, 57-62.

13- Knaus, C., & Zwicker, M. (2014). Progressive image denoising. *IEEE Transactions on Image Processing,* 23, 3114-3125.

14- Lehmussola, A., Ruusuvuori, P., & Yli-Harja, O. (2006). Evaluating the performance of microarray segmentation algorithms. *Bioinformatics*, 22(23), 2910-2917.

15- Linker, R., Cohen, O., & Naor, A. (2012). Determination of the number of green apples in RGB images recorded in orchards. *Computers and Electronics in Agriculture*, 81, 45–57.

16- Malamas, E.N., Petrakis, E.G.M., Zervakis, M., Petit, L., & Legat, J.D. (2003). A survey on industrial vision systems, applications and tools. *Image and Vision Computing,* 21,171–188.

17- Mizushima, A., & R. Lu. (2013). An image segmentation method for apple sorting and grading using support vector machine and Otsu's method. *Computers and Electronics in Agriculture,* 94, 29–37.

18- Moradi, G., Shamsi, M., Sedaghi, M. H., & Alsharif, M.R. (2011, April). Fruit defect detection from color images using ACM and MFCM algorithms. *International Conference on Electronic Devices, Systems and Applications (ICEDSA),* 2011 International Conference on (pp. 182-186). IEEE.

19- Moradi. G., Shamsi, M., Sedaaghi, M.H. Moradi, S., & Alsharif, M. R. (2012). Apple defect detection using statistical histogram based EM algorithm. *19th Iranian Conference on Electrical Engineering (ICEE).*

20- Payne, A.B. Walsh, K.B., Subedi, P.P., & Jarvis, D. (2013). Estimation of mango crop yield using image analysis– Segmentation method. *Computers and Electronics in Agriculture,* 91, 57–64.

21- Rakun, J., Stajnko, D., & Zazula, D. (2011). Detecting fruits in natural scenes by using spatial-frequency based texture analysis and multiview geometry. *Computers and Electronics in Agriculture,* 76, 80-88.

22- Romano, G. Argyropoulos, D., Nagle, M., Khan, M.T., & Muller, J. (2012). Combination of digital images and laser light to predict moisture content and color of bell pepper simultaneously during drying. *Journal of Food Engineering,* 119, 438–448.

23- Saberkari, H., & Shamsi, M. (2012). Comparison of different algorithms for ECG signal compression based on transfer coding. *IEEE Symposium on Industrial Electronics and Applications (ISIEA 2012)*, 23-26.

24- Saberkari, H., Bahrami, S., Shamsi, M., Amoshahy, M.J., Badri Ghavifekr, H., & Sedaaghi, M.H. (2015). Fully automated complementary DNA microarray segmentation using a novel fuzzy-based algorithm. *Journal of Medical Signals and Sensors*, 5(3), 182-191.

25- Huynh-Thu, Q., & Ghanbari, M. (2008). Scope of validity of PSNR in image/video quality assessment. *Electronics Letters*, 44(13), 800-801.

6

Technology Acceptance Model (TAM) as a Predictor Model for Explaining Agricultural Experts Behavior in Acceptance of ICT

Amir Alambaigi [1] and Ismael Ahangari [2*]

Abstract

This study aimed to develop Technology Acceptance Model (TAM) model to explain adoption of information technologies process. a Descriptive – correlation study was conducted and data were collected through a survey. Statistical population was West Azerbaijan Agricultural Extension agents who 120 of them were selected randomly using the Krejcie and Morgan table. A questionnaire was employed to measure the variables in the model. Its validity was confirmed by a panel of experts. The Cronbach's alpha coefficient ranged between from 0.704 to 0.816 show satisfied reliability. For data processing, partial least squares (PLS) method as a new approach to structural equation modeling was used. The results showed that among three variables for development of technology acceptance model including Job relevance, experience and organization willingness to invest, the first and second show significant effects. Thus, Job relevance and experience as an external variable was added to the basic TAM. Other relations between variables in basic technology acceptance model in current study were also seen significant. Our developed TAM can explain 64% of the actual behavior of employee in information technology utilization. TAM is one of the most influential extensions of Ajzen and Fishbein's theory of reasoned action (TRA) in the literature. The theories behind it assume that when a person forms an intention to act, that s/he will be free to act without limitation. While In the real world there will be many constraints, such as limited freedom to act. For example, people in organized working environments are forced to use most of the relevant applications irrespective of their opinion or attitude. In this research mentioned model was used as a strong model to predict actual use behavior that affected by three variables namely Job relevance, experience and organization willingness to invest.

Keywords:
Technology acceptance model, Job relevance, IT, Partial least squares

[1] *Assistant Professor, Department of Agricultural Extension and Education, Faculty of Agricultural & Natural Resources, University of Tehran, Tehran, Iran*

[2] *PhD Student, Department of Agricultural Extension and Education, Faculty of Agricultural & Natural Resources, University of Tehran, Tehran, Iran*

* *Corresponding author's email: ahangari.tarvij@ut.ac.ir*

INTRODUCTION

In recent decades technological innovation has become an integral part of modern life and the global economy and Identify the dominant patterns on adoption of this innovation has become the main goal of many researchers in academia and industry. In this respect, at least during the last twenty years many theoretical models to identify behavioral factors influencing the adoption of information and communication technologies have been proposed in which the technology acceptance model has been interested by researchers more than any other models in various areas of acceptance. Technology acceptance is important issues to agricultural extension systems at the world and many of extension studies have been directed to study resistance to change and technologies acceptance. The models with theatrical basis that often have been provided based on change theory or the overall based on the Theory of Planned Behavior (TPB), considered that adoption and implementation of a new technology is from an attitude, or a behavioral plan, and ultimately a behavior led to the adoption of change. Study of factors influencing the intention is one of the growing areas in the field of Motivating factors of adoption (Ahangari, 2014).

Technology Acceptance Model (TAM) was developed in the 1980s, and researchers such as Davis argued that the key of increasing use of information technology is in increase the rate of its adoption (Melas *et al.,* 2011). Information technology can be knowledge, skills and tools that helps to collect, manage, use and share information related activities of the organization and can be cause of promote them. The application of information processing involving both computer hardware and software that deals with the storage, retrieval, sharing, and use of health care information, data, and knowledge for communication and decision-making (Holden and Karsh, 2010).

In this regard, many methods and models has been applied to studying effective factors on information technology acceptance and the most reliable of them is technology acceptance model that studies factors at the individual level. This model has been applied through many researches in different countries and its applicability has been studied (Dillon and Morris, 1996). At recent decades, several models have been proposed in the context of technology adoption but what is indisputable is that the main and underlying concept of all models of technology acceptance by the user is presented in Figure 1 (Yaghoobi and Shakeri 2010).

Among the models of Technology Acceptance Model (TAM), Theory of Reasoned Action (TRA) and Theory of Planned Behavior (TPB) have been most widely used in studies of information technologies adoption (Yaghoobi and Shakeri 2010). The personal and social factors have been considered as two important constructs and their effects has been analyzed on intention or plan to use different technologies. According to the Logical action theory of Fishbein and Ajzens (1975), person's behavior is follower of intention or plan that affected on their attitudes are formed due to the normative influences. In fact acceptance technology model has been used to explanation this event that why some people accept one technology but some of them reject that and two variable, perceive of usefulness and perceive of ease of use have been introduced as basic elements of this process (Cothran, 2011). However at the recent decades high efforts have been noticed to introduce the other effective variables, that Kulviwat *et al.* (2009) by studying this process have saturated Technology Accept-

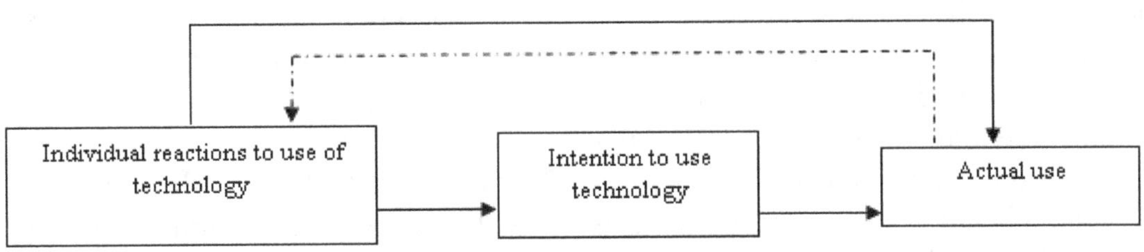

Figure 1: Underlying factors of the technology acceptance model by the user (Venkatesh *et al.*, 2003)

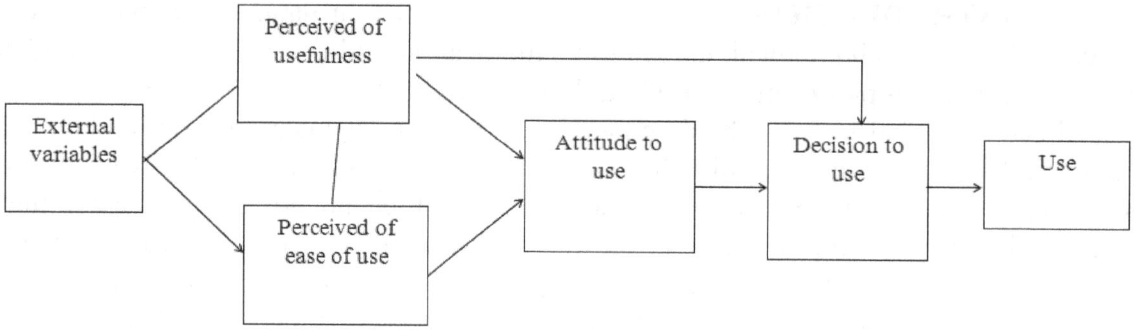

Figure 2: Technology acceptance model (Davis *et al.,* 1989)

ance model by adding variables such as trust, experience and job relevance.

Technology acceptance model is an influencing techno-social model that aimed at describe the adoption of an information system by user (Letho and Lee, 2013). Also this model is applying to explain technology acceptance decisions among a large number of different technologies, different contexts and different samples users (Hu *et al.*, 2003). Generally, the model has been analyzed relationships of effective factors on computer acceptance by users and it's a model on individual level. The efforts taken to develop TAM has been noticed Mainly in the form of three research approach include derived from similar models and complete core of basic model, Introducing alternative or added important factors of view and finally, using some of the intermediate factors those effect on perceive of usefulness and perceive of ease of use (Wixom and Todd, 2005). Figure 2 shows process and performance of the model.

As is seen, two factors (perceive of usefulness and perceive of ease of use) comprise the base of model. These two factors affect the attitude of people toward using technology, because of the decision to use the technology, and ultimately using is done. Also, perceived ease of use effects on the perceived of usefulness. The external factors could include any factor such as organizational factors; social factors, how to teach, helping from other people and etc. (Davis *el al.*, 1989).

The results of their researches showed that TAM could predict about 40 percent of influencing factors on information technology acceptance and this model has been recognized as a useful theorical model to understanding and explaining behavior of IT using. Perceived of ease of use is defined as the extent to which a person believes that using a system would be free of mental effort. Perceived of usefulness is defined the extent that a person believes that using a system would his or her job performance (Davis, 1989). Behavioral intention or purpose is the intensity of individual's intention and will to perform the target behavior (Morris and Dillon, 1997). The model of Davis has this advantage than the before descriptive model that is based on a theory. This model has elements of many previous studies in itself (Lucas and Spilter, 2000). Gyampah and Salam (2004) showed that both training and project communication influence the shared beliefs that users form about the benefits of the technology and that the shared beliefs influence the perceived usefulness and ease of use of technology. The perceived usefulness of electronic services in general is the main determinant of the intention to use e-government services. Risk perception, personal experience, perceived behavioral control and subjective norm were found to significantly predict the perceived usefulness of electronic services in general, while trust in e-government was the main determinant of the perceived usefulness of e-government services (Kuttschreuter *et al.*, 2007). System attributes, perceived usefulness and perceived ease of use, indeed, positively engender motorists' attitudes towards ETC (Electronic Toll Colection) service adoption (Chen *et al.*, 2007).

Theoretical framework development

As was expressed in addition to the basic core model, various researchers to develop original technology acceptance model, studied other variables under the title external variables and have been analyzed its effect on the variables of the original model. In this study we use of three variables (job relevance, company's willingness to fund to establishment IT and individual experience factor) to reach the organizational information technology acceptance model among users based on Kim's pattern. Company's willing to fund is defined here as the degree to which a company believes that funding would enhance an individual's adoption of technology. Without adequate funding, it is virtually impossible for individuals, employees, or organizations to adopt and use new technologies.

Experience refers to the effect of Company's willing to fund might change with prior experience, the level of an individual's satisfaction when adopting prior technology. Experience has been regarded as important in identifying individual differences (Kim, 2008). For example, prior experience with a similar technology was found to be a major factor influencing an individual's attitude in adoption decisions (Dabholkar, 1996). Job relevance means that individuals have different perceptions of outcomes they expect obtain from technology because of the different nature of their job; also they are exposed to external information, which may affect them in choosing which technology they need. This means that organization in its tasks has defined use of intended technology to everyone explicitly and also its effects on improving the performance of job duties is absolutely clear. In other words, being dependent job tasks to a particular technology under the title defined communication technology of career and indicates the intensity of employees need to a particular technology to increase performance. Two dimensions of fitness of the technology to perform job duties and being common in organization are two key dimensions (Kim, 2008).

Ozkan and Kanat (2011) noticed that the model of e-Government was an improvement over TAM in terms of predictive power. The constructs trust, perceived behavioral control and attitudes successfully explained the intention to use an e-Government service. Nasri and Charfeddine (2012) examined empirically the factors that affect the adoption of Internet banking by Tunisian bank customers. They used the Technology Acceptance Model (TAM) and Theory of Planned Behavior (TPB) as base model. The results showed that perceived of usefulness is a significant determinant influencing factor to predict intention of Internet banking using. Yi *et al.* (2006) explained 57% of the physician's intention to accept an innovation, is dependent on status of variables included in the model. Lee and Wu (2011) in a study in Taiwan paid to understanding aspects of technology acceptance model on e-service quality formation. The results illustrate that perceived trust and usefulness, which are the factors of the technology acceptance perspective, positively moderate the relationship between e-service quality, perceived service value, and service satisfaction. Holden and Karsh (2010) believe improving job performance, easier to perform job duties, improving the quality of job operations and related technologies with some important aspects of job responsibilities all are determinants factors of technology usefulness feeling that can affect employee behavior.

Lehto and Lee (2013) arrived at this result that behavioral intention was significantly influenced by both perceived usefulness and user satisfaction. The results of study of Yen *et al.* (2011) indicated that TAM, either original or parsimonious, is successful in explaining user intention to use wireless technology in organizations. Lin *et al.* (2011) presented that the core constructs of the TAM have strong influences on user-intention towards e-Government products. Kim *et al.* (2009) showed that path magnitudes were significantly changed by technology features and complexity. Perceived usefulness had more influence on feature acceptance when basic features were used, and perceived ease of use had more impact on feature acceptance when advanced features were used. According to the study of Melas *et al.* (2011) in Greece, TAM predicts a substantial proportion of the intention to use clinical information systems. Also they presented that public knowledge has a significant positive effect on perceived ease of use. Their

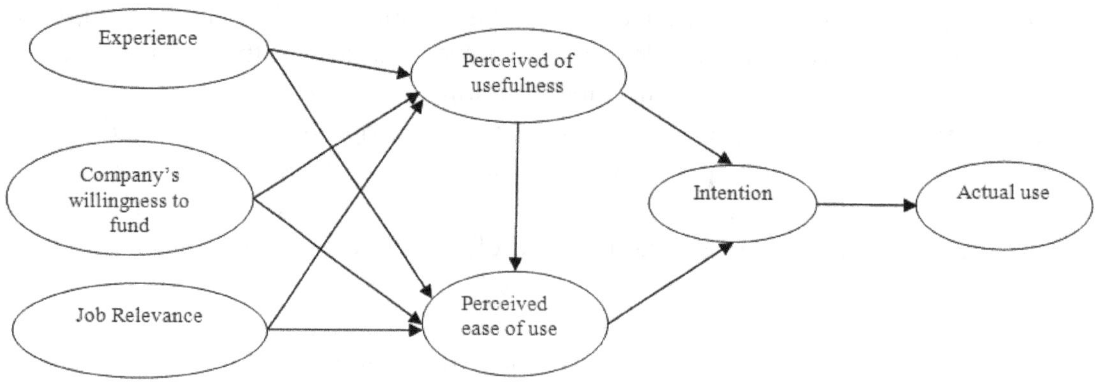

Figure 3: Research framework

result showed that job relevance, prior experience and knowledge are influencing variables toward perceived ease of use and perceived usefulness of information and communication technologies that finally, led to attitude to use of that. Based on the literature review this study aimed at investigating the role of external variables in technology acceptance with organizational context that its conceptual model is as follows.

Research hypotheses

1) Company's willingness to fund has a significant effect on perceived usefulness of information technologies.

2) Company's willingness to fund has a significant effect on perceived ease of use of information technologies.

3) Job relevance has a significant effect on perceived usefulness of information technologies.

4) Job relevance has a significant effect on perceived ease of use of information technologies.

5) Perceived usefulness has a significant effect on perceived ease of use of information technologies.

6) Perceived usefulness has a significant effect on using intention of information technologies.

7) Perceived ease of use has a significant effect on using intention of information technologies.

8) Experience has a significant effect on perceived usefulness of information technologies.

9) Experience has a significant effect on actual use of information technologies.

10) Intention has a significant effect on actual use of information technologies.

MATERIALS AND METHODS

Research method was descriptive – correlational. Data have been collected through a survey. Statistical population was West Azerbaijan Agricultural Extension agents that 120 of them were selected randomly using the Krejcie and Morgan table. A questionnaire was designed to measure the variables in the model and the views of individuals were collected in a five pints scale. In which library studies, questionnaire were designed. Its validity was confirmed by a panel of experts. The Cronbach's alpha coefficient was used to assess the reliability and values ranged between from 0.704 to 0.816. The variables have been investigated in this study were contains three types of variables; independent external variables (job relevance, company's willingness to fund and experience), mediator variables (perceived ease of use, perceived of usefulness and behavioral intention) and dependent variable (actual use). Partial least squares method was dominant statistical method to test research hypotheses. This method is prediction causal method that considering a multivariate approach. When the problem is multi-faceted and complex, and knowledge about that is little, this method has high efficiency. The Partial Least Squares Method as a variance-based approach in structural equation modeling that is considered a second-generation method and has overcome on the weaknesses of the first generation of multivariate techniques (e.g, including multiple linear regression, diagnostic analysis, logistic regression, factor analysis and cluster analysis), such as simple look at the linear stage, especially in linear regression, exist the default based on visibility of the studied variables and ignoring measurement error of variables. The Partial Least Squares Method provides possibility of structures factor formation to measurement latent traits by relevant markers and the same time modeling of relationships between independent and dependent latent traits (Haenlein and Kaplan, 2004). New approach of structural equation modeling in Partial Least Squares format generally, is used when the sample size is not sufficient to estimate the significant of parameters and researcher without concern about the number of parameters and their fit with sample size can evaluate the model. To estimate structural equation models by partial least squares method, the sample size for estimating a model should be five to ten times than number of existing paths in the model (Majchrzak *et al.*, 2005). Given that the present sample size was 120 and 10 paths has been studied in the Model, then the sample size is sufficient to estimate the model. In this study to estimate the significant of parameters of the model, especially path coefficients the bootstrap method used that tests significant of parameters through the simulation and by resampling method. In order to exact estimating of models, minimum number of resampling equivalent to 500 provides reliable results (Newsted, 1996). In this study, is also used 800 beside 500 to increase the accuracy of estimate.

RESULTS AND DISCUSSION

According to findings 65.8 percent (n=79) of respondents had a bachelor's degree. Also 81.7 percent (n = 98) were in formal employment status. The average of respondent's age was 43.14 years with 8.04 years standard deviation. Minimum age was 22 years, maximum age was 58 years. Average of job experience was 18.87 years with 7.74 years standard deviation and minimum and maximum of experience was between one to 30 years. Based on the conducted literature review to measure the variables of the model relevant markers were used. As is discussed in structural equation modeling approach, in the first should investigate the validity of selected measures to measuring latent variables and confirmatory factor analysis is performed for this purpose that the results of observations are shown in Table 1.

Composite reliability coefficients indicator and Cronbach's alpha coefficient for all the structures of the study were higher from 0.6 and 0.7. It explains that structures reliability was satisfied. To calculate the validity was used of structures validity method. In format of structure validity this issue is studying that do selected markers have sufficient accuracy to measure the desired structures or not? Confirmatory Factor Analysis is used. If critical value of t-value for factor loading of each indicator be higher than 1.96, satisfied constructing validity otherwise drop this indicator. In this case, this structure has sufficient accuracy to measure that construct or latent trait. The results of factor loading show that all indicators have a significant p-level at 0.01. In structural equation modeling, in addition to structure validity to examine the importance of the selected markers, the diagnostic validity is intended in the sense that markers of each structure finally provide an appropriate separation in terms of measurement in related to the other structures of the model. In simple terms, each markers only measure its structure and combining them must be in a manner that all structures are well separated. To investigation diagnostic validity, square roots of Average Variances Extracted (AVE's) has been used (Hair *et al.*, 1998). To this purpose, square roots of average variances extracted should be more than other correlations of the other factors with this structure. The results of discriminate validity shows that selected markers to measure the structures have a good diagnostic validity because the AVE's for each structure that in the diameter of matrix in table 2 has been showed is higher than all of the Correlations of other factors with that factor.

After ensuring that the structures are measured correctly, then by using the structural model, causal relations between structures will be reviewed. In fact considering the results of relationship between dependent and independent structures, we can study significant relation between the structures of research by using the re-

Table 1: Actor loading of each indicator of model structures

Latent variables	Indicators	Factor loading	CR	CA	AVE	VIF
Actual behavior	If technology be available I will actual use of it	0.845	0.833	0.799	0.714	1.435
	If technology not available I will ask from managers or Supervisor to take it	0.844				
	According to the IT's benefits I'll use it to perform my job duties	0.820				
Behavioral intentions	I'll recommend use of IT to my colleagues	0.922	0.902	0.801	0.755	1.335
	Always I'll search to finding new capabilities of IT to perform my job duties.	0.863				
Perceived usefulness	Using of IT in my job would increase my productivity	0.867	0.889	0.815	0.619	1.354
	I would find a smartphone useful in my job	0.861				
	Using of IT in my job be able me to perform my duties rapidly	0.636				
Perceived ease of use	Using of IT would enhance my effectiveness on the job		0.901	0.816	0.698	1.225
	Using of IT improve my job efficiency	0.806				
	I would find IT is easy to use	0.743				
Experience of using	The prospect of using this technology and what it could offer me is conceivable for me	0.799 0.875	0.775	0.730	0.633	1.204
	Learning to operate IT would be easy for me					
	The process of IT using is tangibles for me	0.714				
	I want to see the benefits of IT before I apply it	0.939				
	The IT provides me a more efficient and organized tool in my job	0.806 0.786				
	If i have access to the IT, i will use it always	0.519				
	My organization with having experts in case of any problem with the technology urgently take action to fix it	0.853				
Company's willingness to fund	To update an existing and related needs technology, my organization has taken appropriate measures	0.928	0.896	0.704	0.642	2.073
	In my job, if a company pays for any cost to use a smartphone, I will definitely use it	0.784				
	When I need to particular help in related to IT, my organization is willing to meet that through investment	0.859				
	Applying IT in our organization is tangibles	0.852				
	application of IT is appropriate to the nature and duties of my job	0.797				
Job relevance	I feel this technology could facilitate my job processes	0.717	0.902	0.762	0.609	2.551
	For those who wish to enter into the organization as a partner, it is necessary to be familiar with the IT	0.663				
	In my job, usage of a IT is high	0.733				
	In my job, usage of a IT is relevant	0.898				

Table 2: The matrix of the square roots of average variances extracted and correlation

	1	2	3	4	5	6	7
Actual behavior (1)	**0.844**						
Behavioral intentions (2)	0.345	**0.880**					
Perceived usefulness (3)	0.252	0.334	**0.786**				
Perceived ease of use (4)	0.158	0.237	0.356	**0.835**			
Experience of using (5)	0.264	0.350	0.261	0.078	**0.795**		
Company's willingness to fund (6)	0.228	0.224	0.295	0.279	0.083	**0.801**	
Job relevance (7)	0.457	0.267	0.362	0.321	0.125	0.706	**0.780**

* Diagonal is square roots of average variances extracted

lated coefficient. To study significant of path coefficients, bootstrap method has been used as a simulation statistical method. To this mean, the resembling method was used at two statuses 500 and 800 of sample that recommended in partial least squares method. The result show in both cases, being meaningful or not of the parameters, was no change and the results have strong validity (Table 2 and Table 3)

The first hypothesis of research indicates the Company's willingness to fund has a significant effect on perceived usefulness of information technologies from statistical aspect. Studying path coefficients is explaining that effect of company's willing to fund on perceived usefulness is calculated 0.08 that has significant level higher than 0.05. Therefore, there was not sufficient evidence to reject the null hypothesis. In other words, enough evidence to prove that the company's willing to fund has significant effect on perceived usefulness of information technologies were not presented. Studying significant of coefficient to this path in different moods of statistical simulation is not reported.

The second hypothesis of research explains the Company's willingness to fund has a significant effect on perceived ease of use of information technologies from statistical aspect. Studying path coefficients is explaining that effect of company's willing to fund on perceived ease of use is calculated 0.08 that has significant level higher 0.05. Therefore, there was not sufficient evidence to reject the null hypothesis. In other words, enough evidence to prove this the company's willing to fund has significant effect on perceived ease of use of information technologies were not presented. Studying significant of coefficient to this path in different moods of statistical simulation is not reported.

The third hypothesis of research explains the job relevance has a significant effect on perceived usefulness of information technologies from statistical aspect. Studying path coefficients is explaining that effect of job relevance on perceived usefulness is calculated 0.28 that has significant level higher 0.02. Therefore, there was sufficient evidence to reject the null hypothesis. In other words, enough evidence to prove this the job relevance has significant effect on perceived usefulness of information technologies were presented. According to the above findings is characterized that the job relevance is the reason to perceived usefulness of IT application and whatever scope of the use of information technology be more relevant to the desired job, the staff will evaluate using of it more advantageous.

The fourth hypothesis of research explains the job relevance has a significant effect on perceived ease of use of information technologies from statistical view. Studying path coefficients is explaining that effect of job relevance on perceived ease of use is calculated 0.26 that has significant level higher 0.02. Therefore, there was sufficient evidence to reject the null hypothesis. According to the above findings is characterized that the job relevance is the reason to perceived ease of IT application and whatever area of the use of information technology be more relevant to the desired job, the staff will evaluate using of it easier. Thus we can know that perceived ease of technology using in organizational studies as a function of job relevance.

The fifth hypothesis of research explains the

Table 3: Path coefficients and significant of direct path

Studied path		Path coefficient	p-value	S.E	R^2	VIF	Resampling rate BT = 500	BT = 800
Company's willingness	Perceived usefulness	0.08	0.26	0.124		1.995	0.645	0.501
Job Relevance	Perceived usefulness	0.28	0.02	0.136	0.58	2.013	2.05	2.18
Experience of using	Perceived usefulness	0.72	< 0.01	0.085		1.016	8.47	8.68
Job Relevance	Perceived ease of use	0.26	0.02	0.117		2.104	2.22	2.34
Company's willingness	Perceived ease of use	0.08	0.25	0.126	0.27	2.002	0.634	0.689
Perceived usefulness	Perceived ease of use	0.37	< 0.01	0.076		1.155	4.86	5.05
Perceived usefulness	behavior intention	0.29	< 0.01	0.092		1.145	3.15	3.23
Experience of using	Actual behavior	0.36	0.03	0.088	0.43	1.139	4.09	4.17
Perceived ease of use	behavior intention	0.27	0.03	0.102		1.145	2.64	2.84
behavior intention	Actual behavior	0.79	< 0.01	0.078	0.64	1.139	10.12	10.25

perceived usefulness has a significant effect on perceived ease of use of information technologies from statistical aspect. Studying path coefficients is explaining that effect of perceived usefulness on perceived ease of use is calculated 0.37 that has significant level lower 0.01. Therefore, there was sufficient evidence to reject the null hypothesis. According to the above findings is characterized that the perceived usefulness is the reason to perceive ease of IT application and whatever perceived usefulness be higher, the staff will evaluate using of it easier. Thus we can know that perceived ease of technology using in organizational studies as a function of staff's perception from perceived usefulness.

The sixth hypothesis of research explains the perceived usefulness has a significant effect on using intention of information technologies from statistical aspect. Studying path coefficients is explaining that effect of perceived usefulness on using intention is calculated 0.29 that has significant level lower 0.01. Therefore, there was sufficient evidence to reject the null hypothesis. According to the above findings is characterized that the perceived usefulness is the reason to using intention of IT application and whatever perceived usefulness be more, the staff have higher intention to use. Thus we can know that intention of technology using in organizational studies as a function of staff's perception from perceived usefulness.

The seventh hypothesis of research explains the perceived ease of use has a significant effect on using intention of information technologies from statistical aspect. Studying path coefficients is explaining that effect of perceived ease of use on using intention is calculated 0.27 that has significant level 0.03. Therefore, there was sufficient evidence to reject the null hypothesis. According to the above finding is characterized that the perceived ease of use is the reason to using intention toward IT application and whatever perceived ease of use be more, the staff have higher intention to use. Thus we can know that using intention as a function of staff's perception from perceived ease of use.

The eighth hypothesis of research explains the experience has a significant effect on perceived usefulness of information technologies from statistical aspect. Studying path coefficients is explaining that effect of experience variable on perceived usefulness is calculated 0.72 that has significant level lower 0.01. Therefore, there were sufficient evidence to reject the null hypothesis. According to the above finding is characterized that the experience is the reason to perceived usefulness toward IT application and whatever experience be more, the staff will evaluate using of that more useful.

The ninth hypothesis of research explains that experience has a significant effect on actual use of information technologies from statistical aspect. Studying path coefficients is explaining that effect of experience on actual use is calculated 0.36 that has significant level 0.03. Therefore, there were sufficient evidence to reject the null hypothesis. In other words, enough evidence to prove the experience has significant effect on actual use of information technologies were presented. According to the above findings is characterized that the experience is the reason to actual use of IT application and whatever experience be more, the staff will use that more.

The tenth hypothesis of research explains that intention has a significant effect on actual use of information technologies from statistical aspect. Studying path coefficients is explaining that effect of the intention variable on actual use is calculated 0.79 that has significant level lower 0.01. Therefore, there were sufficient evidence to reject the null hypothesis. In other words, enough evidence to prove the intention has significant effect on actual use of information technologies were presented. According to the above findings is characterized that the intention is the reason to actual use of IT application and whatever user intention toward technology using be more, the staff will have more actual use (Figure 4).

To demonstrate the validity of the results of research model, specifically fitness indexes of structural equation models was used in format of partial least squares method. Average Variance Inflation Factor index was 1.485. This result shows multiple alignment of the model is controlled well and the model accuracy to predict

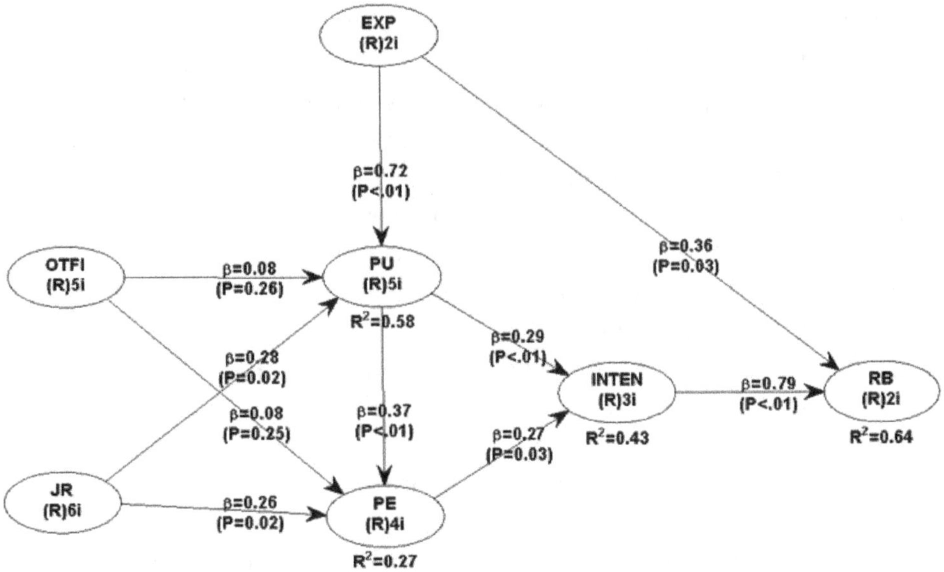

Figure4: The path model of effects of research variables based on partial least squares estimation

the dependent variable has a good reliability. In other words, each of the independent structures that effect on independent variables, account unique part of the variance of the dependent variable associated with it. The Variance Inflation status for predictive independent variables of study shows each of the independent variables has their unique role and in formation of amount of variance accounted by the model, there is no a significant interferes. Because the VIF index for all structures in the direction having predictor variables have been measured below the critical 5 that this critical values is the maximum coefficient of the overlap tolerance of explained variance and lower values of that indicate no replacement property among independent variables and that is the main assumptions of simultaneous equations systems. Values of explained variance mean and path coefficients mean show that the relationship between variables is well known. The amount of explained variance average index was 0.482 and significance level of that was 0.01 and Suggests that a significant proportion of the variance in the data expressed by existing relationships. Also the path coefficients average index with 0.351 value and significance level of that was 0.01. These shows existing coefficients to explain related causal relationship are repeatable and they have required accuracy to explain relations. Eventually it became clear that this model can estimate 64 percent of real behavior changes, 43 percent of use intention changes, 58 percent of perceived usefulness changes and 27 percent of perceived ease of use.

CONCLUSIONS

Although different models for technology acceptance have been proposed, the Theory of Planned Behavior as a heart of TAM continues to be the one most widely used and a generally accepted theory to explain intention to technology acceptance. Nevertheless, new questions have emerged with respect to its specific configuration in the case of actual behavior. There is a call by some researchers to advance our understanding of how the TAM may be improved or modified to better reflect the full complexity of acceptance processes.

Results of this study showed among the three added variables to TAM the organization's willingness to fund, job relevance and previous experience in using the technology, the company's willingness to fund's variable did not show a significant role on perceived usefulness and perceived ease of use. However, job relevance variable showed a significant role on perceived usefulness and perceived ease of use. These results indicate that one of the suggested variables in organizational studies, which can help to develop technology acceptance model, is job relevance variables that directly have effects on perception of usefulness and ease of use, and indirectly on the actual application of information technology in doing the responsibility of the job. Therefore determine the technology abilities to staff and demonstrate the capabilities of technology to doing job duties better can be one solutions of breaking down resistance to technology acceptance by the employees.

The third added variable to the technology acceptance model as previous experience in using technology has a very substantial effect on perceived usefulness and actual behavior variable too. This result shows that any organizational attempt to providing learning experience opportunities for employees and job rotation in addition that can also provide the opportunity to learning experience, will increases employees readiness to technology acceptance from their perceptions of usefulness aspect. The above result is obtained in studies such as Kim (2008). Therefore step by step implementation of the organization required technologies it will be appropriate approach for the promotion of its basic functionality and mentally prepared formation to accept. This finding is consistent with Ecceles theory, stating that attitude mediates relationship between past performance and future success (Ecceles, 1990). In fact, this theory explains what one gets is a function of past experience and shaping his related attitude. The study also showed that experiencing is a major factor among the studied factors of technology acceptance model. Other results of studied paths those were also considered in other models of technology acceptance have significant path coefficient, this would mean that proposed relationships in the primary model of technology acceptance also in this study had significant path coefficients, in this face that perceived usefulness has a direct effect on ease of use and also perceived usefulness and perceived ease of use have a significant effect on acceptance intention. On the other it was found using intention is one of the important predictor variables of behavior in real condition. This finding corresponded with the results of its previous similar studies (Holden and Karsh, 2010; Hu *et al.,* 2003; Kim *et al.,* 2009; Legris *et al.*, 2003; Melas *et al.*, 2011; Ozkan and Kanat, 2011). Based on the research findings can provide the following suggestions:

- It is suggested that while providing opportunities related to job rotations to this mean that staff be familiar with backgrounds of application of IT, the extent of its application to staff be more transparent. Also in providing the technology, need context in the staff be considered and in inadequacy of that by introducing aspects of the new technology those are more tangible and has a greater proximity with job context of each employee's be gained need experience for them.
- On the other hand before providing any technology, in the first be investigated the point of view of staff and if there are negative perceptions provide appropriate strategies to solve them. Being IT departments active in the organization can will be an important role to gain this positive experience and also one of the responsibilities of this unit will be identifying the amount and type of information technology use among staff.
- Since in the present study, external variables are in the organization level it suggests that can be use variables such as trust, perceived self-efficacy of computer and self-image as an individual variable to predicting behavior in other studies.
- Standardization scales for measuring the shape and usage amount of information technology in regards to job relevance variable can be noted as a research context in the area of technologies acceptance.
- As regards to significant role of experience and job relevance should be used variables such as age, work experience and education level as moderator variables in the relationships of acceptance model.
- As regards company's willingness variable did not show significant effect the pathology in the studied organizations should be done. It seems that the trend that organization is ahead in terms of investment in information technology is Contrary to the employees perceptions of usefulness and ease of use and finally having appropriate intention to necessary using in job context.

ACKNOWLEDGEMENT

The study was conducted with the help of Department of Agricultural Extension and Education of Tehran University and Agricultural Extension

Management of West Azerbaijan Province. We extend special thanks to all those who gave their time to participate in this study.

REFERENCES

1- Ahangari, E. (2014). *Providing a model of Environmental Friendly technologies acceptance among the West Azerbaijan province's wheat farmers.* Unpublished thesis, University of Tehran, Iran. (In Persian)

2- Charfeddine, L., & Nasri, W. (2012). Factors affecting the adoption of Internet banking in Tunisia: An integration theory of acceptance model and theory of planned behavior. *Journal of High Technology Management Research*, 23, 1-14.

3- Chen, C., Fan, Y., & Farn, C. (2007). Predicting electronic toll collection service adoption: An integration of the technology acceptance model and the theory of planned behavior. *Emerging Technologies,* 15(5), 300-311.

4- Cothran, T. (2011). Google scholar acceptance and use among graduate students: A quantitative study. *Journal of Library and Information Science Research,* 33(4), 293–301.

5- Dabholkar, P. A. (1996). Consumer evaluations of new technology-based self-service options: An investigation of alternative models of service quality. *International Journal of Research in Marketing*, 13 (1), 29-51.

6- Davis, F. D. (1989). Perceived usefulness, perceived ease of use, and user acceptance of information technology. *Management Information Systems Journal*, 13, 319- 339.

7- Davis, F. D., Bagozzi, R. P., & Warshaw, P. R (1989). User acceptance of computer technology: A comparison of two theoretical models. *Management Science,* 35(8), 982-1003.

8- Dillon, A., & Morris, M. G. (1996). User acceptance of information technology: Theories and models. In M. E. Williams (Ed.), *Annual Review of Information Science and Technology* (ARIST): Vol. 31 (pp. 3-32).

9- Eccles, J. S. (1990). Predictors of math anxiety and its influence on young adolescents, course enrollment intention and performance in mathematics. *Journal of Educational Psychology*, 82, 60-70.

10- Elahi M. M. Cagampang, F. R. Mukhtar, D. Anthony F.W. Ohri, S. K & Hanson, M. A. (2009). Long-term maternal high-fat feeding from weaning through pregnancy and lactation predisposes offspring to hypertension, raised plasma lipids and fatty liver in mice. British. *Journal of Nutrition,* 10 (2), 514–519.

11- Fishbein, M., & Ajzen, I. (1975). Belief, Attitude, Intention, and Behavior: An Introduction to Theory and Research. Reading, MA: Addison-Wesley

12- Gyampah, K. A., & Salam, A. F. (2004). An extension of the technology acceptance model in an ERP implementation environment. *Information & management*, 41 (6), 731-745.

13- Haenlein, M., & Kaplan, A. (2004). A Beginner's Guide to Partial Least Squares Analysis. *Understanding Statistics*, 3, 283-297.

14- Hair, J.F. Jr., Anderson, R.E., Tatham, R.L., & Black, W.C. (1998). Multivariate Data Analysis, *(5th Edition)*. Upper Saddle River, NJ: Prentice Hall.

15- Holden, R. J., & Karsh, B. T. (2010). The technology acceptance model: Its past and its future in health care. *Journal of Biomedical Informatics*, 43, 159–172.

16- Hu, P.J.H., Clark, T.H.A., & Ma, W.W. (2003). Examining technology acceptance by school teachers: a longitudinal study. *Information and Management,* 41, 227–241.

17- Kim, S.H. (2008). Moderating effects of job relevance and experience on mobile wireless technology acceptance: Adoption of a smartphone by individuals. *Information and Management,* 45, 387–393.

18- Kim, H.J., Mannino, M., & Nieschwietz, R.J. (2009). Information technology acceptance in the internal audit profession: Impact of technology features and complexity. *International Journal of Accounting Information Systems*, 10, 214–228.

19- Kulviwat, S., Bruner, C.G., & Al-Shuridah, O. (2009). The role of social influence on adoption of high tech innovations: The moderating effect of public/private consumption. J*ournal of Business Research*, 62, 706–712.

20- Kuttschreuter, M., Horst, M., & Gutteling, J. M. (2007). Perceived usefulness, personal experiences, risk perception and trust as determinants of adoption of e-government services in The Netherlands. *Computers in Human Behavior,* 23, 1838–1852.

21- Lee, F.H., & Wu, W.Y. (2011). Moderating effects of technology acceptance perspectives on e-service quality formation: Evidence from airline websites in Taiwan. *Journal of Expert Systems with Applications*, 38, 7766–7773.

22- Legris, P., Ingham, J., & Collerette, P. (2003). Why do people use information technology? A critical review of the technology acceptance model. *Information and Management,* 40(3), 191–204.

23- Lehto, M. R., & Lee, D. U. (2013). User acceptance of YouTube for procedural learning: An extension of the technology acceptance model. *Computers and Education,* 61, 193-208.

24- Lin, F., Fofana, S. S., & Liang, D. (2011). Assessing citizen adoption of e-Government initiatives in Gambia: A validation of the technology acceptance model in information systems success. *Government Information Journal*, 28, 271-279.

25- Lucas Jr., H. C., & Spitler, V. (2000). Implementation in a world of workstation and networks-. *Information and Management,* 38, 119- 128.

26- Majchrzak, A., Lim, R., & Chin, W. W. (2005). Managing Client Dialogues during Information Systems Design to Facilitate Client Learning. *MIS Journal*, 29(4), 653-672.

27- Melas, C. D., Zampetakis, L. A., Dimopoulou, A., & Moustakis, V. (2011). Modeling the acceptance of clinical information systems among hospital medical staff: An extended TAM model. *Journal of Biomedical Informatics*, 44, 553–564.

28- Morris, M.G., & Dillon, A. (1997). How user perceptions influence software use. IEEE Software, 14(4), 58-65.

29- Nasri, W., & Charfeddine, L. (2012) Factors Affecting the Adoption of Internet Banking in Tunisia: An Integration Theory of Acceptance Model and Theory of Planned Behavior. *The Journal of High Technology Management Research*, 23, 1-14.

30- Newsted, P. R., Chin, W. W., & Marcolin, B. L. (1996). A Partial Least Squares Latent Variable Modeling Aproach for Measuring Interaction Effects: Results from a Monte Carlo Simulation Study and Voice Mail Emotion/Adoption Study. Proceedings of the seventeenth international conference on information systems, Cleveland, Ohio, December 16-18, 1996.

31- Ozkan, S., & Kanat, I.E. (2011). E-Government adoption model based on theory of planned behavior: Empirical validation. *Government Information Journal,* 28, 503–513.

32- Yaghoobi, N.M., & Shakeri, R. (2010). Comparative analysis of the technology acceptance model.

7

Evaluation of Palm Groves Technical Efficiency Using Bootstrap Data Envelopment Analysis: A Case Study of Roodkhanehbar Area, Iran

***Gholamreza Zamanian* [1] *and Mostafa Khajeh Hassani* [2*]**

Keywords:
bootstrap, Data Envelopment Analysis, Hormozgan Province, Keriteh date, Roodkhanehbar, technical efficiency

Abstract

Roodkhnehbar area, having approximately 111 thousands of Keriteh palm trees, is one of the most important areas of date production in the Rudan County[1] and the source of peoples' income in this area, directly or indirectly. As a result, its production efficiency has a critical importance to the orchardists in this region. This study aims to evaluate technical efficiency of palm groves in this area using input-oriented bootstrap data envelopment analysis and sampling 50 palm groves of Keriteh date producers of Roodkhanehbar area in 2013. The results suggested that 64% of date producers operate with less than 50% efficiency and only 14% of them operate efficiently. The study, then, carries the implication that it is recommended to train the orchardists, providing a chance for successful orchardists to share their experiences with others in an attempt to optimize allocation of inputs.

[1] Assistant Professor, Economics, University of Sistan and Baluchestan, Zahedan, Iran
[2] PhD Candidate, Agricultural Economics, University of Sistan and Baluchestan, Zahedan, Iran
* Corresponding author's email: zamanian@eco.usb.ac.ir

[1] Rudan County is a county in Hormozgan Province in Iran

INTRODUCTION

One of the most important economic goals and policies implemented in Iran is to decrease dependency on oil revenues and increase non-oil products revenues. Considering oil market depression in recent years and harmful effects which single-product export has on the country, the importance of this policy becomes further egregious. Taking this into account and because of high potentials in agricultural sector, the country policy makers and economists should consider these facts in their planning. Agricultural sector of Iran includes a wide range of activities and the most important of them is date production, which as a potential substitute, can provide more work opportunities and thereby support the development of the country. As shown below (Figure 1), producing 1.061 million tons of dates in 2012, Iran had the world's second rank in date production after Egypt. In addition to meeting domestic demands andexporting 112030 tons of dates in 2011 Iran is ranked as the third country after Iraq and Pakistan,.

Despite Iran's rank in production and export of dates, this country was ranked twelfth among 37 producing countries in 2012, and has had not great performance per unit area compared to other competitors such as Egypt, China, Sudan, Colombia, and America, because of inappropriate and traditional production methods in producing chain in nation groves (FAO, 2012). For this reason, to maintain the position and increase comparative advantages, and subsequently, increase the competition power in foreign markets, it is always necessary that the performance should be improved by studying efficiency and productivity indexes of groves of different regions of the country.

Hormozgan Province is one of the most important producers of dates in the country. There are over 100 varieties of dates in the country, 80 of which are planted in Hormozgan palm gardens. The most important and favorable varieties include Mazafati, Khanizi, Shahani, Piaram, Khasoei, Keriteh, Mordasang, Zarak,and red Kolk. This province is located in fourth and seventh ranks, respectively in terms of palm cultivation and production. It takes five months to harvest dates in this and some other provinces. Roodkhanehbar area, located in Rudan County, is one of the palm growing areas where this activity has so much influence on the quality of residents' lives. This area has the biggest area under Keriteh date cultivation with about 100 groves and over one hectare and 111 thousands palm trees. Despite being very nutritious, Keriteh variety is not considered among other expensive varieties such as Piaram and Mazafati and is produced as human, livestock and industrial food supply in foreign markets of China, India, and Afghanistan. The majority of the people in this region are laborers at palm groves, and any improvement in the allocation of resources which leads to the reduction of production costs, will consequently increaselabor profit, which will considerably accelerate the development of the region. What is of great importance is evaluation of technical efficiency of the region palm grovesas well as identification of inefficient units; theselead to improvement in resource allocation. Given the perceived necessities, the present study aimed to evaluate the technical efficiency of Roodkhanehbar regionKeriteh palm groves in 2013 using bootstrap input oriented data envelopment analysis. Despite the fact that extensive studies have already been conducted

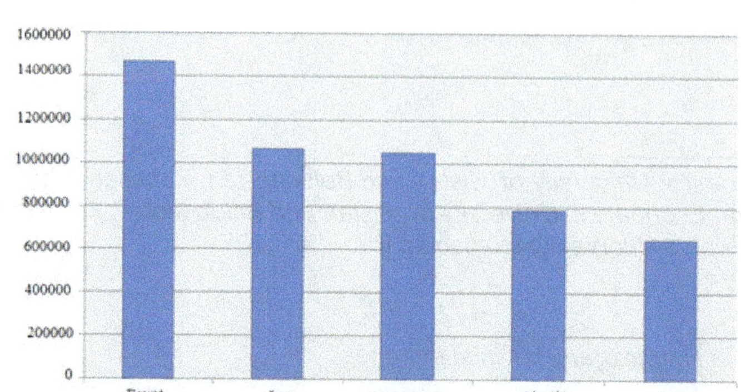

Figure 1. The first five date producers of the world

in the field of evaluating agricultural sector activities efficiency inside and outside the country, a few of them have used bootstrap data envelopment analysis, the most important of them include (Balcombe et al., 2008; Brümmer, 2001; Dong & Featherstone, 2006; Gocht & Balcombe, 2006; Odeck, 2009). A theoretical foundations and methodology of study;brief description of the data used and the results of model estimation, are presented here.

MATERIALS AND METHODS

Theoretical foundations

In 1957 Farrell calculated efficiency of American agricultural sector based on Economic theories using nonparametric methods. Citing five principles, he constructed a collection called production possibility and considered a part of its frontier as an estimation of production function. Every Decision Making Unit (DMU), which is placed on this frontier, is efficient and inefficient otherwise (Farrell, 1957). Because of scientific problems in measuring and limitations raised in the Farrell method, it did not find much practical application and remained silent for years until 1978. To remedy this problem, Charnes, Cooper, and Rhodes (CCR) introduced the method of Data Envelopment Analysis (DEA) by universalizing the Farrell method such that it includes the characteristic of production process with several factors and outputs. In this method, it is not required to follow a certain default order so as to estimate the production function, and efficiency of a firm (decision making unit) is measured relative to the efficiency of other firms (Cooper et al., 2006). Unfortunately, in actual situation the maximum amount of output produced from a particular input cannot be calculated by using a sample, because the studied sample is a sample of an unknown population, and efficient frontier of the population is unknown. As shown in the following figures, the efficiency of unit A in Figure 2 significantly changes compared to Figure 3 (Bahadori et al., 2013).

This efficiency change is a consequence of nonparametric nature of DEA model. As can be seen, DEA frontier depends on and is sensitive to the samplesuch that by changing the sample, the previous frontier crumbles down. Of course, not all the shortcomings of this model are due to its nonparametric nature; however, it also depends on the sample size (Bahadori et al., 2013). To overcome this problem, Simar (1996) designed a method entitled Bootstrap DEA approach to evaluate the variability of efficiency for every sampling (Löthgren & Tambour, 1996; Simar, 1996). In this method, the Bootstrap technique is used to demonstrate ranking and sensitivity of efficiency scores obtained by DEA method relative to the variation of sample combination. Bootstrap is a resampling technique proposed by Efron and Tibshirani and is applied to estimate sampling distribution characteristics of an estimator (in cases when it is difficult to obtain it with other methods) (Ebadi, 2011; Efron & Tibshirani, 1993). In the simplest format, Bootstrap is a random selecting among thousands of pseudo samples by using simple random sampling,which replace a set of series of actual samples (original sample);by using each pseudo sample a pseudo estimation of pseudo efficiency score is obtained. These thousands pseudo estimates form an empirical distribution for estimator which is used as an estimate of the sampling distribution of the main population.

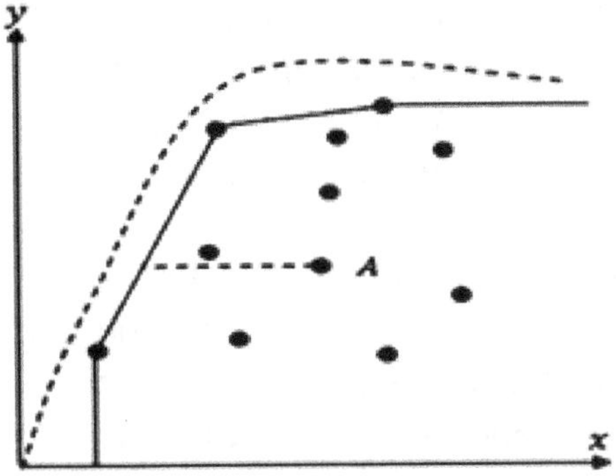

Figure 2: The first sample

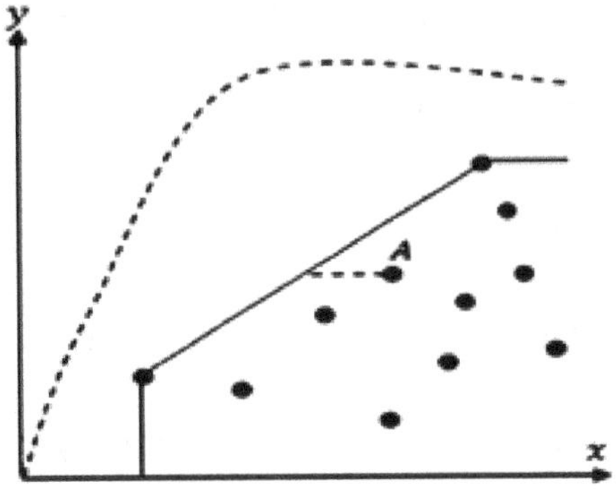

Figure 3: The second sample

Overall data generation process (DGP) and Bootstrap

The activity of a production unit produces q output $(y \in R_{+}^{q})$ using p. inputs $(x \in R_{+}^{q})$ can be shown with production set of ψ consisting of feasible and physical set (x, y), as the following equation:

$$\psi=\{(x, y) \in R_{+}^{p+q} | x \rightarrow y\} \quad \text{Input x can produces output y} \tag{1}$$

This set which may be shown as its two components, includes input set and output setso that the input set for $\forall y \in \Psi$ is shown as equation (2) and the output set for $\forall x \in \Psi$ is shown by equation (3) as below:

$$X(y)=\{X \in R_{+}^{P} | (x,y) \in \psi\} \tag{2}$$

$$Y(x)=\{Y \in R_{+}^{P} | (x,y) \in \psi\} \tag{3}$$

The relation between two sets of (2) and (3) can be explained by a set of standard assumptions provided by Shephard including the convexity assumption $X(y)$ for all y (Y(x) for all x) and the disposability assumption of inputs and outputs and so on (Shephard, 1970). The efficient frontier expressed by Farrell can be shown as a subset of $X(y)$ or $Y(x)$ which are shown as $\partial X(y)$ and $\partial Y(x)$, respectively, and in equations (4) and (5):

$$\partial X(y)=\{x | x \in X(y),\ \theta x \notin X(y),\ \forall 0<\theta<1\} \tag{4}$$

$$\partial Y(x)=\{y | y \in Y(x),\ \beta y \notin Y(x),\ \forall \beta>1\} \tag{5}$$

The already mentioned equations can be used to define as calculation indexes to obtain input- and output- oriented efficiencies for k^{th} firm (x_k, y_k) as equations (6) and (7). In the following just the input oriented model is described [2] (Färe & Grosskopf, 2006; Simar & Wilson, 1998).

$$\theta_k=Min\{\theta | \theta x_k \in X(y_k)\} \tag{6}$$

$$\beta_k=Min\{\beta | \beta y_k \in X(x_k)\} \tag{7}$$

If $\theta_k=1$, the firm (x_k, y_k) operates efficiently. Yet, if efficiency level is lower than unit $(\theta_k<1)$, this production unit (x_k, y_k) operates inefficiently and can produce using less input. For further description, it will be beneficial to definefunction level of input oriented kth firm at production level y_k, by the following equation:

$$X^{\partial}(x_k | y_k)=\theta x_k \tag{8}$$

Note that $X^{\partial}(x_k | y_k)$ is intersection point of the efficient frontier $\partial X(y)$ and radius θx_k, and in order to calculate input oriented efficiency, the ratio of radial distance of firm is to be measured (x_k, y_k) to its equivalent point on the efficient frontier of $\partial X(y)$, $X^{\partial}(x_k | y_k)$. Since the production set (ψ) and thus input set $(X(y))$ and production efficient frontier $(\partial X(y))$ are unknown, the efficiency level of kth production unit (θ_k) will be unknown as well. Supposedly, using data generation process *(DGP)* as represented by ρ, we can construct a random sample set $\chi=\{(x_i,y_i) | i=1,2,...,n\}$; $\hat{X}(y)$ and $\partial\hat{X}(y)$ can be obtained by using the equation method M and estimation of desired sets in the form of $\hat{\Psi}$. Therefore, efficiency of production unit (x_k, y_k) can be estimated via the following equation (Simar & Wilson, 2000).

$$\hat{\theta}_k = Min\left\{\theta | \theta x_k \in \widehat{\partial X(y_k)}\right\} \tag{9}$$

Note that sampling properties of $\hat{\Psi}$, $\hat{X}(y)$, $\partial\hat{X}(y)$ and subsequently $\hat{\theta}_k$ all depend on data generation process of ρ, which is unknown. Furthermore, even if ρ is known, achieving them by M method is very difficult, especially when M is a non-parametric method. In circumstances like thiswhere partial specification and analysis of sampling properties of estimators are very difficult or impossible, the Bootstrap method can be the most appropriate and practical for this aim. Assume data generation process of ρ is known and can be reached to an acceptable estimation like $\hat{\rho}$ by the main sample (χ), $\hat{\rho}$ is used to generate data set $\chi^* = \{(xi^*, yi^*) | i=1, 2, ..., n\}$; by using M method and this pseudo sample, an estimation of respected sets of $\hat{\Psi}^*$, $\hat{X}^*(y)$ and, $\partial\hat{X}^*(y)$ corresponded to the pseudo

[2] *The statements presented for input oriented model can be easily rewritten for output oriented model as well*

sample χ^*; and thus an estimation of the efficiency level $\hat{\theta}_k^*$ of kth firm under the study (x_k, y_k) is obtained as relation (10).

$$\hat{\theta}_k^* = Min\left\{\theta \mid \theta x_k \in \partial\widehat{X^*(y_k)}\right\} \tag{10}$$

It should be noted that sample properties of estimators $\hat{\Psi}^*$, $\hat{X}^*(y)$, $\partial\hat{X}^*(y)$ and consequently $\hat{\theta}_k$ are totally depended on production method ρ which is an unknown method. Additionally even ρ was known, obtaining them using *M* method would be too difficult from the original sample χ only if $\hat{\rho}$ can be fully known; In this case, they may be difficult to calculate analytically. Yet, by using *Mont Carlo method*, an approximation of sample distributions can be simply achieved. Using $\hat{\rho}$, B pseudo samples χ_{b^*} *(b= 1, 2,...B)* are generated, and then by *M* method, pseudo estimates of $\hat{\Psi}_b^*, \hat{X}^*(y)_b$, and $\partial\hat{X}^*(y)_b$, (b=1, 2, ..., B) for each pseudo sample are obtained, and ultimately efficiency level $\{\hat{\theta}_{k,b}^*\}_{b=1}^{B}$ for each studied unit is calculated. The empirical density function $\{\hat{\theta}_{k,b}^*\}_{b=1}^{B}$ is Mont Carlo approximation of distribution $\hat{\theta}_k^*$ subject to $\hat{\rho}$. The Bootstrap method is based on the idea that if $\hat{\rho}$ is an acceptable approximation of ρ, the known Bootstrap distribution will simulate sample distribution of estimators of *Ψ, X(y), ∂X(y)* and θ_k which are of interest but unknown. Accordingly, in order to measure the efficiency level θ_k of the firm (x_k, y_k) the equation (11) must be satisfied.

$$\left(\hat{\theta}_k^* - \hat{\theta}_k\right) \mid \hat{\rho} \sim \left(\hat{\theta}_k - \theta_k\right) \mid \rho \tag{11}$$

Where, θ_k, $\hat{\theta}_k$ and $\hat{\theta}_k^*$ are defined by equations (6), (9), and (10), respectively. The above equation is valid and true if $\hat{\rho}$ is a consistent estimate of ρ. As suggested by to the equation (11), bias of $\hat{\theta}_k$ can be obtained from the main estimator of the population θ_k as follows:

$$bias_{\rho,k} = E\left(\hat{\theta}_k\right) - \theta_k \tag{12}$$

The equivalent of the above equation in the Bootstrap space may be stated as the following equation.

$$bias_{\hat{\rho},k} = E\left(\hat{\theta}_k^*\right) - \hat{\theta}_k \tag{13}$$

The expected value for $\hat{\theta}_k^*$ may be substituted by its *Mont Carlo approximation* as follows:

$$E\left(\hat{\theta}_k^*\right) = \frac{1}{B}\sum_{b=1}^{B}\hat{\theta}_{k,b}^* = \overline{\theta}_k^* \tag{14}$$

Therefore:

$$\widehat{bias}_k = \frac{1}{B}\sum_{b=1}^{B}\hat{\theta}_{k,b}^* - \hat{\theta}_k = \overline{\theta}_k^* - \hat{\theta}_k \tag{15}$$

As suggested by the equation (11), one can obtain an estimate of (12) using (15). Thus:

$$bias_{\rho,k} \approx \frac{1}{B}\sum_{b=1}^{B}\hat{\theta}_{k,b}^* - \hat{\theta}_k = \overline{\theta}_k^* - \hat{\theta}_k \tag{16}$$

By correcting the bias of the main estimator ($\hat{\theta}_k$), the bias-corrected estimator is obtained as equation (17) bellow:

$$\begin{aligned} \tilde{\theta}_k &= \hat{\theta}_k - bias_{\rho,k} \approx \hat{\theta}_k - \left(\frac{1}{B}\sum_{b=1}^{B}\hat{\theta}_{k,b}^* - \hat{\theta}_k\right) \\ \tilde{\theta}_k &\approx 2\hat{\theta}_k - \frac{1}{B}\sum_{b=1}^{B}\hat{\theta}_{k,b}^* \\ \tilde{\theta}_k &\approx 2\hat{\theta}_k - \overline{\theta}_k^* \end{aligned} \tag{17}$$

For this estimator, the term of bias-corrected estimator which is applied as the obtained bias, is not its exact value but an approximate one. Hence, the bias of estimator is not eliminated but just modified. In addition, the standard deviation of estimator $\hat{\theta}_k^*$ is shown as equation (18).

$$\widehat{se}_{\rho,k} = \left\{\frac{1}{B-1}\sum_{b=1}^{B}\left(\hat{\theta}_{k,b}^* - \overline{\theta}_k^*\right)\right\}^{\frac{1}{2}} \tag{18}$$

At the end of this section, the confidence

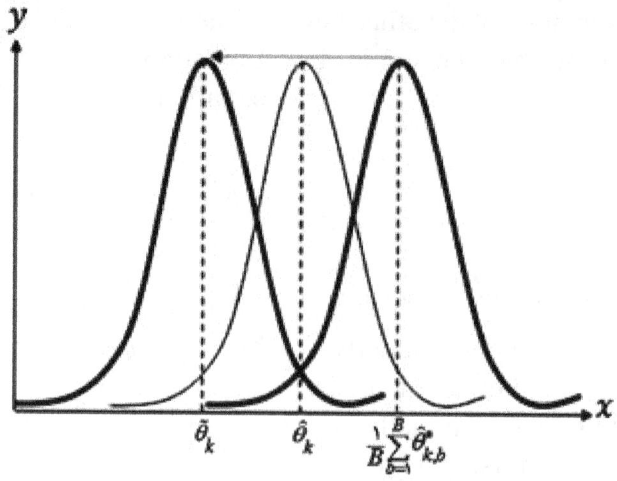

Figure 4. Bias- correction of distribution function

interval of θ_k is determined and the empirical distribution functions of $\hat{\theta}^*_{k,b}$, *(b=1, 2,..., B)* is provided after correcting the bias. We need a corrected empirical density distribution function with the centrality of bias-corrected estimator ($\tilde{\theta}_k$) from θ_k. Thus, we should move the empirical density function $\hat{\theta}^*_{k,b}$ to the left by $2\widehat{bias}_k$. As shown in figure 4, if it moves to the left by $1\widehat{bias}_k$, the empirical density function will be centralized at $\hat{\theta}_k$ rather than $\tilde{\theta}_k$ (Bahadori et al., 2013).

Therefore, the empirical density function can be shown as $\hat{\theta}^*_{k,b}$ and the confidence interval θ_k with coverage level (1-2α) can be shown by the equations (19) and (20), respectively.

$$\tilde{\theta}^*_{k,b} = \hat{\theta}^*_{k,b} - 2\widehat{bias_k} \tag{19}$$

$$\left(\hat{\theta}_{k,low}, \hat{\theta}_{k,up}\right) = \left(\tilde{\theta}_k^{*(\alpha)}, \tilde{\theta}_k^{*(1-\alpha)}\right) \tag{20}$$

In which α^{th} percent (critical value) in phrase $\tilde{\theta}_k^{*(\alpha)}$ is applied to determine confidence interval and so empirical density function $\tilde{\theta}^*_{k,b}$, *(b=1, 2,..., B)*. If the empirical density function is $\tilde{\theta}_k^{*(\alpha)}$ unbalanced, it would be preferred to select the median as the distribution centrality of $\tilde{\theta}_k$. What which has not been answered in this section is how $\hat{\rho}$ should be selected. Since the answer depends on the method of estimating *M*, the second subsection briefly describes the data envelopment analysis method (*M* method), and third one addresses various algorithms of $\hat{\rho}$ selection (Bahadori et al., 2013; Simar & Wilson, 1998, 2000).

Data envelopment analysis

The *data envelopment analysis approach* measures the efficiency of decision making unit *(x_i, y_i)* based on determination of production collection resulting from the sample $\chi = ((x_i, y_i), i=1, 2, ..., n)$ and its general form is shown as the following equation:

$$\hat{\Psi}_{DEA} = \left\{(x_i, y_i) \in R^{p+q} \mid y_i \le Yz, x_i \ge Xz, \sum_{i=1}^{n} z_i = 1, z \ge 0, i = 1, 2, .., n\right\} \tag{21}$$

Given the above equation, the input set for the production level of *y* is estimated as follows:

$$\widehat{X(y)} = \left\{x \in R_+^p \mid (x, y) \in \hat{\Psi}_{DEA}\right\} \tag{22}$$

The estimated efficient input oriented frontier for the above input set at the output level of *y*, is shown by $\partial\widehat{X(y)}$; this frontier is a subset of $\widehat{X(y)}$ and obtained from definition of $\hat{\Psi}_{DEA}$. Considering the above-mentioned definitions, for each *DMU(x_i, y_i)*, the estimated efficiency level ($\hat{\theta}_i$) of ith*DMU* with variable returns to scale assumption, may be computed by solving the equation (23), using the linear planning:

$$\hat{\theta}_i = \min\left\{\theta \mid y_i \le Yz, \theta x_i \ge Xz, \sum_{i=1}^{n} z_i = 1, \right. \\ \left. z > 0, 0 \le \theta_i \le 1, i = 1,2,..,n\right\} \quad (23)$$

Indeed, in order to estimate the efficiency level of ith firm ($\hat{\theta}_i$), the ratio of radial distance between the points in which the *DMU* operates (x_i, y_i) and its equivalent point on the efficient input-oriented frontier $(\hat{x}^{\partial}(x_i \mid y_i), y_i)$ is calculated via the above equation. In this equation, $(\hat{x}^{\partial}(x_i \mid y_i)$ is a level of consumed input that the *(DMU)* with a specific production level y_i should achieve to function efficiently. (By moving from x_i to $(\hat{x}^{\partial}(x_i \mid y_i)$ along with radius (θx_i) (Cooper et al., 2006; Schmidt, 2008).

$$\hat{x}^{\partial}(x_i \mid y_i) = \hat{\theta}_i x_i \quad (24)$$

The Bootstrap data envelopment analysis estimator

The Bootstrap method is a statistical resampling method which it effectiveness has been proven in terms of its caliber to perform statistical inference on complex issues. The most important step in the bootstrap method is proper determination of method ρ or *DGP* from data sample of population. in The assumption underlying the use of the Bootstrap method is to estimate sample distribution of estimator using empirical distribution of estimations obtained from resampling. Bootstrap data envelopment analysis algorithms proposed by Simar, Wilson (1998) *(SW)*, and Lothgren and Tambour (1998) *(LT)*, are based on the same model of data generation process *(DGP)*, and it is assumed that to generate data at a specific level of output, the input amount required for generating pseudo sample is obtained from random radial deviations of isoquant curve of input set. Each input in the observed sample of input- output, is shown as (25):

$$(x_i, y_i) = \left(\frac{x^{\partial}(x_i \mid y_i)}{\theta_i}, y_i\right) \quad (25)$$

Where $x^{\partial}(x_i \mid y_i) \in \partial X(y_i)$ is an unobserved point (on constructed efficient frontier) equivalent radial of firm (x_i, y_i) on the efficient frontier $\partial X(y)$ which is compared with firm location to calculate the firm efficiency. It is assumed the actual efficiency scores are taken from a similar distribution like θ_i F_θ, $i=1, 2, ..., n$, therefore it can be said that the Data Generation Process *(DGP)* model is an idea subject to output and input ratios and that random elements of production processes can be completely shown with the random input-oriented efficiency index. The basic idea of Bootstrap simulation is to imitate data generation process.

Bootstrap algorithms of *LT* and *SW* in each sampling are as follows: Subject to the observed ratio of inputs and outputs, resembling data are generated within two steps. In the first step, the input frontier curve is estimated using the observed sample; then using input frontier and pseudo efficiency values taken from some estimates of distribution F_θ, the amounts of Bootstrap pseudo inputs are generated by *DGP* iteration introduced in (25). This step in the algorithm provided by *LT* is established based on a simple resembling from empirical distribution of estimated efficiency scores used in generating pseudo efficiencies. Unlike the *LT* method, *SW* algorithm in the first step uses smoothed resembling process, and is based on the argument of compatibility. In the second step, the estimation of Bootstrap efficiency is performed through calculating the radial distance of efficient frontier produced by pseudo sample from pseudo input (in *LT* algorithm) or the original input amounts (in *SW* algorithm). In what follows, *LT* and *SW* algorithms are described together with their differences.

***LT* algorithm**

Steps of bootstrap algorithm provided by *LT* are as follow:

1. Using the original estimated efficiency values $\{\hat{\theta}_i, i = 1,2,...,n\}$, the input-output vectors

are changed to the form of equation (26).

$$\left(\hat{x}^{\partial}(x_i \mid y_i), y_i\right) = \left(x_i . \hat{\theta}_i, y_i\right) \quad (26)$$

2. resampling is performed from *n* estimated technical efficiencies $\{\hat{\theta}_{in}\}$ and independent replacement. The efficiencies obtained from resampling are represented as δ_i^*, $i=1, 2, ..., n$.

3. Pseudo Bootstrap data is generated from equation (27).

$$\left(x_i^*, y_i^*\right) = \left(\frac{\hat{x}^{\partial}(x_i \mid y_i)}{\delta_i^*}, y_i\right) \quad (27)$$

4. Estimating Bootstrap efficiency scores using pseudo data is obtained by solving the model (28) with linear programming.

$$\hat{\theta}_{i,b}^{LT*} = \min_{\theta,z}\left\{\theta \mid y_i \le Yz, \theta x_i^* \ge X^* z, \sum_{i=1}^{n} z_i = 1, z > 0\right\} \quad (28)$$

5. Second to fourth steps of this algorithm are repeated in order to make B times of Bootstrap efficiency $(\hat{\theta}_{i,b}^{LT*}, b=1,...,B)$ for the specific firm.

In this algorithm, the estimated Bootstrap frontier and Bootstrap efficiency scores are resembled based on the resampling technical efficiency scores obtained from empirical distribution of efficiency scores that in turn are resulted from the original sample. In addition, Bootstrap iterations of efficiency based on the resampled data are based on the original data, as the main estimations does (Lothgren, 1998; Lothgren & Tambour, 1999).

SW algorithm

The algorithm introduced by Simar and Wilson (*SW*) is different from Lothgren and Tambour's algorithm with respect to the second to fourth steps. In the second step of smoothing process, the main values of estimated efficiencies according to the empirical distribution core smoothing, are used to produce smoothed-resamples of pseudo efficiency scores. Application of smoothing process is based on reflective modification of *Gaussian kernel density* of estimation function, already mentioned by Silverman (1986). This process has been discussed by Simar and Wilson (1998) in detail. Suppose that is a non-smooth resample taken independently with replacement from empirical distribution of main values of technical efficiency $\{\hat{\theta}_{in}\}$. The smoothing process is formed in two steps: First, a small disturbance is added to δ_i^* and then the correction in resampling sequence is applied.

In order to produce smoothed pseudo efficiency first, a small disturbance equal to $h\varepsilon_i^*$ (where, *h* indicates Bandwidth and ε_i^* has been taken from a normal identical independent standard distribution) $\tilde{\delta}_i^*$ is added to δ_i^* in order to make pseudo efficiency. Considering the fact that efficiency scores are bounded within unit of distance, (the input-oriented efficiency scores resulting from *DEA*), reflective process is used in the equation (29) to generate $\tilde{\delta}_i^*$ as

$$\tilde{\delta}_i^* = \begin{cases} \delta_i^* + h\varepsilon_i^*, if \left(\delta_i^* + h\varepsilon_i^*\right) \le 1 \\ 2 - (\delta_i^* + h\varepsilon_i^*), otherwise. \end{cases} \quad (29)$$

If it is determined $\delta_i^* + h\varepsilon_i^* > 1$, $\tilde{\delta}_i^*$ changes to a symmetric image of $\delta_i^* + h\varepsilon_i^*$ reflection, on the frontier point $\tilde{\delta}_i^* = 2 - (\delta_i^* + h\varepsilon_i^*))$. One of the most important issues in use of smoothing process is selecting bandwidth parameter (*h*). As Silverman (1986) showed in his study, there are several approaches to select bandwidth (*h*). In his study, Lothgern (1998), using Mont Carlo simulation, shows that value of *h* can be calculated using a strong, automated bandwidth selection law for a variable proposed by Silverman (1986) as follows:

$$h = \circ.9 \circ n^{-\frac{1}{5}} \min\left\{\hat{\sigma}_{\hat{\theta}}, \frac{R_{13}}{1.34}\right\} \quad (30)$$

where, $\hat{\sigma}_{\hat{\theta}}$ represents anestimation of estimated internal standard deviation of efficiency values $\{\hat{\theta}_{in}\}$ and *R13* is interquartile range of empirical

Table 1
Summary of Statistics

	Date(kg)	Date palm(N)	Water(h)	Labor(p/d)	Land(h)
Maximum	10000	500	3085	103	197100
Mean	2241.96	170.34	540.46	28.22	21152.14
Minimum	300	10	25	2	1230
Variance	3842300	19885.94	380502.5	472.1751	903295263
Standard deviation	1960.179	143.1673	627.9034	22.12528	30611.141

distribution of $\{\hat{\theta}_{in}\}$. After above steps, smoothed resampling efficiency values (γ^*) is determined via modifying $\tilde{\delta}_i^*$ as follows:

$$\gamma_i^* = \frac{\bar{\delta}_i^* + \left(\tilde{\delta}_i^* - \bar{\delta}_i^*\right)}{\dfrac{\sqrt{1+h^2}}{\hat{\sigma}_{\hat{\theta}}^2}} \tag{31}$$

where, $\bar{\delta}_i^* = \sum_{i=1}^{n} \frac{\delta_i^*}{n}$ is resampling main efficiency scores mean value. The second fundamental difference between the two algorithms provided by *LW* and *SW* is in the fourth step. In *SW* algorithm, the estimated Bootstrap input-oriented efficiency scores of ith production firm is obtained according to the ratio of radial distance (at fixed output level) of consumed input of ith firm; resulting from original data to its corresponding point located on Bootstrap pseudo production is quant curve, which is in contrast with *LT* algorithm in which the pseudo efficiency scores are produced from the pseudo data associated with Bootstrap pseudo frontier.

SW algorithm can be briefly described as below:

1. Input-output vectors used in the calculation of main efficiency scores (calculated from the original sample) $\{\hat{\theta}_i, i = 1, 2, \ldots, n\}$, are replaced into the following equation:

$$\left(\hat{x}^{\partial}(x_i, y_i), y_i\right) = \left(x_i.\hat{\theta}_i, y_i\right) \tag{32}$$

2. In this stage, the smoothed resampling efficiency scores (γ^*) are as follows:

2.1 Using the equation (30), estimated efficiency scores are applied to determine bandwidth (h).

2.2 The values of $\{\delta_i^*\}$ are produced by resampling along with replacement from empirical distribution of estimated efficiency scores $\{\hat{\theta}_{in}\}$.

2.3 Using equation (29), a string of $\{\tilde{\delta}_i^*\}$ is produced.

2.4 Bootstrap pseudo data are generated as the following equation:

$$\left(x_i^*, y_i^*\right) = \left(\frac{\hat{x}^{\partial}(x_i, y_i)}{\gamma_i^*}, y_i\right) \tag{33}$$

4. The estimator of Bootstrap efficiency scores is obtained using pseudo data and by solving the model (34) with linear planning:

$$\hat{\theta}_{i,b}^{SW*} = \min_{\theta,z}\left\{\theta \mid y_i \le Yz, \theta x_i \ge X^* z, \sum_{i=1}^{n} z_i = 1, z > 0\right\} \tag{34}$$

5. The second to fourth steps of this algorithm are repeated B times to make a B-member set of Bootstrap efficiencies $(\hat{\theta}_{i,b}^{LT*}, b = 1, \ldots, B)$ of the specified firm (Lothgren, 1998; Silverman, 1986;Simar & Wilson, 1998).

Data and Specification of Variables

With about 100 palm groves and more than one-hectare cultivation, RoodkhanehbarRegion (including the villages of MiyanChilan, Bagh-Narges, Regab, Hizbandegan, Delbanan, Chah-Saifullah, and Baghshah) is one of the date producers of Hormozgan Province. Among the varieties planted in Hormozgan Province, the Keriteh variety is the dominant palm,and accordingly, in this study, the efficiency of Keriteh palm production has been evaluated. Random

Table 2
Bias- Corrected Technical Efficiency Scores and the Formation of Confidence Intervals

P- G	Es-Bi	Bi-Corr	S-Dev	Conf-int		Es-Bi	Bi-Corr	S-Dev	Conf-int		S-E
		CRS		L	U		VRS		L	U	
1	0	1	0	1	1	0	1	0	1	1	1
2	-0.0006	0.0703	0.0055	0.0607	0.0786	-0.0291	0.1274	0.0058	0.0601	0.0793	0.551805
3	-0.0002	0.073	0.0061	0.0628	0.0827	-0.0045	0.0808	0.006	0.0622	0.082	0.903465
4	0.0005	0.0662	0.0075	0.0551	0.0786	-0.0282	0.1929	0.0115	0.1199	0.1575	0.343183
5	0	0.0268	0.0028	0.0224	0.0311	-0.0618	0.1504	0.0028	0.0223	0.0311	0.178191
6	0.0001	0.0133	0.0016	0.0109	0.0157	-0.0534	0.12	0.0015	0.011	0.0158	0.110833
7	-0.0003	0.0669	0.0058	0.0567	0.0756	-0.017	0.1141	0.0062	0.07	0.0903	0.586328
8	-0.0002	0.035	0.0032	0.0296	0.0396	-0.0825	0.1995	0.0031	0.0297	0.0397	0.175439
9	0.0002	0.1336	0.015	0.1108	0.1564	-0.2659	0.6656	0.015	0.1104	0.1564	0.200721
10	-0.001	0.1868	0.0162	0.1594	0.2115	-0.0957	0.3794	0.0154	0.1611	0.212	0.492356
11	0	0.1537	0.0164	0.1278	0.1786	-0.034	0.2215	0.0166	0.1273	0.1791	0.693905
12	0	1	0	1	1	0	1	0	1	1	1
13	-0.0016	0.2122	0.0172	0.1806	0.2379	-0.7914	1	0.0176	0.1802	0.2396	0.2122
14	-0.0005	0.1138	0.0137	0.0953	0.1376	0.0002	0.1161	0.0137	0.0949	0.1376	0.980189
15	-0.0002	0.0853	0.0099	0.0696	0.1004	-0.209	0.5031	0.01	0.0697	0.1005	0.169549
16	-0.0001	0.018	0.0016	0.0153	0.0205	-0.0775	0.1727	0.0016	0.0153	0.0204	0.104227
17	-0.0014	0.2967	0.0249	0.2566	0.3383	-0.1297	0.5543	0.0244	0.2548	0.3372	0.53527
18	-0.0011	0.2971	0.0308	0.2477	0.3422	-0.705	1	0.031	0.2467	0.3423	0.2971
19	0.0002	0.0932	0.0109	0.0763	0.1104	-0.106	0.306	0.0108	0.0769	0.1106	0.304575
20	-0.0002	0.181	0.0214	0.1491	0.2151	0.0004	0.1814	0.0212	0.1487	0.2145	0.997795
21	-0.0012	0.2586	0.0213	0.2207	0.2923	-0.1567	0.5936	0.0196	0.2474	0.3118	0.435647
22	-0.0023	0.2642	0.0231	0.2226	0.2978	-0.1589	0.6097	0.0197	0.2581	0.3247	0.433328
23	-0.0017	0.2553	0.0222	0.2169	0.2888	-0.1462	0.5541	0.0208	0.2246	0.2919	0.460747
24	-0.0008	0.1206	0.01	0.1028	0.136	-0.0461	0.2158	0.0108	0.1059	0.1408	0.558851
25	0.0024	0.3147	0.0347	0.262	0.371	-0.2429	1	0.0114	0.7396	0.7753	0.3147
26	-0.0006	0.4087	0.0331	0.3543	0.4646	-0.1095	0.6544	0.0345	0.378	0.4928	0.624542
27	0	1	0	1	1	0	1	0	1	1	1
28	-0.0006	0.2165	0.0207	0.1827	0.249	-0.0323	0.2849	0.0182	0.1909	0.2504	0.759916
29	-0.0011	0.2712	0.0253	0.2298	0.311	-0.0566	0.4494	0.0255	0.2961	0.3799	0.603471
30	-0.0006	0.2707	0.0253	0.2297	0.3084	-0.0402	0.3563	0.0225	0.2372	0.3117	0.759753
31	-0.0012	0.1834	0.0173	0.1544	0.2086	-0.101	0.3825	0.017	0.1539	0.2077	0.479477
32	-0.0001	0.1297	0.012	0.1102	0.148	-0.1961	0.5218	0.0121	0.1101	0.1487	0.248563
33	-0.0008	0.1359	0.0125	0.1146	0.1542	-0.2714	0.6769	0.0125	0.1141	0.1535	0.200768
34	-0.0006	0.1348	0.0134	0.1125	0.1546	-0.1162	0.3664	0.0136	0.1126	0.1546	0.367904
35	-0.0019	0.2511	0.0234	0.2109	0.2842	-0.0539	0.4164	0.0239	0.2724	0.3519	0.603026
36	0.0002	0.3237	0.0307	0.2755	0.3725	-0.0694	0.5406	0.0303	0.3549	0.4537	0.598779
37	-0.0046	0.5443	0.0515	0.457	0.6163	-0.1049	1	0.0221	0.8546	0.9241	0.5443
38	-0.003	0.4347	0.042	0.3635	0.4972	-0.0629	0.5682	0.0371	0.3816	0.5017	0.765048
39	-0.0007	0.1673	0.0191	0.1366	0.1965	-0.0835	0.3335	0.019	0.1369	0.196	0.501649
40	-0.0009	0.1324	0.0142	0.1083	0.1528	-0.0431	0.3012	0.0176	0.1886	0.2455	0.439575
41	-0.0004	0.3403	0.0346	0.2857	0.392	-0.0672	0.4956	0.0266	0.3173	0.4047	0.686642
42	0	0.018	0.0021	0.0147	0.0212	-0.0682	0.1544	0.0021	0.0147	0.0213	0.11658
43	-0.0002	0.0449	0.0047	0.0372	0.0519	-0.1079	0.2603	0.0046	0.0371	0.0513	0.172493
44	0.0001	0.0943	0.0109	0.0776	0.1116	-0.0883	0.2711	0.0106	0.0775	0.111	0.347842
45	-0.0004	0.1188	0.0121	0.0992	0.1371	-0.0093	0.1367	0.0121	0.0993	0.1369	0.869056
46	-0.002	0.3041	0.0275	0.2567	0.3434	-0.0674	0.4999	0.0279	0.319	0.4122	0.608322
47	-0.0023	0.2061	0.017	0.1744	0.2311	-0.1138	0.4398	0.0172	0.1825	0.2385	0.468622
48	-0.0005	0.1152	0.0105	0.0974	0.1311	-0.0287	0.2098	0.0125	0.1337	0.1741	0.549094
49	-0.0002	0.0821	0.0077	0.0695	0.0937	-0.1584	0.3985	0.0078	0.0697	0.0943	0.206023
50	-0.0001	0.1769	0.0156	0.1517	0.2022	-0.0411	0.286	0.016	0.1776	0.2305	0.618531
Average	-0.00065	0.228828	0.0161	0.202316	0.253398	-0.11064	0.44126	0.014796	0.232942	0.280748	0.5036088

sampling and questionnaire were used to determine the sample size of the studied population, and Bartlett's table was used to collect data. The Table was applied to determine minimum returned sample size of a given population size for continuous and categorical data. Based on this table, for a population of 100 palm groves over a hectare, at probability level of 10% error,

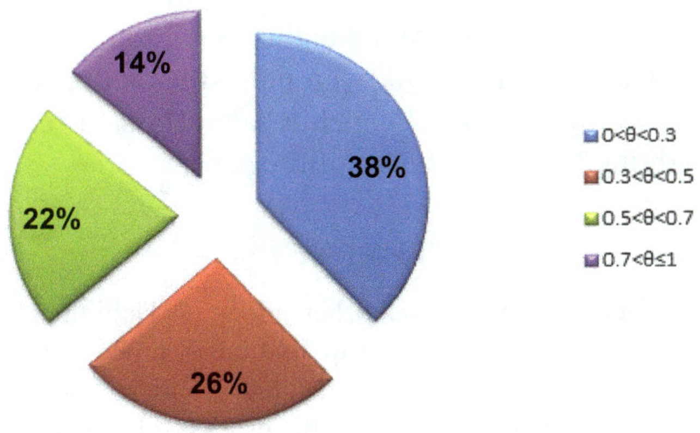

Figure 5. Distribution of pure technical efficiency scores for date producers

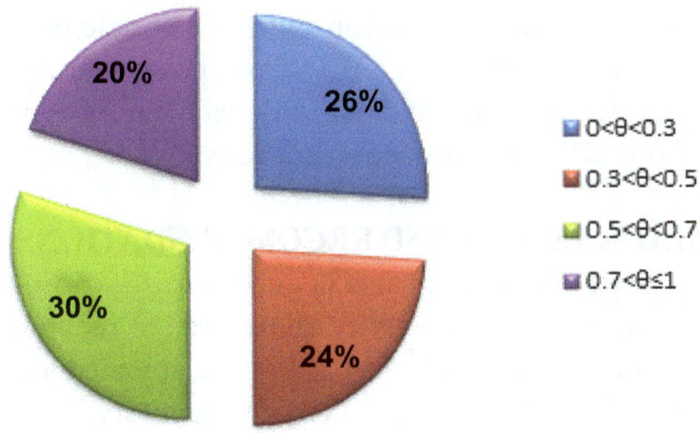

Figure 6. Distribution of pure scale efficiency scores for date producers

a sample with a population of at least 46 palm groves should be collected (Barlett et al., 2001). As such, in this paper, from producer population with over a hectare cultivation of palm in Roodkhanehbar region in 2013, a sample with the size of 50 palm groves was taken. Information regarding the amount of palm product (Kilograms) and four basic inputs including labor (person/Labor Day), land (hectares), water (h), and the number of date palms (N) were used to evaluate the efficiency of palm production in the region. A summary of the statistics is listed in Table 1.

RESULTS AND DISCUSSION

Using smoothed Bootstrap, the input-oriented data envelopment analysis provided by Simar and Wilson, the technical efficiency scores and related confidence intervals with 1000 iterations, (B=1000) at 95% confidence level were calculated in this survey, assuming VRS[3] and CRS[4]. CRS assumption is only appropriate when all firms are operating at an optimal scale, and it implies the total technical efficiency (TE). Using CRS specification when all firms are not operating at an optimal scale results in gross measures of TE by scale efficiencies (SE). Additionally, using VRS specification permits the calculation of TE devoid of these SE effects and refers to pure technical efficiency. Therefore, TE is decomposed to pure efficiency and scale efficiency; therefore, scale efficiency can be calculated by the ratio of technical efficiency and pure technical efficiency scores.

The results are listed in Table 2 and Figure 2. The bandwidth used in this study was computed via the equation (30) proposed by Silverman and was equal to 0.0005.

Abbreviations used in the above table are as listed:

P-G: Palm groves

Es-Bi: Estimated bias

Bi-Corr: Bias-corrected technical efficiency

S-Dev: Standard Deviation

Conf-Int (L-U): Confidence interval (lower-bound-upper bound)

S-E: Scale Efficiency

As can be seen in the table above, the average pure technical efficiency scores resulting from Bootstrap input-oriented DEA under VRS for Keriteh palm cultivation were about 44%, which indicates that growers in this region are able to produce the same amount of product with about an average of 66% saving in resource consumption, and therefore, reduce their production costs. Figure 5 also demonstrates that 38% of studied palm groves had efficiency less than 30% and that only 14% of growers operated efficiently.

---

[3] *Variable Returns to Scale*

[4] *Constant Return to Scale (here, restrictions [$\sum_{i=1}^{n} z_i = 1$] in CRS model is not applied.)*

Similarly, as shown in the Figure 6 below, the scale efficiency score for only 20% of date producers was more than 0.7. This shows that the activity of most of them has been far distant from efficient production capacity.

CONCLUSION AND RECOMMENDATIONS

Since producing dates is either directly or indirectly the main activity and source of income for people in Roodkhanebar Region, any increase in technical efficiency (TE) of the activity leads to saving product inputs consumption, reduces production costs, and increases output level with existing inputs, and finally, increases both profit and welfare of the people of the region.

Due to the importance of the issue, this study aimed to evaluate the management (pure) and scale efficiency of date production in this region. As shown in Table 2, date producers of this region, having average management (pure) and scale efficiencies scores equal to 0.44 and 0.50 respectively, operated weakly, and significant savings will be possible for given quantity of outputs by efficient management and operating in optimum scale. Furthermore, as listed in Figures 2 and 3, the distribution of scores of management and scale efficiencies showed 64% of producers had management efficiency less than 0.50 and 50% of producers had scale efficiency less than 0.50. Given the low efficiency scores and high number of inefficient producers (in terms of both management and optimum scale efficiencies), and alsobecauseincreasing lower levels of efficiency is more convenient and easier than increasing it for higher levels, (e.g., increasing efficiency from 20 to 50% is easier than increasing it from 70 to 100%), the efficiency of producers in this region can be remarkably increased by taking appropriate promoting and educating actions to improve the general knowledge of optimum resource consumption andto facilitate proper input access. In this regard, it is recommended to establish a center for training and experience share, where successful orchardists could share their experiences with others in an attempt to optimize efficient use of inputs and production as well.

ACKNOWLEDGEMENT

We would like to take this opportunity to thank Mr. Omid Tajik who helped us withdata collection and Ms. Pooneh Khajeh Hassani for editing the final version of the paper.

REFERENCES

1- Bahadori, A., Hosseini Nahad, S., & Habibinia, G. (2013). Use of Bootstrap simulation process for estimating non parametric efficient frontier "Investigating problems existing in the process provided in Saeed Ebadi's paper. *Research in Operation And Its Applications, 37*(2), 113-135.

2- Balcombe, K., Fraser, I., Latruffe, L., Rahman, M., & Smith, L. (2008). An application of the DEA double bootstrap to examine sources of efficiency in Bangladesh rice farming. *Applied Economics, 40*(15), 1919-1925.

3- Barlett, James E, Kotrlik, Joe W., & Higgins, Chadwick C. (2001). Organizational research: Determining appropriate sample size in survey research. *Information Technology,Learning,and Performance Journal, 19*(1), 43-50.

4- Brummer, B. (2001). Estimating confidence intervals for technical efficiency: the case of private farms in Slovenia. *European Review of Agricultural Economics, 28*(3), 285-306.

5- Cooper, W., Seiford, L., & Tone, K. (2006). *Introduction to data envelopment analysis and its uses: with DEA-solver Software and References*: Springer Science and Business Media.

6- Dong, Fengxia, & Featherstone, Allen M. (2006). Technical and scale efficiencies for chinese rural credit cooperatives: a bootstrapping approach in data envelopment analysis. *Journal of Chinese Economic and Business Studies, 4*(1), 57-75.

7- Ebadi, S. (2011). A method for ranking efficiency scores using Bootstrap. *Research in Operation and its Applications, 29*(2), 29-44.

8- Efron, B., & Tibshirani, R. (1993). An Introduction to the Bootstrap, Monographs on Statistics and Applied Probability, Vol. 57. *New York and London: Chapman and Hall/CRC*, 321-335.

9- FAO, STAT. (2012). FAOSTAT-Statistical Database, 2012.

10- Fare, R., & Grosskopf, . (2006). *New directions:efficiency and productivity* (Vol. 3): Springer Science and Business Media.

11- Farrell, M. (1957). The measurement of productive efficiency. *Journal of the Royal Statistical Society. Series A (General), 120*(3), 253-290.

12- Gocht, Al., & Balcombe, K. (2006). Ranking efficiency units in DEA using bootstrapping an applied analysis for Slovenian farm data. *Agricultural Economics, 35*(2), 223-229.

13- Lothgren, M. (1998). How to bootstrap DEA estimators: a Monte Carlo comparison. *WP in Economics and Finance* (223).

14- Lothgren, M., & Tambour, M. (1999). Testing scale efficiency in DEA models: a bootstrapping approach. *Applied Economics, 31*(10), 1231-1237.

15- Lothgren, M., & Tambour, M. (1996). *Scale Efficiency and Scale Elasticity in DEA-models: A Bootstrapping Approach*: Economic Research Inst.

16- Odeck, J. (2009). Statistical precision of DEA and Malmquist indices: a bootstrap application to Norwegian grain producers. *Omega, 37*(5), 1007-1017.

17- Schmidt, S. (2008). *The measurement of productive efficiency and productivity growth:*Oxford: Oxford University Press.

18- Shephard, RW. (1970). *The theory of cost and production functions, Princeton.* New Jersey: Princeton Univ. Press (1st edition, 1953).

19- Silverman, B. (1986). *Density estimation for statistics and data analysis* (Vol. 26): CRC press.

20- Simar, L. (1996). Aspects of statistical analysis in DEA-type frontier models. *Journal of Productivity Analysis, 7*(2-3), 177-185.

21- Simar, L., & Wilson, P. (1998). Sensitivity analysis of efficiency scores: How to bootstrap in non-parametric frontier models. *Management Science*, 44(1), 49-61.

22- Simar, L., & Wilson, P. (2000). A general methodology for bootstrapping in non-parametric frontier models. *Journal of Applied Statistics, 27*(6), 779-802.

8

Measuring the Technical Efficiency of Canola Farmers and Determining the Effective Factors in Tabriz County, Iran

Sepideh Rostami Dolatabadi [1] and Mohammad Ghahremanzadeh [2]*

Abstract

The purpose of this study was to measure the technical efficiency of canola farmers and determine the underpinning factors. Required data were collected using stratified random sampling method and questionnaire survey. In this context, 157 canola farmers from Tabriz County, Iran, were asked to fill out the questionnaire in 2012-2013 growing season. The technical efficiency was evaluated using stochastic frontier approach. The results of inputs production elasticity indicated that all inputs were consumed in economic area. In addition, the highest (0.504) and lowest (0.095) elasticity rate was related to water consumption and education level, respectively. The estimated technical efficiency demonstrated that the highest and lowest technical efficiencies were 25 and 95% with an average of 80%. It should be noted that the highest efficiency was observed in the farms with the area covering 3-4 hectares. Furthermore, the estimated inefficiency model indicated that education level, training course number, and cultivated area negatively affected the inefficiency. By contrast, age of farmer had a positive effect on inefficiency. As the results show, educating and training farmers for the optimum use of inputs as well as improving their knowledge as to promote agricultural products should be taken into account.

Keywords:
Canola, Stochastic Frontier Function, Tabriz County, technical efficiency

[1] *M.A., Agricultural Management, Department of Agricultural Extension, Education and Management, Tabriz Branch, Islamic Azad University, Tabriz, Iran.*
[2] *Associate Professor, Agricultural Economics, Department of Agricultural Economics, University of Tabriz*
* *Corresponding author's email: Ghahremanzadeh@tabrizu.ac.ir*

INTRODUCTION

Self-sufficiency in agricultural products is one of the most important concerns in each country. Since food security is closely associated with the country's political stability, increase in production might be a good solution. It is clear that crop production can be increased in two ways: by increasing the cultivated area or by improving the efficiency. It is obvious that although increasing the cultivated area is the easiest way, there are some limitations; for instance, increasing the cultivated area of a certain crop means stopping the cultivation of other crops. Furthermore, due to water scarcity, especially during the recent years, increasing cultivated area is not a good solution. Therefore, the most logical solution is to increase the efficiency. Generally, increasing efficiency and also identifying its effective factors can be considered as suitable supplement for the established policies that encourage and conserve domestic production (Dashti et al., 2011). Identifying effective factors, implementing proper policies and making careful plans play a crucial role in increasing production through improving efficiency of canola growers in Tabriz County. Consequently, improving efficiency is inevitable.

As oilseeds have high energy, they play an important role in supplying energy in developing countries. Amongst oil seeds, canola contains approximately 40-45% oil, which is extracted for use as a premium edible vegetable oil. The remaining canola meal is widely used as protein source (38-43%) in animal feeds (Ahmadi & Javidfar, 1998). Canola cultivation in rotation with wheat reduces pest and disease as well as weed population. Canola cultivation in East Azerbaijan dates back to 1993. It has been estimated that 810 hectares are allocated to canola production in Tabriz County that produces 70% of province total production (East Azerbaijan Jihad-Agriculture Organization, 2014). Accordingly, the aim of this study was to measure the technical efficiency of canola growers and determine its affecting factors.

Several studies have been carried out on agricultural units' efficiency in Iran and different parts of the world. Rafaty et al. (2011) studied the technical, locative and economic efficiencies of cotton growers using stochastic frontier approach. The average technical, allocative and economic efficiencies were85, 90 and 77%, respectively. The results indicated that the cultivated area, machinery, work force, chemical fertilizer amount and irrigation schedule hadpositive and significant effect on efficiency. Similar results were found by Dashti et al. (2011) who studied technical efficiency of broiler farmers in Songhor and Kolyaee County. They used stochastic frontier approach for estimating technical efficiency. Average technical efficiency was found to be 82.17. In addition, the results showed that day-old-chick-numbers, level of equipment and training courses per year hadsignificant and positive impact on technical efficiency. Pishbahar and Nasiri (2012) measured technical efficiency in strawberry farms using stochastic frontier approach in Kurdistan County. Average technical efficiency was found to be 83%. Inefficiency model indicated that education level, sponsorship, experience and weeding number had negative effect on inefficiency. By contrast, plot number, plant age and plant distance had positive effect on inefficiency. Doorandish et al. (2013) studied technical efficiency in dairy farms using Cobb-Douglas function. Average technical efficiency was 93%. Their results showed that experience, farmer's main job and subsidy management hadpositive and significant effect on technical efficiency. Esfandyari et al. (2013) examined technical efficiency of rice production using stochastic frontier approach and indicated that average technical efficiency was about 83% and that the factors affecting inefficiency were age, education level, farmer's experience, land size, land ownership, and membership in the cooperative production.

Otieno et al. (2012) focused on technical efficiency in beef cattle production in Kenya. The average technical efficiency was found to be 69%, suggesting that there is considerable scope to improve beef production in Kenya. Effective institutional support is also necessary in order to improve efficiency, including improved access to market contracts, better farm management skills and off-farm income opportunities. Agom et al. (2012) measured technical efficiency of cocoa farms in Nigeria using stochastic frontier approach.The average technical efficiency was found to be 69%. There were some variables affecting inefficiency negatively such as farmer age, farm size, education level, farmer gender and history of land use. Karthic et al. (2013) investigated technical efficiency in turmeric reduction using stochastic frontier approach. The results indicated that the average technical efficiency was more than 80% and that the education level and farmers experience negatively affected inefficiency. In addition, variables such as nitrogen and potassium amount, irrigation frequency and machinery showed positive and significant effect on turmeric yield. Adenuga et al. (2013) studied technical and economic efficiency in potato production using stochastic frontier approach and found the average efficiency as to be about 80%. The results showed that farmers' age, their training and credit accessibility hadnegative effect on inefficiency. Trujillo and Iglesias (2014) assessed technical efficiency of pineapple farms using stochastic frontier approach and revealed that education level, farmers' experience and credit accessibility had negative influence on inefficiency. Mohammad and Saghaian (2014) measured technical efficiency in rice farms located in eight different provinces in South Korea from 1993 to 2012 using stochastic frontier approach. Average technical efficiency was 77% and there was no significant difference in technical efficiency between provinces. In addition, family labor was suggested to be replaced by wage labor. Based on the reviewed literature, there are afew studies on the efficiency facets of canola productionwhereas efficiency improvement is the best approach to develop canola cultivation. Thus, the aim of this study is to measure the technical efficiency of canola farmers and determine its affecting factors.

MATERIALS AND METHODS

Efficiency is the ratio of the output to the input value; in fact, efficiency is an inclusive concept consisting of three different areas: engineering, management, and economy. In economy, efficiency means optimal allocation of resources (Mehrgan, 2004). Efficiency is optimum use of resources to produce a specific outcome with a minimum amount or quantity of waste expense, or unnecessary effort. Finally, the unit or units which have the highest output in a certain degree of technology with proper management would have the highest efficiency (Zeranejad & Hajiabad, 2009). Farrell (1957) divided the general concept of efficiency into three parts: technical efficiency, allocative (price) efficiency, and economic efficiency. Technical efficiency is a measure of a farmer's success in producing maximum output from a given set of input. Allocative efficiency is the extent to which a farmer equates the marginal value product of a factor of production to its price. Economic efficiency is calculated by multiplying technical and allocative efficiency. In order to measure efficiency in a certain unit, two common approaches are used: (1) Stochastic Frontier Production Function (parametric), and (2) Data Envelopment Analysis Approach (nonparametric). Since the agricultural production process is a random process, stochastic frontier production function is more common in agricultural studies. Stochastic frontier analysis (SFA) is a method of economic modeling. It has its starting point in the stochastic production frontier models introduced by Aigner et al. (1977). Unlike data envelopment analysis, stochastic frontier production function acknowledges the fact that factors outside the control of the farmer can considerably affect unit's efficiency. The production frontier model proposed by Battese) 1991 (can be expressed as:

$$Yi = f(X_i,)\ exp\ (V_i - U_i) \quad i = 1, 2 \ldots\ldots, N. \qquad (1)$$

where, Y_i represents the possible output level of *ith* farm, *fis* an appropriate production function (e.g., Cobb-Douglas or translog), *Xi* is a (1×k) vector of production inputs and, β is a (k×1) vector of parameters to be estimated. The term V_i is a two-sided $(-\infty < V_i < +\infty)$ normally distributed random error $V \sim N[0, \delta_v^2]$ that represents the systematic error which accounts for random variation in output due to factors beyond the control of farmers. The term U_i is a one-sided $(U_i \geq 0)$ efficiency component that captures the inefficiency in production relative to the stochastic frontier. The two components V_i and U_i are assumed to be independent of each other.

The technical efficiency of an individual producing unit is defined in terms of the ratio of the observed output to the maximum production of the corresponding frontier output, given the best available technology (Onyenweaku & Effiong, 2006). Thus, the technical efficiency of ithunit in the context of the stochastic frontier production function is given in the form of equation 2.

$$TEi = Y_i/Y_i^* = F(x_i;\beta)exp(V_i-U_i)/F(x_i;\beta)exp(V_i) = exp(-U_i) \quad (2)$$

where, Y_i is an observed output and Y_i^* is the frontier output. Yi achieves its maximum value of $F(x_i;\beta)$ exp (V_i) if TE_i is 1. Otherwise, a value of TEi less than 1 provides a measure of the shortfall of the observed output from maximum feasible output in an environment characterized by stochastic elements that vary across producers.

The technical inefficiency effect model, U_i, proposed by Batteseand Coelli (1995) is described by

$$U_i = \delta_0 + \delta_1 Z_i + \varepsilon_i \quad (3)$$

where, U_i is non-negative random variable representing inefficiency in production relative to the stochastic frontier in the *ith* form, Z_i is vector of explanatory variables associated with the technical inefficiency effects in the *ith* form, δ is vector of unknown parameters to be estimated, and ε_i is error term. $U_i=0$ means thatthe production is on the frontier and is technical while $U_i > 0$ means that production is inefficient since it will lie below the frontier.

Empirical Model

In this study, the Translog and Cobb-Douglas production functions of frontier model specification for the data were carried out for testing the functional form, inefficiency effect, determinants of coefficients and model best fit to the data. After comparing two functional forms, the results showed that the Cobb-Douglas model fitted the data and the Translog model was rejected. Therefore, the production technology of canola farmers in the studied area was assumed to have been specified by the Cobb-Douglas frontier production function. The Cobb-Douglas functional form is preferable to other forms if there are three or more independent variables in the model (Hanley & Spash, 1993). Thus, the Cobb-Douglas functional form has been widely used in farm efficiency analysis for both developing and developed countries.

The experimental model for investigating stochastic frontier production function is

$$LnTC_i = \beta_0 + \beta_1 LnSa_i + \beta_2 LnLL_i + \beta_3 LnWi + \beta_4 LnKA_i + \beta_5 LnSeed_i + V_i - U_i \quad (4)$$

where, *Sa* is the pesticide cost (herbicide and insecticide) based on *IRR*[1], *LL* is the work force number (person/day), *W* is the irrigation frequency, *KA* is the amount of consumed fertilizer (kg), Seed is the amount of consumed seed (kg), *Ln* is the natural logarithm, β_i are regression parameters to be estimated while V_i and U_i are as defined earlier.

In order to study the factors influencing technical inefficiency, stochastic frontier production function and the factors affecting technical inefficiency are estimated simultaneously.

$$U_i = \delta_0 + \delta_1 Edu_i + \delta_2 AC_i + \delta_3 Se_i + \delta_4 Amozesh_i \quad (5)$$

where, U_i is the inefficiency term, δ_0 is the constant term, *Edu* is the canola growers education level, *AC* is the cultivated area, *Se* is the canola growers' age and *Amozesh* is the training course number.

Data Sources

Tabriz County is the most populated and historic city in the north west of Iran, and the present capital of East Azerbaijan Province. Agriculture is considered as one of the main pillars of the economy in this city. Tabriz has about 810 hectares (46% of total area) under canola cultivation, with total production of 1,710 tons (70% of total production in province). In this study, the required information and data were collected with a questionnaire filled out by canola growers in 2012-2013 growing season.The statistical population of the study included all canola growers in Tabriz city. According to data published by Ministry of Jihad-Agriculture, there were 198 canola growers

[1] *IRR: Iranian Currency ($1= 25000 IRR)*

in 2012-2013 growing season in Tabriz. In addition, there were reportedly four regions in Tabriz where growers were producing canola including Kojabad (75 farmers), Khosroshah (62 farmers), Mayan-Sofla (51 farmers) and Gharamlak and Khajedizak (10 farmers). Considering the four different regions of canola cultivation, sampling was performed using stratified sampling method. Cochran's formula was used to determine the sample size. The sample number for each region was found as follow: Kojabad 53 samples, Khosroshah 44 samples, Mayan-Sofla 36samples and Gharamlak and Khajedizakseven samples.

RESULTS AND DISCUSSION

According to Table 1, most canola growers (31%) are in the age range of 51-60 years. There is just 3% older than 71 years old. These results indicate that younger farmers are not interested in canola production. The education level results showed 45.2% were illiterate, 29.3% were lowly educated and 25.5% were moderately educated, suggesting that education level among the farmers is low. In case of canola cultivation experience, 14%of the farmers had 2 years or less experience of canola production. In addition, 26% of the farmers had 5-6 years of experience, 19% of the farmers had 7-8 years of experience and 2% of the farmers had 3-4 years of experience (most frequent). There were only 8% of the farmers with more than 8 years of experience. Therefore, we can conclude that there is short history in canola cultivation in the region. Regarding promotion courses, 5-9 sessions per course showed the highest frequency (23%).

Statistical properties of the consumed inputs and canola production in the farms are given in Table 2. The results indicated that average cultivated area was 2.3 hectares. However, the highest frequency (64%) was related to 1.5-3.5 hectares. Therefore, we can conclude that average cultivated area in the region was low. For instance, 20% of farmers were working on lands smaller than 1 ha and 16% were working on lands larger than 3.5 hectares.

Stochastic frontier model as Cobb-Douglas form with, inefficiency model was estimated and the results are shown in Table 3. The results indicated that λ value was significant at 0.01 probability level, suggesting that it is possible to estimate linear inefficiency model and stochastic frontier approach model at the same time. Likelihood ratio test was calculated for

Table 1
Summary Statistics of Qualitative and Quantitative Characteristics of Farmers

Variable	Class	Frequency	Percentage
Age (year)	≤ 30	3	1
	31-40	28	17
	41-50	47	29
	51-60	50	31
	61-70	23	14
	≥71	6	3
Family size(person)	≤ 2	10	6
	3-4	64	40
	5-6	59	37
	7-8	18	11
	≥ 8	6	3
Canola cultivation experience (year)	≤ 2	23	15
	3-4	47	29
	5-6	41	26
	7-8	31	19
	≥ 8	15	9
Number of passed training courses	≤ 1	4	2
	1-4	30	19
	5-9	37	23
	10-14	34	21
	15-19	22	14
	≥ 20	30	19

Table 2
Summary Statistics of the Quantitative Variables in Canola Production

Variable	Unit	Minimum	Maximum	Average	Standard
Irrigation before seed sowing	H	0	69	10.5	2.3
Irrigation during seed sowing	H	8	240	61.63	42.63
Irrigation during plant growth	H	25	765	154.29	129.43
Cattle manure	1000 kg	0	4	0.05	0.405
P fertilizers	Kg	0	600	157.8	105.10
N fertilizers	Kg	0	1000	212.4	155.93
Herbicides	liter	0	10	1.82	1.34
Herbicides cost	IRR	0	2700000	778152.8	5.32
Insecticide	liter	0	5	0.99	0.89
Insecticide cost	IRR	0	2500000	486178.3	4.36
Seed	Kg	6	120	25.8	15.4
Work force for fertilizing during seed sowing	Person/day	0.25	6	1.31	1.04
Work force for irrigation during crop growth	Person/day	3	75	8.5	7.5
Work force for pesticides and fertilizing	Person/day	0.5	8	2.9	1.16
Family work force during seed sowing	Person	0	2	0.96	0.308
Family work force during crop growth	Person/day	0.5	2	1.05	60.38
Seed cost	IRR	17000	17500	17445.8	155.86
Canola yield	1000 kg	0.4	28	5.9	4.63
Cultivated area	Ha	0.5	10	2.3	1.5

Notes: H: hour, Ha: hectare, IRR: Iranian Currency ($1= 25000 IRR) Source: Derived from field survey data 2013

zero tests based on independent variables of inefficiency model *($H_0 = \delta_1 = \delta_2 = \ldots \delta_3 = 0$)*. In fact, technical inefficiency effects were assayed by

$$LR=-2\ln\lambda=-2(\log L_{UR}-10gL_R)=-2(-234.83-(-21.96)=425.74$$

$LR= 425.74$ $\quad x^2=0.351$ $\quad LR>x^2$

Since the calculated LR value was greater than the critical value (0.351) and it was significant at 5 % probability level, the null hypothesis is rejected suggesting that thewhole regression is significant. Therefore, it can be concluded that there are technical inefficiency effects among farmers so that some part of inefficiency is affected by inefficiency variables.

According to Table 3, in inefficiency model,

Table 3
The Results of Simultaneous Estimation of Stochastic Frontier Production Function, and Inefficiency Affects Models

Variable	Parameter	Coefficients	t-statistic
Stochastic frontier production:			
Intercept	β_0	-1.733	-1.56*
N fertilizer amount	β_{KA}	0.163	-1.72*
Work force number	β_{LL}	0.303	1.94*
Pesticide cost	β_{sa}	0.074	5.01***
Canola cultivated area	β_{AC}	0.15	0.38***
Inefficiency effects model:			
Intercept	δ_0	-1.760	-0.70***
Education level	δ_1	-0.866	-2.09**
Number of training courses	δ_2	-2.37	-2.74***
Farmer's age	δ_3	0.122	2.69***
Canola cultivated area	δ_4	-1.76	-3.58***
Variance parameters:			
Lambda	λ	1.35	22.46***
E(Sigma U)	$E(\delta_U)$	4.32	
E(Sigma V)	σ_V	0.27	
Log-Likelihood	*-234.83*		

Notes: *p<0.1, **p<0.05, ***p<0.01

Table 4
Technical Efficiency Frequency of Canola Growers in Tabriz County

Class (efficiency %)	Absolute frequency	Relative frequency
40	1	0.00636
40.01-50	8	0.05732
50.01-60	9	0.057
60.01-70	9	0.057
70.01-80	30	0.19108
80.01-90	75	0.47770
90.01-100	25	0.15923
	Average 80	
	Standard deviation 0.131	
	Minimum 25	
	Maximum 95	

Table 5
Technical Efficiency Frequency of Canola Growers in Tabriz County Based On Cultivated Area

Absolute frequency	Frequency	Relative frequency	Absolute frequency		
			Min	Average	Max
Less than 1	31	19.78	0.44	0.7786	0.93
1.1-2	68	43.58	0.25	0.7933	0.94
2.1-3	30	12.32	0.42	0.8319	0.95
3.1-4	11	4.51	0.73	0.8510	0.94
4.1-5	8	3.69	0.75	0.8244	0.90
Greater than 5	7	2.87	0.68	0.7770	0.84

there are some significant factors with negative coefficients such as education level. As expected, utilization increases with education level. Therefore, we can use advanced equipment and use inputs properly. Training course number had significant and negative coefficient. Therefore, it can be concluded that inefficiency effect decreases with increasing training courses. These results reveal the effect of educational programs held by Jihad-Agriculture Organization. Cultivated area had significant and negative effect on inefficiency, stating that increases in cultivated are increases canola grower's efficiency. In addition, farmers' age showed a significant and positive effect on inefficiency. The results indicated that older farmers are more conservative and do not prefer to use new technologies which results in lower efficiency.

As shown by Table 4, the average technical efficiency was 80%. On average, canola growers' efficiency is augmentable up to 20%. The lowest and highest technical efficiency of produces was found to be 25 and 95%, respectively. In other words, the difference between minimum and maximum efficiency is 70%. Based on results presented in Table 4, more than 82% of canola growers have technical efficiency of more than 70%.

As mentioned in Table 5, almost 48% of canola growers have efficiency of 80-90%. Therefore, it can be concluded that technical efficiency improvement is one of the efficient methods for increasing production. Technical efficiency frequency in relation to cultivated area is shown in Table 5. Average efficiency of 85.1% was related to cultivated area of 3-4 hectares, and then 2-3 hectares showed average efficiency of 83.19%. Furthermore, average efficiency of farmers working on 4-5 hectares lands was 82.44%. According to these results, average efficiency increases with cultivated area.Theseresults suggest that efficiency in larger units is more than small units. Finally, it can be concluded that with the current technology, management factors and optimum use of inputs can improve technical efficiency among farmers, especially in small scales.

CONCLUSIONS

The results indicated that the difference between the most efficient and inefficient canola growers is 70%. This clearly suggests that there is a considerable gap which can be removed by proper management. In fact, management factors are derived from socio-economic factors, affecting production directly or indirectly. Average technical efficiency of canola growers was 80%. Accordingly, it can be said that if canola growers use current technologies more efficiently, the canola production can increase up to 20%. The results demonstrated that education level, training course number and cultivated area have significant and negative effect on inefficiency, while farmers' age showed a significant and positive effect on inefficiency. Finally, based on the results the highest average efficiency was related to 3-4-halands with average efficiency of 85%. Considering the results, authors suggest that Jihad-Agriculture Organization should use educational and promotional programs to promote farmers' skills. In addition, farmers should be trained in order to choose the best sowing date, irrigation and fertilizing, weed management, harvest time and foliar application of nutrients such as zinc. Considering the gap between canola growers in the region, it is necessary to convey knowledge and information from efficient farmers to other farmers. Considering this fact that the highest technical efficiency was observed in 2-4-ha lands, it is recommended that canola growers use such lands to produce canola. In addition, they should enterprise to integrate small farms.

REFERENCES

1- Adenuga, H., Muhammad-Lawal, A., &Rotimi, O. A. (2013). Economics and technical efficiency of dry season tomato production selected areas in Kwara State, Nigeria. *Agris On-line Papers in Economics and Informatics, 5*(1), 11-19.

2- Agom, I. D., Ben Ohen, S., Okoiltam, K., & Nyambilnya, N. (2012). Analysis of Technical efficiency of small holder cocoa farmers in Cross River State, Nigeria. *International Journal of Agricultural Management and Development,2*(3), 177-185.

3- Ahmadi, M.R., & Javidfar, F. (1998). *Canola nutrition, Joint Stock Company of oilseeds cultivation.* Unpublished Report.

4- Aigner, D.J., Lovell, C.A.K. & Schmidt, P. (1977).

5- Battese, G.E., & Coelli, T. J. (1995). Frontier production functions, technical efficiency and panel data: With application to paddy farmers in India. *Journal of Productivity Analysis, 3* (1), 153–169.

6- Battese, G.E. (1991). Frontier production functions and technical efficiency: A survey of empirical applications in agricultural economics. *Agricultural Economics, 7*,185-208.

7- Dashti, G., Yavari, S., Pishbahar, E., & Hayati, B. (2011). Factors affecting technical efficiency of broiler chicken units in Songhor and Kolyaee city. *Iranian Journal of Animal Science Research, 21*(3), 83-95 (In Persian).

8- Doorandish, A., Nikokar, A., Hoseinzadeh, M., & Looshabi, A. (2013). Multi-product technical performance of dairy farms in North Khorasan province. *Agricultural Economics and Development, 2*, 112-122 (In Persian).

9- East Azerbaijan Jihad-Agriculture Organization. (2014). *Agricultural Statistics Year Book,* Department of Planning and Economic, Tabriz, Iran.

10- Esfandyari, M., Shahraki, J. Z., & Karbasi, A. (2013). Determining the technical efficiency along with rice production risk in Marvdasht city. *Agronomy Journal (Pajouhesh and Sazandegy) 100*, 76-91 (In Persian).

11- Farrell, M. J. (1957). The measurement of productive efficiency. *Journal of the Royal Statistical Society, 120*(3), 253-290.

12- Hanley, N., & Spash, C. L. (1993). *Farm management research for small farmer development.* Food and Agriculture Organization of the United Nations, Rome.

13- Karthic, V. K., Alagumani, T., & Amarnath, J. S. (2013). Resource–use efficiency and technical efficiency of turmeric production in Tamil Nadu, Department of Agricultural Economics, Agricultural College and Research Institute, Madurai-641003. Tamil Nadu. *Agricultural Economics Research Review, 26*(1), 109-114.

14- Mehrgan, M. R. (2004). *Evaluation of the organizations performance, quantitative approach using data envelopment analysis,quantitative approach using Data Envelopment Analysis*. Tehran University Press, Tehran, Iran.

15- Mohammed, R., & Saghaian, S. (2014). *Technical efficiency estimation of rice production in South Korea.* Selected paper prepared for presentation at the 2014 Southern Agricultural Economics, Association (SAEA) Annual Meetings in Dallas, 1-16.

16- Onyenweaku, C. E., & Effiong, E. O. (2006). *Technical efficiency in pig production in Akwa-Ibom State, Nigeria.* Paper presented at the 40thAnnual

Conference of the Agricultural Society of Nigeria held at NRCRI, Umudike.

17- Otieno, D. J., Hubbard, L., & Ruto, E. (2012). *Determinants of technical efficiency in beef cattle production in Kenya.* Selected Paper prepared for presentation at the International Association of Agricultural Economists (IAAE) Triennial Conference, Foz do Iguacu, Brazil, 18-24 August.

18- Pishbahar, E., & Nasiri, A. (2012). Study on the technical performance of strawberry producers in Sanandaj city. *Journal of Sustainable Agriculture and Production Science, 22*(4), 125-134 (In Persian).

19- Rafati, M., Azarinfar, Y. A., Zad, M., Barabari, A., & Kazemnejad, M. (2011). Study of technical, allocative and economic efficiency in cotton farms in Gorgan, Golestan Province. *Journa of Agricultural Economics, 3*(1), 121-141 (In Persian).

20- Trujillo, J. C., & Iglesias, W.J. (2014). Measurement of the technical efficiency of small pineapple farmers in Santander, Colombia: A stochastic frontier approach. *Brazilian Journal of Rural Economics and Sociology, 51*(1), 49-62.

21- Zeranejad, M., &Hajiabad, R. (2009). Assessment of the technical efficiency of wheat production in Iran. *Journal of Economic Research, 9*(2), 147-151 (In Persian).

9

On the Comparison of the Applications of Conventional Ranking Techniques in Determining the Priority Factors Affecting Seed Production of Medicinal Plants: Case of Guilan Province, Iran

Seyyed Ali Noorhosseini[1], Esmaeil Fallahi[2] and Sepideh Gholinezhad[3]*

Abstract

Aimed at identifying and prioritizing promoters and deterrent affecting seed production of medicinal plants, the present study was conducted in 2014. The Delphi method was conducted using a panel of 13 experts in Guilan Natural Resources and Agriculture Organization. In the first round of the study, multiple-response techniques were used for content analysis. Based on the results of the first round, the second round questionnaire was featuring a Likert type scale that was then filled out by the first round respondents. Next, by comparing the conventional techniques, data were analyzed for investigation. The use of ranking techniques led largely to the same results or results with slight differences. The study recommends devoting due attention to the technical and environmental conditions such as proper storage, to avoid the wasteful harvesting of seeds at risk, and to provide favorable conditions for the domestication of medicinal plants. In addition, the role of research, training, and information dissemination of the economic benefits would receive high importance. Furthermore, designing and implementing a legal monitoring mechanism and an active and comprehensive policy making process, as well as providing facilities and equipment necessary for the preparation, cultivation, and harvest of seeds by decision-making, executive institutions, and organizations would notably be effective to tackle the present challenges.

Keywords:
Medicinal plants, Delphi method, Shannon entropy, triangular fuzzy numbers, seed production

[1] *Young Researchers and Elite Club, Rasht Branch, Islamic Azad University, Rasht, Iran*
[2] *Assistant Professor, Department of Agricultural Economics, University of Shiraz, Shiraz, Iran*
[3] *Young Researchers and Elite Club, Qaemshahr Branch, Islamic Azad University, Qaemshahr, Iran*
* *Corresponding author's email: noorhosseini.sa@gmail.com*

INTRODUCTION

Development of pharmaceutical, cosmetic, food, and the other related industries have increased the demand for raw materials. One of the major categories of raw materials is a medicinal plant. Although, large amounts of chemicals and synthetic ingredients with similar to medicinal plants are produced and traded, natural compounds still have a special place in the production of qualified products, and hence, many companies prefer to use raw materials of these plants. Limitations of the World Health Organization (WHO) play a key role in the use of natural ingredients in the manufacture of these products as stricter rules have been adopted since 2000. According to reports from credible sources, about 70 thousand plant species in traditional medicine and supplements are in use around the world. A study conducted by the World Health Organization found that about 80 percent of the population in developing countries relies on herbal medicine. At least 25 percent of the drugs listed in the reputable world pharmacopoeias is derived from plants. Currently, 121 types of active herbal ingredients are used in the pharmaceutical industry worldwide. Out of total 252 essential medicines introduced by the World Health Organization, 11.1% are herbal medicines and a significant number are synthetic medicines that are derived from natural materials (Ebadi & Abbasian, 2014). Overall, while demand for herbal medicines in developing countries is increasing, there are signs that consumers in developing countries have lost interest in modern medicines and started using herbal medicines (Kumar, 2008).

Medicinal plants are important natural resources that have been used by humans since the ancient times. Given the growing world population, health and human health issues have surfaced as one of the important concerns of human society. As regards plant diversity, especially medicinal plants, Iran are a unique place in the geography of the plant world. According to botanists and scientists of natural resources, there are approximately 8,000 species, and Iranian researchers' studies have shown that 2,000 species have medicinal, aromatic, spiced, and cosmetic properties. In addition, from all recognized plant species, nearly 1,850 plant species are native to Iran (endemic), which are grown exclusively in the territory of Iran and are considered as an exclusive capacity (Irvani et al., 2013; Jasemi et al., 2016). Based on the foregoing cases, there is an enormous economic potential in the production and trade of medicinal plants in the country. Exploiting these potential and managing problems in this field, jobs could be created by medicinal plant production as well as in strengthening the national economy.

Seed production is involved as one of the most important technical and economic issues in this field. Seed is the most important and basic part of the plant that has a vital role in the reconstruction, preservation, and transfer of plant genetic material as well as the mechanisms of distribution, reproduction, and survival of the plant in very difficult circumstances. Moreover, seed production, not only reduces costs, but also decreases depending on the import economically and provides the possibility for a raise in economic returns for producers (Noorhosseini et al., 2011). In addition, the seeds have multiple uses, such as pharmaceutical, industrial, and commercial. Therefore, managing medicinal plant seed production and identifying factors affecting these crops should be considered as research priorities for the planners. Despite the stated importance of the medicinal plant seed production, very few studies in the field of medicinal plant seed production have been carried out, particularly in the country.

Inspired by the studies that had been carried out in Bangladesh, Sharmin (2004) showed that factors such as education level, history of planting, cultivation of medicinal plants and the degree of satisfaction from the income of growing these plants have a positive effect on the development of cultivation.

Yadav and Misra (2012) examined the role of market information systems in the development of forest plants. The results showed that applying these systems in the market solved the problems existing in development and helped suppliers, producers, and consumers. Providing information on quality standards, packaging, and use, method of production, and detection methods increased the quality of plants and helped consumers and suppliers to recognize medicinal plants.

Sher et al. (2013) examined medicinal plants in northern Pakistan from an economic standpoint. The study results showed that lack of awareness of medicinal properties and economic importance of these plants among local people caused damages to vegetation and led to inadequate production of herbal medicines.

Rassam et al. (2012) prioritize the factors affecting the cultivation, and development of medicinal plants in North Khorasan province. The results showed that among the four factors affecting the development of cultivation, economic criteria have the greatest impact. Support policies, training extension services, and social and cultural factors were ranked in the next priority.

Ameri et al. (2013) analyzed consumers' statistics, modes of consumption, production, and awareness of medicinal plants. The results showed that 71.5% of people in Iran use medicinal plants. 70% of men and 73.5% of women are medicinal plant users.

As noted above, although studies have been done on the medicinal plants, little research has been performed on the management of medicinal plant seed production, while based on the above discussions; this is one of the most important issues faced by planners. Accordingly, the present study aims to rank promoter and deterrents affecting medicinal plants seed production in Guilan province as one of the potential regions of medicinal plants in the country. To achieve the research objectives and to achieve more accurate responses, this study compared the common ranking techniques including average, ratio variation, Shannon entropy weighting, and triangular fuzzy numbers.

MATERIALS AND METHODS

The present study capitalized on Delphi as a qualitative research method for data collection. Thirteen experienced and skillful experts in medicinal plant cultivation and production in Guilan, north of Iran constituted the professional panel's members in this study. It should be noted that the experts were identified after numerous interviews and then invited to participate. Each of these experts was active more than a decade in the species identification, cultivation, and production of medicinal plants in Guilan Province.

In order to rank the promoter and deterrents affecting seed production of medicinal plants, this study was conducted in two rounds: In the first phase of the research, multiple response analysis and in the second round, conventional ranking techniques (average, variation ratio, Shannon entropy weighting, and triangular fuzzy numbers) were used. In addition for evaluating experts' agreement about the promoters and deterrents affecting seed production of medicinal plants, Kendall W coefficient was used.

Multi-response analysis

Multi-response analysis is a suitable method to analyze the data from the questionnaires with open questions. Therefore, data analysis and statistical methods were used in the first round of the research. In this round, two questions were raisedregarding to the promoters and deterrents affecting seed production of medicinal plants. After analyzing the content of responses provided in the first round, a questionnaire was designed for the second round where participants were asked to explain their agreement rate regarding each problem and strategies suggested in terms of five-point Likert type scale (1= very disagreed to 5= very agreed). Factors are coded and proposed as analyzed variables presented in Table 1.

Variation ratio

Variation ratio is used to study the distribution of data measured on the nominal or ordinal

Table 1
Type and Number of Responses Raised by the Experts

Variables	Total responses	Number of respondents
Promoting factors affecting the seed production	25	11
Inhibiting factors affecting the seed production	30	16
Total	55	27

scale. In fact, this variable is used to study the consistency or non-consistency of responses. Equation (1) used for its calculation (Kalantari, 2012).

$$VR = 1 - \frac{\textit{Maximum frequency}}{\textit{Total frequency}} \quad (1)$$

Shannon entropy

The concept of Shannon's entropy has a dominant role in information theory. This concept has been developed in different scientific fields, such as physics, social sciences, and so on. The researchers use this formula to estimate the degree of importance of alternatives in the following rounds (Soleimani-damaneh, 2009; Ghorbani, 2012):

- Conversion of decision matrix to the unscaled matrix:

$$R_{ij} = \frac{x_{ij}}{\sqrt{\sum_{i=1}^{m} x_{ij}^{2}}} \quad (3)$$

$$P_{ij} = \frac{R_{ij}}{\sum_{i=1}^{m} R_{ij}} \quad (4)$$

In the above formulas: n= number of indices, m = number of existing options, x= mean

- Calculate the amount of indices entropy:

$$E = -K\sum_{i=1}^{m}\left[P_i \times \ln P_i\right] \quad (5)$$

In the above formula: $K= 1/ln(m)$

- Calculate the amount of uncertainty or deviation degree:

$$d_j = 1 - E_j \quad (6)$$

- Calculate index weight:

$$W_i = \frac{d_i}{\sum_{i=1}^{n} d_j} \quad (7)$$

Fuzzy sets

A fuzzy set is a continuum of grades of membership. Such a set is characterized by a membership function that assigns to each object a grade of membership ranging from zero to unity. Fuzzy sets and fuzzy logic are powerful mathematical tools for modeling. Triangular fuzzy number is a special kind of fuzzy sets. It is possible to use different fuzzy numbers according to the situation. In applications, it is often convenient to work with triangular fuzzy numbers (TFNs) because of their computational simplicity. Moreover, they are useful in promoting representation and information processing in a fuzzy environment. Therefore, in this paper, the researchers used triangular fuzzy numbers (Nazari et al., 2012).

After the data were collected, the fuzzy triangular numbers were converted into absolute fuzzy numbers by means of the Minkowski formula. The following formula is used for converting fuzzy triangular numbers into absolute fuzzy numbers:

$$\textit{Absolute fuzzy numbers} = m + b\text{-}a/4 \quad (8)$$

In the above formula (a) is expressed as low limit (bound), (b) is expressed as the top limit (bound) and (m) means the biggest membership degree.

Fuzzy triangular numbers and absolute fuzzy numbers of the linguistic variables are recorded in Table 2.

Table 1
Conversion Triangular Fuzzy Numbers to a Certain Number

Number of qualitative	Triangular fuzzy number	Final fuzzy number
Very low Importance (Very disagree)	(0,0,0.25)	0.075
Low importance (disagrees)	(0,0.3,0.5)	0.275
The medium (intermediate)	(0.2,0.5,0.8)	0.5
High-importance (I agree)	(0.5,0.7,1)	0.725
High importance (Very agree)	(0.7,1,1)	0.925

After collecting the data, the weight of each indicator is achieved by multiplying the number of acquired responses for each indicator with the deterministic numbers and dividing the product by the number of respondents. By dividing the weight of each indicator on the total weights of indicators, normalized weight is achieved.

Indicator weight =

$$\frac{very\ low\ impor\tan t\ frequency \times 0.075 + \ldots}{the\ number\ of} \frac{++ very\ high\ imporant\ frequenxy \times 0.925}{respondants} \quad (9)$$

Normalized weight =

$$\frac{weight\ \ of\ \ each\ \ indicator}{total\ \ weight\ \ of\ \ each\ \ indicator} \quad (10)$$

Kendall's *W* coefficient

In this study, Kendall's *W* consistency coefficient was used to define the consensus r`ate of panel members. The Kendal consistency coefficient is a measure for defining consistency and the agreement rate among several series of scores related to N objects or persons. In fact, by using this measure, it is possible to find the score correlation between K series of scores. Such a measure is useful, particularly in the studies relating to the "relevance among the judges". Kendall's W consistency coefficient indicates that individuals arranging several items based on their importance fundamentally applied similar measures to judge of the importance of each item thus in this regard agree with each other. This scale is calculated using following equation (Kalantari, 2012):

$$W = \frac{s}{\frac{1}{12}K^2(N^3 - N)} \quad (11)$$

$$S = \sum\left(R_j - \frac{\sum R_j}{N}\right)^2 \quad (12)$$

S = sum of square of standard deviations of R_j from means

R_j Total score relating to a factor.

K= the number of scores sets (the numbers of turns).

N= the number of ranked factors.

1/12 $K^2(N^3-N)$ = Maximum sum of squares of deviations from mean R_j.

i. e. sum of S observing if there is c

ii.

iii.

iv. Complete agreement between K rankings.

The amount of this scale is equity when there is complete consistency or agreement and is zero when there is no complete consistency. Thus, if Kendall's W consistency coefficient rate doesn't change or if it is slightly raised, it indicates that there is no increase in the agreement and opinion asking may be terminated.

Statistical calculations

Performing the calculation, using Shannon entropy weight and triangular fuzzy numbers, and acquiring unreliable result taking a number of indicators to be tested using this technique, is very time consuming and complex due to their specific structure. To facilitate achieve dependable results and reduce common errors in these calculations, SPSS, and Microsoft Excel were used.

RESULTS AND DISCUSSION

Results prioritization promoter affecting seed production of medicinal plants using the multi-response analysis indicated that items of implementation of appropriate cropping patterns for seed production of medicinal plants on the basis of scientific findings (P1) and creation of appropriate conditions for the cultivation and domestication of medicinal plants (P9) that individually accounted for 20% of cases. Also, in the first rank, as well as the items creating research and pattern farms for cultivation of medicinal plants in the province of Guilan (P2) and enhance the quality and vigour of medicinal plants seed through proper storage and environmental conditions favorable (P7), each of which separately accounted for 16 percent of cases are in second rank. Other items were in third rank (Table 3).

Results based on content analysis of promoter affecting seed production of medicinal plants in the second phase using three methods: ratio variation, Shannon entropy weight, and fuzzy triangular are summarized in Table 3 below.

Prioritization of each promoter affecting seed production of medicinal plants in Guilan, the ratio variation show that the items of implementation of appropriate cropping patterns for seed production of medicinal plants on the basis of scientific findings (P1) and creating research and pattern farms for cultivation of medicinal plants in the province of Guilan (P2) were ranked first and second, respectively among promoter factors affecting production seeds of medicinal plants. The items farmers' awareness of the economic advantages of seed production of medicinal plants (P3) and access to suitable facilities getting seeds and seed storage of medicinal plants for planting season (P4) were in ranks last.

Prioritization with triangular fuzzy numbers showed that implementation of appropriate cropping patterns for seed production of medicinal plants on the basis of scientific findings (P1) in the first rank and items planning and special attention to seed of medicinal plants in the pharmaceutical industry (P5) and farmers' awareness from the economic advantages of seed production of medicinal plants (P3) were second in importance. Moreover, the items of enhance the quality and vigor of medicinal plants seed through proper storage and environmental conditions favorable (P7) and preventing the indiscriminate harvesting of medicinal plant seeds of rare species and under threat in Guilan province (P6) were in the third rank.

The results of the Shannon entropy method to benchmark the coefficient of variation for the prioritization of promoter factors affecting seed production of medicinal plants indicated that items enhance the quality and vigor of medicinal plants seed through proper storage and environmental conditions favorable (P7), planning and special attention to seed of medicinal plants in the pharmaceutical industry (P5), preventing the indiscriminate harvesting of medicinal plants seeds of rare species and under threat in Guilan province (P6), creation of appropriate conditions for the cultivation and domestication of medicinal plants (P9), training processing and separating seed of medicinal plants (P10) and creating research and pattern farms for cultivation of medicinal plants in the province of Guilan (P2) were ranks of first to third respectively.

Despite the fact that some minor differences in the ranking of the three methods were proposed, the common feature of all three methods was most important items in the ranks of one to five. In general, weaknesses of different methods of ranking made the compilation of the methods used in this study. The average ranking confirms the importance of the issues raised in the study is presented in Table 4. Accordingly, items of farmers' awareness from the economic advantages of seed production of medicinal plants (P3), planning and special attention to seed of medicinal plants in the pharmaceutical industry (P5), preventing the indiscriminate harvesting of medicinal plants seeds of rare species and under threat in Guilan province (P6), enhance the quality and vigour of medicinal plants seed through proper storage and environmental conditions favorable (P7), implementation of appropriate cropping patterns for seed production of medicinal plants on the basis of scientific findings (P1) and access to suitable facilities getting seeds and seed storage of medicinal plants for planting season (P4) were more important, so that all of the items in each of the three methods with the rating of one to five.

The results prioritization deterrents affecting seed production of medicinal plants using the Multi-response analysis indicated that item lack of facilities and equipment for the harvesting and preparation of seeds (D12) that individually accounted for 20% of cases, in the first rank, as well as the item high humidity and seasonal rainfall in the province to prevent the seed production of medicinal plants (D8), each of which separately accounted for 16.67 % of cases are in second rank. Items of lack of laws prohibiting the harvesting of natural species in danger of extinction (D2) and lack of refrigerator equipment and ultracold required for seed storage of medicinal plants (D15) were in third rank and

Table 3

Prioritize of Promoter Factors Affecting Seed Production of Medicinal Plants

Factors	Multi-response analysis (First round)			Ratio of variations (Second round)			Prioritization 1
	Number of Responses	Percent of Responses	Percent of Cases	Mean	SD	VR	
Implementation of appropriate cropping patterns for seed production of medicinal plants on the basis of scientific findings (P1)	5	20.00	38.46	4.385	0.870	0.154	1
Creating research and pattern farms for cultivation of medicinal plants in the province of Guilan (P2)	4	16.00	30.77	4.539	0.660	0.231	2
Farmers' awareness from the economic advantages of seed production of medicinal plants (P3)	1	4.00	7.69	4.615	0.870	0.231	3
Access to suitable facilities getting seeds and seed storage of medicinal plants for planting season (P4)	1	4.00	7.69	4.615	0.870	0.231	3
Planning and special attention to seed of medicinal plants in the pharmaceutical industry (P5)	1	4.00	7.69	4.615	0.650	0.308	4
Preventing the indiscriminate harvesting of medicinal plants seeds of rare species and under threat in Guilan province (P6)	1	4.00	7.69	4.615	0.650	0.308	4
Enhance the quality and vigor of medicinal plants seed through proper storage and environmental conditions favorable (P7)	4	16.00	30.77	4.769	0.599	0.385	5
Buy ensure production of medicinal plants including the seed (P8)	1	4.00	7.69	4.308	0.947	0.385	6
Creation of appropriate conditions for the cultivation and domestication of medicinal plants (P9)	5	20.00	38.46	4.615	0.650	0.462	7
Training processing and separating seed of medicinal plants (P10)	1	4.00	7.69	4.539	0.660	0.462	7
Access to mechanization and agricultural equipment associated with the production seed of medicinal plants (P11)	1	4.00	7.69	4.308	1.109	0.462	8
Total	25	100.00	192.31	-	-	-	-

Continued Table 3

Factors	Triangular fuzzy numbers (Second Round)			Prioritization 2	Shannon entropy (Second Round)			Prioritization 3
	Importance	Normalized Weight	Normalized Weight Rating		E_j	D_j	W_j	
Enhance the quality and vigor of medicinal plants seed through proper storage and environmental conditions favorable (P7)	0.8442	0.0932	9.3246	3	0.6009	0.3991	0.0745	1
Planning and special attention to seed of medicinal plants in the pharmaceutical industry (P5)	0.8462	0.0935	9.3458	2	0.5645	0.4355	0.0813	2
Preventing the indiscriminate harvesting of medicinal plants seeds of rare species and under threat in Guilan province (P6)	0.8442	0.0932	9.3246	3	0.5647	0.4353	0.0813	2
Creation of appropriate conditions for the cultivation and domestication of medicinal plants (P9)	0.7827	0.0864	8.6449	6	0.5647	0.4353	0.0813	2
Training processing and separating seed of medicinal plants (P10)	0.8308	0.0918	9.1759	4	0.5547	0.4453	0.0831	3
Creating research and pattern farms for cultivation of medicinal plants in the province of Guilan (P2)	0.7731	0.0854	8.5387	8	0.5547	0.4453	0.0831	3
Farmers' awareness from the economic advantages of seed production of medicinal plants (P3)	0.8769	0.0969	9.6856	1	0.4767	0.5233	0.0977	4
Access to suitable facilities getting seeds and seed storage of medicinal plants for planting season (P4)	0.7981	0.0881	8.8148	5	0.4767	0.5233	0.0977	4
Implementation of appropriate cropping patterns for seed production of medicinal plants on the basis of scientific findings (P1)	0.8308	0.0918	9.1759	4	0.4620	0.5380	0.1005	5
Buy ensure production of medicinal plants including the seed (P8)	0.7808	0.0862	8.6236	7	0.4335	0.5665	0.1058	6
Access to mechanization and agricultural equipment associated with the production seed of medicinal plants (P11)	0.8462	0.0935	9.3458	2	0.3919	0.6081	0.1136	7

Table 4
Comparison of Prioritization Methods for Promoter Factors Affecting Seed Production of Medicinal Plants

Factors	VR	Triangular fuzzy	Shannon entropy	Mean
Farmers' awareness from the economic advantages of seed production of medicinal plants (P3)	3	1	4	2.67
Planning and special attention to seed of medicinal plants in the pharmaceutical industry (P5)	4	2	2	2.67
Preventing the indiscriminate harvesting of medicinal plants seeds of rare species and under threat in Guilan province (P6)	4	3	2	3.00
Enhance the quality and vigor of medicinal plants seed through proper storage and environmental conditions favorable (P7)	5	3	1	3.00
Implementation of appropriate cropping patterns for seed production of medicinal plants on the basis of scientific findings (P1)	1	4	5	3.33
Access to suitable facilities getting seeds and seed storage of medicinal plants for planting season (P4)	3	5	4	4.00
Creating research and pattern farms for cultivation of medicinal plants in the province of Guilan (P2)	2	8	3	4.33
Training processing and separating seed of medicinal plants (P10)	7	4	3	4.67
Creation of appropriate conditions for the cultivation and domestication of medicinal plants (P9)	7	6	2	5.00
Access to mechanization and agricultural equipment associated with the production seed of medicinal plants (P11)	8	2	7	5.67
Buy ensure production of medicinal plants including the seed (P8)	6	7	6	6.33

Warehousing and inappropriate treatment and unscientific of medicinal plant seed (D13) was in fourth (Table 5).

Results based on content analysis of deterrents affecting seed production of medicinal plants in the second phase using three methods: ratio variation, Shannon entropy weight and fuzzy triangular are summarize in Table 5.

The results of the ratio of variations method for the prioritization the deterrents factors affecting the seed production of medicinal plants indicated that items the increasing destruction of forests and pastures and destruction of various species of medicinal plants (D1), lack of laws prohibiting the harvesting of natural species in danger of extinction (D2), lack of organizational policy on development of medicinal plants products (D3), lack of awareness from sale market and the demand-supply of medicinal plants seeds (D4), lack of support from farmers for the production of medicinal plants (D5) and lack of expertise for the seed production of medicinal plants (D6) were ranks of first to fifth respectively.

The two methods of ratio variation and triangular fuzzy numbers showed more or less similar results with regard to prioritize deterrent affecting seed production of medicinal plants. In Shannon entropy method based on coefficient of variation, the displacement only between ranks one to six. It seems that the coefficient of variation is not a sufficient basis for ranking in statistical methods. It only is measure the distribution of statistical data to be handled as a normal benchmark.

In total, according to Table 6, items of the increasing destruction of forests and pastures and destruction of various species of medicinal plants (D1), lack of organizational policy on development of medicinal plants products (D3), lack of laws prohibiting the harvesting of natural species in danger of extinction (D2), lack of expertise for the seed production of medicinal plants (D6), lack of awareness from sale market and the demand-supply of medicinal plants seeds (D4) and lack of support from farmers for the production of medicinal plants (D5) were more important, so that all of the items in each of the three methods with the rating of one to five.

After performing the first and second rounds of study, given the fact that Kendall's W coefficient rates for promoter and deterrents were 0.172 and 0.159, respectively, ($p<0.05$ and

Table 5
Prioritize of deterrents factors affecting seed production of medicinal plants

Factors	Multi-response analysis (First round)			Ratio of variations (Second round)			Prioritization 1
	Number of Responses	Percent of Responses	Percent of Cases	Mean	SD	VR	
The increasing destruction of forests and pastures and destruction of various species of medicinal plants (D1)	1	3.33	7.69	4.769	0.832	0.077	1
Lack of laws prohibiting the harvesting of natural species in danger of extinction (D2)	3	10.00	23.08	4.539	0.967	0.231	2
Lack of organizational policy on development of medicinal plants products (D3)	1	3.33	7.69	4.539	0.776	0.308	3
Lack of awareness from sale market and the demand-supply of medicinal plants seeds (D4)	1	3.33	7.69	4.385	0.961	0.385	4
Lack of support from farmers for the production of medicinal plants (D5)	1	3.33	7.69	4.385	0.961	0.385	4
Lack of expertise for the seed production of medicinal plants (D6)	1	3.33	7.69	4.385	0.870	0.385	5
Lack of sufficient information available to farmers for seed production of medicinal plants (D7)	1	3.33	7.69	4.385	0.961	0.385	6
High humidity and seasonal rainfall in the province to prevent the seed production of medicinal plants (D8)	5	16.67	38.46	4.846	1.463	0.462	7
Control of livestock and wildlife to prevent the attack on the seed of medicinal plants (D9)	1	3.33	7.69	4.231	0.927	0.538	8
Lack of encourage and promote ranging from public and private sectors for the seed production of medicinal plants (D10)	1	3.33	7.69	4.231	0.927	0.538	8 9
Increased importing seed of medicinal plants and a lack of willingness to invest in seed production in the country (D11)	1	3.33	7.69	3.846	1.281	0.538	10
Lack of facilities and equipment for the harvesting and preparation of seeds (D12)	6	20.0	46.15	4.308	0.630	0.615	11
Warehousing and inappropriate treatment and unscientific of medicinal plants seed (D13)	2	6.67	15.38	4.000	1.000	0.615	12
Lack of demand for buying seeds of medicinal plants (D14)	1	3.33	7.69	4.000	1.080	0.615	13
Lack of refrigerator equipment and ultra-cold required for seed storage of medicinal plants (D15)	3	10.00	23.08	4.000	1.155	0.615	14
Lack of the gene bank of medicinal plants in Guilan (D16)	1	3.33	7.69	3.846	0.987	0.692	-
Total	30	100.00	230.77	-	-	-	

Continued Table 5

Factors	Triangular fuzzy numbers (Second Round)			Prioritization 2	Shannon entropy (Second Round)			Prioritization 3
	Importance	Normalized Weight	Normalized Weight Rating		E_j	D_j	W_j	
Lack of facilities and equipment for the harvesting and preparation of seeds (D12)	0.7846	0.0642	6.4212	5	0.616	0.384	0.045	1
Lack of organizational policy on development of medicinal plants products (D3)	0.8288	0.0678	6.7831	2	0.562	0.438	0.052	2
The increasing destruction of forests and pastures and destruction of various species of medicinal plants (D1)	0.8750	0.0716	7.1608	1	0.555	0.445	0.053	3
Lack of expertise for the seed production of medicinal plants (D6)	0.7962	0.0652	6.5156	4	0.513	0.487	0.058	4
Lack of laws prohibiting the harvesting of natural species in danger of extinction (D2)	0.8269	0.0677	6.7674	3	0.490	0.510	0.060	5
Lack of awareness from sale market and the demand-supply of medicinal plants seeds (D4)	0.7962	0.0652	6.5156	4	0.481	0.519	0.061	6
Lack of support from farmers for the production of medicinal plants (D5)	0.7962	0.0652	6.5156	4	0.481	0.519	0.061	6
Lack of sufficient information available to farmers for seed production of medicinal plants (D7)	0.7962	0.0652	6.5156	4	0.481	0.519	0.061	6
Control of livestock and wildlife to prevent the attack on the seed of medicinal plants (D9)	0.7654	0.0626	6.2638	6	0.482	0.518	0.061	6
Lack of encourage and promote ranging from public and private sectors for the seed production of medicinal plants (D10)	0.7654	0.0626	6.2638	6	0.482	0.518	0.061	6
Warehousing and inappropriate treatment and unscientific of medicinal plants seed (D13)	0.7154	0.0585	5.8546	8	0.442	0.558	0.066	7
Lack of the gene bank of medicinal plants in Guilan (D16)	0.6827	0.0559	5.5870	9	0.435	0.565	0.067	8
Lack of demand for buying seeds of medicinal plants (D14)	0.7154	0.0585	5.8546	8	0.420	0.580	0.069	9
Lack of refrigerator equipment and ultra-cold required for seed storage of medicinal plants (D15)	0.7173	0.0587	5.8703	7	0.402	0.598	0.071	10
Increased importing seed of medicinal plants and a lack of willingness to invest in seed production in the country (D11	0.6788	0.0556	5.5556	10	0.365	0.635	0.075	11
High humidity and seasonal rainfall in the province to prevent the seed production of medicinal plants (D8)	0.6788	0.0556	5.5556	11	0.333	0.667	0.079	12

Table 6

Comparison of Prioritization Methods for Deterrents Factors Affecting Seed Production of Medicinal Plants

Factors	VR	Triangular fuzzy	Shannon entropy	Mean
The increasing destruction of forests and pastures and destruction of various species of medicinal plants (D1)	1	1	3	1.67
Lack of organizational policy on development of medicinal plants products (D3)	3	2	2	2.33
Lack of laws prohibiting the harvesting of natural species in danger of extinction (D2)	2	3	5	3.33
Lack of expertise for the seed production of medicinal plants (D6)	5	4	4	4.33
Lack of awareness from sale market and the demand-supply of medicinal plants seeds (D4)	4	4	6	4.67
Lack of support from farmers for the production of medicinal plants (D5)	4	4	6	4.67
Lack of sufficient information available to farmers for seed production of medicinal plants (D7)	6	4	6	5.33
Lack of facilities and equipment for the harvesting and preparation of seeds (D12)	10	5	1	5.33
Control of livestock and wildlife to prevent the attack on the seed of medicinal plants (D9)	8	6	6	6.67
Lack of encourage and promote ranging from public and private sectors for the seed production of medicinal plants (D10)	8	6	6	6.67
Warehousing and inappropriate treatment and unscientific of medicinal plants seed (D13)	11	8	7	8.67
Lack of demand for buying seeds of medicinal plants (D14)	12	8	9	9.67
Lack of refrigerator equipment and ultra-cold required for seed storage of medicinal plants (D15)	13	7	10	10.00
Increased importing seed of medicinal plants and a lack of willingness to invest in seed production in the country (D11	9	10	11	10.00
High humidity and seasonal rainfall in the province to prevent the seed production of medicinal plants (D8)	7	11	12	10.00
Lack of the gene bank of medicinal plants in Guilan (D16)	14	9	8	10.33

Table 7

Kendall's W coefficient rate

Variable	Kendall's Wa	p-value
Promoter factors affecting the seed production	0.172*	0.027
Deterrents factors affecting the seed production	0.159**	0.009

** , * = significant at 5% and 1% levels respectively

p<0.01), it represents the agreement rate between the experts which removes the need to perform next Delphi stage (Table 7).

CONCLUSIONS AND RECOMMENDATIONS

Nowadays, medicinal plants due to the important role that community health are the attention focus of scientific, research and business centers. Iran is one of the few countries with a wide range of climates. Therefore, it is very diversity of vegetation and some native plant species, as well as Iran (endemic), which are grown exclusively in Iranian territory as a monopoly on the country's capacity is. Directing the massive capabilities and management problems in this area could also create jobs and self-sufficiency in the production of many products derived from these plants contributes significantly to the national economy. Meanwhile, seed production as one of the most important technical and economic issues involved in this area which should be managed as one of the priorities of the research, should be planned. Accordingly, this study sought to identify and prioritize promoter and deterrent factors affecting seed production of medicinal plants in Guilan province as one of the potential production of medicinal plants in the country. To achieve the research objectives and to achieve more precise answers, ranking study common techniques were compared. In the first stage of research, Multi-Response analysis, eleven as promoter factor affecting seed production and seed production as well as barriers to sixteen of medicinal plants established in Guilan Province. In the next round is to determine the priority of these factors, common techniques were compared ranked.

According to the results of common techniques of prioritization, factors of planning and special attention to seed of medicinal plants in the pharmaceutical industry (P5), preventing the indiscriminate harvesting of medicinal plants seeds of rare species and under threat in Guilan province (P6), enhance the quality and vigor of medicinal plants seed through proper storage and environmental conditions favorable (P7), implementation of appropriate cropping patterns for seed production of medicinal plants on the basis of scientific findings (P1) and access to suitable facilities getting seeds and seed storage of medicinal plants for planting season (P4) as the promoter affecting seed production of medicinal plants was in top priority. The important deterrents factors affecting seed production of medicinal plants were also include items such as The increasing destruction of forests and pastures and destruction of various species of medicinal plants (D1), lack of organizational policy on development of medicinal plants products (D3), lack of laws prohibiting the harvesting of natural species in danger of extinction (D2), lack of expertise for the seed production of medicinal plants (D6), lack of awareness from sale market and the demand-supply of medicinal plants seeds (D4) and lack of support from farmers for the production of medicinal plants (D5), respectively.

According findings of this study with regard to the appropriate warehousing as one of the most important factors affecting seed production of medicinal plants known, recommended the establishment of appropriate warehouses and the creation of a favorable the environmental conditions to improve the production of seed in the venture. In addition, the design and implementation of an active and comprehensive policy and legal regulatory system by relevant organizations helps to prevent improper harvesting seeds of rare and threatened species and will target the development of different species. Also suggested by the members of various committees, to take advantage of the benefits of research, education, and information are necessary proper interaction between the Ministries of Agriculture Jihad, Health, and Medical Education, Research and Technology. In addition, executive institutions support for the design and construction of machinery, preparation, cultivation, and harvesting of medicinal plants and seeds.

ACKNOWLEDGEMENT

The authors gratefully acknowledge the Dr. Mohammad Sadegh Allahyari for friendly helps and comments. All the experts are being thanked for their patience in answering our questions.

REFERENCES

1- Ahmadi, F., Nasriani, Kh., & Abazari, P. (2008). Delphi technique: a tool in research. *Iranian Journal of Medical Education,* 8(1), 175-186.

2- Ameri, F., Vahabi, M.R., Khatoon Abadi, A. & Andalibi, L. (2013). On the relevance of medicinal plants consumers in Iran: investigating statistics for consumers, states of consumption, informative and source area. *Teb va Tazkiyeh,*22(3), 37-42. (In Persian)

3- Ebadi, M., & Abbasian, J. (2014). Evaluation of world trade of Medicinal and Aromatic Plants. The second national festival and exhibition of medicinal plants, natural and traditional products in Tehran, Iran. 6-9 Sep 2014, 7 p.

4- Ghorbani, M., Arabzad, S. M., & Bahrami, M. (2012). Implementing shannon entropy, swot and mathematical programming for supplier selection and order allocation. *International Journal of Supply Chain Management*, 1(1), 43-47.

5- Irvani, H., Famil H.Sh., & Sadatzade, M. (2013). Comparing factors affecting the development of cultivation of aromatic plants between two groups of flower growers and processor at Kashan. *Iranian Journal of Medicinal and Aromatic Plants*, *29*(1), 225-236

6- Jasemi, M. Moradnezhadi, H. & Salavarz, M. (2016). Entrepreneurial Strategies of Enhancing Competitive Advantage of Medicinal Herbs in Ilam Province, Iran. *International Journal of Agricultural Management and Development,6*(4), 459-466.

7- Kalantari, Kh. (2012). *Processing and analysis of socio-economic research data using SPSS software (50th ed). Tehran: Saba Farhang Publications, Iran.*

8- Kumar, B.(2008). *Problems and prospects of medicinal and aromatic plant cultivation and marketing: A case study from sarlahi district.* Unpublished dissertation, University of Tribhuran, Pakhara, Nepal.

9- Mashayekhi, A., Farhangi, A., Momeni, M., & Alidoosti, S. (2005). Study of key factors influencing the use of information technology in government agencies: the application of the Delphi method. *Management* special edition, 191-232.

10- Mohammadi, A. (2011). Application of TOPSIS Methods of linear assignment and taxonomy in the financial evaluation. *Economic Research Review, 11*(1), 302- 327.

11- Nazari, A. Salarirad, M.M., & Bazzazi, A.A. (2012). Landfill site selection by decision-making tools based on fuzzy multi-attribute decision-making method. *Environmental Earth Sciences, 65*(6), 1631-1642.

12- Noorhosseini, S.A., Ashoori Latmahalleh, D., Allahyari, M.S. & Doozandeh Masooleh, P. (2011) socio-economic factors for adoption of medicinal plants cultivation in eshkevarat region, north of iran. *Journal of Medicinal Plant Research, 5*(1), 30-38

13- Rassam, Gh., Dadkhah, A., Khoshnood Yazdi, A., & Moghadasi, R. (2012). *Prioritize factors affecting the cultivation of medicinal plants in North Khorasan Province.*National conference on natural products and medicinal plants.

14- North Khorasan University of Medical Sciences. Bojnoord, Iran. 27-28 Sep 2014, 5p.

15- Sharmin, L. (2004). Cultivation prospect of medicinal plants in Bangladesh: Experiences from Natore. *BRAC-Research and Evaluation Division Dhaka, Bangladesh.* 37p.

16- Sher, H., Elyemeni, M., & Hussain, K. (2013). Ethnobotanical and economic observations of some plant resources from the northern parts of pakistan. *Ethnobotany Research and Aplications,* 9, 27-41.

17- Soleimani-damaneh, M., & Zarepisheh, M. (2009). Shannon's entropy for combining the efficiency results of different DEA models: Method and application. *Expert Systems with Applications, 36*(3), 5146–5150.

18- Soofi, S. (2012). *Assess the readiness of e-Learning at the University of Guilan University study using fuzzy method.* Unpublished dissertation, Faculty of Humanities, Guilan University, Rasht, Iran, 108 p.

19- Yadav, M., & Misra, S. (2012). Sustainable development: a role for market information systems for non-timber forest products. *Sustainable Development, 20*(2), 128-140.

10

Analysis and Modeling of Yield, CO_2 Emissions and Energy for Basil Production in Iran using Artificial Neural Networks

Sajad Rostami [1], Somayeh Choobin [1], Bahram Hosseinzadeh Samani [1], Zahra Esmaeili [1] and Hemad Zareiforoush [2]*

Abstract

The present study attempts to investigate the potential relationship between input energies, performance production of greenhouse basil, and greenhouse gases emitted from this product. The data were collected from 24 greenhouses using a questionnaire and verbal interaction with farmers. Results of the study showed that the total input energy and total output energy for basil production were 119,852.9 MJ/ha and 61,040 MJ/ha, respectively. The highest rate of energy consumption was related to electricity (52,200 MJ/ha), followed by plastic (23,220 MJ/ha) and chemical fertilizers (13,894 MJ/ha). The energy and productivity indices were estimated at 0.45 and 0.21, respectively, which indicated that the efficiency of energy in the agricultural sector was low. In addition, it was found that the pure energy index and total greenhouse gases emitted from basil production were equal to -722,706.9 and 9,595.6 kg (CO_2), respectively. The highest emission of greenhouse gases was attributed to electricity (2,216 kg/CO_2). Results of modeling proved that artificial neural networks can predict basil performance and CO_2 emissions with a high degree of accuracy (R^2=0.99 and MSE= 0.00023).

Keywords:
artificial neural networks, basil, CO_2, energy flow

[1] *Department of Mechanical Engineering of Biosystem, Shahrekord University, Shahrekord, Iran*
[2] *Department of Agricultural Mechanization Engineering, Faculty of Agricultural Sciences, University of Guilan, Rasht, Iran*
* *Corresponding author's email: rostami.sajad@yahoo.com*

INTRODUCTION

In recent years, development and enhancement of advanced agricultural equipment have led to a remarkable increase in the energy use in this sector so that agricultural operations have become heavily dependent on fossil fuel resources. It has been projected that the total human population of the world would reach 10 billion by 2040. Additionally, the global storage of oil is expected to deplete in 40 years. Such conditions would coerce governments into producing larger amounts of food with less resources of energy. Regarding the aforesaid notions, humankind would be forced to supply for 10 billion people. Due to the fact that many arable lands have been filled with crops in recent years, it seems that employing new methods and optimized usage of arable lands would be the only available way to overcome this problem. Food production must be compatible with the population increase; otherwise, friendly living status of humankind would be put at risk. Thus, production methods that can offer higher rates of yield along with lower amounts of energy consumption would be successful. One of the new methods of intensive agriculture is greenhouse cultivation (Alam et al., 2005). This method is capable of boosting production through optimizing the cultivation context as well as is able to consume more energy out of lower level. This method is categorized by some advantages and disadvantages. One of the benefits of the method is more production per unit area so that one can obtain 10 times more production in the same area (Ozkan et al., 2004). Another advantage of this method is cultivated in seasons other than main season. This advantage has led to the fact that the farmer embarks on doing this by exploiting more resources; however, there are disadvantages in the establishment of greenhouse cultivation. The main drawback of this method is excessive consumption of energy by producers so that they make use of higher values of energy for cultivation. Likewise, due to the low cost of energy in Iran, farmers make no efforts to reduce energy consumption rate. Consequently, a large portion of farmers' income is devoted to supply energy. Analyzing the quality of cultivation along with offering solutions for the reduction of energy consumption enables high-level production using lower amounts of energy resources (Canakci et al., 2005). Each method contributing to the reduction of energy consumption decreases total production cost and increases producer's income. The aim of this research was the examination and determination of basil energy consumption using energy indices (energy efficiency, energy ratio, and added pure energy) and the extent of greenhouse gases emitted from this product.

Alam et al. (2005) investigated energy flow in agricultural sector of Bangladesh over the years 1980-2000. Their examinations were focused on human, animal, machinery, electricity, gasoline, fertilizers, and chemical pesticide subsectors. They reported that energy input and output of agricultural products increased from 6.4 to 17.32 GJ/ha and from 72.22 to 130.05 GJ/ha, respectively, during the studied period. It was concluded that the efficiency of energy (the ratio of input to output) was decreased from 11.23% to 8.1%. Thus, energy input increased faster than the output leading to the decrease in efficiency of energy consumption.

Some researchers investigated four types of greenhouse vegetables in Turkey and concluded that the cucumber had the highest energy among four crops (tomato, cucumber, pepper, and eggplant) categorized by 77.134 GJ/ha. The amount of energy consumption of tomato, eggplant, and pepper was 32.127, 68.98, and 80.25 GJ/ha, respectively. The ratio of energy for tomato, pepper, and eggplant was 1.26, 0.99, and 0.61, respectively. This implies that the increase in inputs is not always accompanied by the increase in outputs in case of producing greenhouse vegetables. This causes problems such as global warming, increase in nutrition in soil, and pollution caused by pesticides (Ozkan et al., 2004). Hence, it is necessary to force the producers to determine the extent of using a contributing factor of energy to increase performance without reducing natural resources. The generation and emission of greenhouse gases in the atmosphere were the undesirable repercussions of excessive use of natural resources in the 20th century.

Global warming is concerned with changes in the Earth induced by human-based activities. The Earth has gotten warmer (more than 0.4°C°C) due to non-normal behaviors. Greenhouse gases related to livestock productions in the agricultural sector have been increased by 14%, reaching up to 4.7 billion tonnes. Such an increase is to some extent due to the increase in total agricultural output and is usually observed in developing countries. Agricultural policies tend to enhance systems which are able to produce more energy while consuming lower amounts of energy (Dalgaard, 2000). Several studies have been carried out in relation to energy consumption and environmental effects of agricultural crop production. The main scope of these studies was to evaluate energy consumption in agricultural crop production (Alluvione et al., 2011; Barut et al., 2011). The average energy consumption has been studied in Iran (Mousavi-Avval et al., 2011; Royan et al., 2012; Tabatabaeefar et al., 2009). However, few studies have addressed the environmental effects of greenhouse gases derived from agricultural activities in Iran (Alam et al., 2005). This is the reason why few information is available regarding the amount of fuel and energy consumption as well as the emission of greenhouse gases caused by agricultural operations (Soltani et al., 2013).

Among the issues on which researchers have tended to draw their attention are predicting, estimation, and modeling of energy flow. It is evident that planning and optimizing the inputs would require modeling of this flow. A variety of methods have been employed to model energy flow and the association between farm (input) and farm performance (output). Regression analysis used to be regarded as the main modeling method until Artificial Neural Networks (ANNS) were developed (Safa & Samarasinghe, 2011).

Recently, the number of scientists and engineers who are interested in modeling of energy consumption and related environmental impacts have been increased (Al-Ghandoor et al., 2009). In the energy area, a wide range of models, from geological models in research on natural resources for modeling future energy demand, has been developed (Safa & Samarasinghe, 2011). Several studies have used ANNs for classifying,predicting, and solving problems in the field of energy. Some researchers have analyzed world primary energy resources, including fossil fuels such as coal, oil and natural gas, using feed forward, back propagation ANN (Rahman & Bala, 2010). Application of ANNs to estimate jute production in Bangladesh was reported by Rahman and Bala (2010). They used an ANN model with six input variables, including Julian day, solar radiation, maximum temperature, minimum temperature, rainfall, and type of biomass to predict the desired variable (plant dry matter) (Rahman & Bala, 2010). Pahlavan et al. (2012) developed a network to predict greenhouse basil production. Safa and Samarasinghe (2011) used ANNs for determination and modeling of energy consumption in wheat production. They compared ANNs with Multiple Linear Regression. They found that artificial neural networks can predict energy consumption better than regression models. The artificial neural network is mainly used to predict energy consumption, energy demand, environmental problems, etc. Relative performance of artificial neural networks has been reported by traditional statistical methods.

No study has been carried out in Chaharmahal-Bakhtiary Province on the basil input and output consumed energy. The aim of the present study was to (1) investigate energy consumption for basil production, (2) calculate different indices of energy efficiency of basil product, (3) determine greenhouse gases derived from basil production, (4) predict energy output of basil production, and (5) examine greenhouse gas emissions based on the input energy.

MATERIALS AND METHODS

Data collection and calculation

The scope of production was considered in Chaharmahal-Bakhtiary Province, Iran. Data were collected from 24 greenhouses in the selected region. The data collection was made by questionnaires and the information on the consumed inputs, the employed machines, and the number of workers was gathered using random sampling. Chaharmahal-Bakhtiary Province is

Table 1
Energy Coefficients of Different Inputs and Outputs in Basil Production

Input/output	Unit	Energy coefficient (MJ/unit)	Ref.
A. Input			
1. Human labor	H	1.96	(Mohammad Shirazi et al., 2012)
2. plastic	Kg	90	(Canakci et al., 2005)
3. Irrigation	M3	1.02	(Erdal et al., 2007)
4. FYM	Kg	0.3	(Mousavi-Avval et al. 2011b)
5. Chemical fertilizers	Kg		
N		60	(Chauhan, Mohapatra, & Pandey, 2006)
K_2O		6.7	(Chauhan et al., 2006)
P_2O_5		11.1	(Chauhan et al., 2006)
Fe		120	(Mandal et al., 2002)
6. Chemicals	Kg		
Fungicides		216	(Erdal et al., 2007)
7. Machinery	Kg yr[a]	62.7	(Canakci et al., 2005)
8. Diesel fuel	L	43.99	(Coxworth et al., 1995)
9. Electricity	KWh	3.6	(Ozkan et al., 2004)
B. Output			
1. Basil yield	Kg	2.18	Calculated

Table 2
GHG Emissions Coefficients of Agricultural Inputs

Inputs	Unit	GHG Emissions Coefficient (kg CO_{2eq} unit^{-1})	Ref.
N	Kg	1.3	(Lal, 2004)
K_2O	Kg	0.2	(Lal, 2004)
P_2O_5	Kg	0.2	(Lal, 2004)
Machinery	MJ	0.071	(Dyer & Desjardins, 2006)
Diesel fuel	L	2.76	(Pishgar-Komleh et al., 2012)
Electricity	KWh	0.608	(Khodi & Mousavi, 2009)
Fungicide	Kg	3.9	(Lal, 2004)

located within the latitudes of 31°9' and 32°38' N and longitudes of 49°30' and 51°32 E. The input energies of the current study included labor, seed, irrigation water, fertilizer, chemical pesticides, diesel fuel, machinery, plastics, and electricity. To calculate the input and output energies, energy equal coefficients were used as shown in Table 1.

Having determined the share of each energy resource for supplying different inputs, the amount of greenhouse gas (GHG) emissions by the energy resource consumption was calculated in terms of a coefficient (Table 2).

The input energy was calculated based on the principle that human force energy is calculated in agricultural operations. In this regard, it was assumed that each farmer works 210 days and 8 hours per day.

To calculate the energy required for annual manufacturing and machine maintenance, the following equation was used (Hatirli et al., 2005).

$$ME=(G.E)/(T.Ca) \qquad (1)$$

where, ME is the energy of machine manufacturing (MJ/ha), G is a constant equal to 158.3 (MJ/kg), Ef is the weight of the tractor (kg), T is the economic life of the tractor (h), and C_a is the effective capacity of the farm (ha/h) which is calculated by Eq. (2):

$$C_a=(S.W.E_f)/10 \qquad (2)$$

where, S is the working speed (km/h), W is the effective working width of the machine (m) and Ef is the farm efficiency.

Tractor fuel consumption is obtained in terms of energy by:

Gasoline tractor (lit/h): 0.06 × tractor power in 3.78 PTO

Diesel tractor (lit/h) = 0.73 × gasoline tractor fuel

Power at the PTO is considered as 80% of tractor nominal power.

To calculate the energy of fertilizers, pesticides and seeds, their consumption rate was multiplied by their energy equivalent (Table 1). The input energy of agriculture sector was obtained using the multiplication of a conversion factor by its amount (Table 1).

Indices are powerful tools which enable the comparison of systems and detailed study of them. They are calculated by Equations 3 to 6 (Pishgar-Komleh et al., 2012).

$$\text{Energy uses efficiency} = \frac{Energy\ output(\ MJ\ ha^{-1})}{Energy\ input\ (MJ\ ha^{-1})} \tag{3}$$

$$\text{Energy productivity} = \frac{Basil\ output\ (kg\ ha^{-1})}{Energy\ input(MJ\ ha^{-1})} \tag{4}$$

$$\text{Specific energy} = \frac{Energy\ input(MJ\ ha^{-1})}{yield\ \ (kg\ ha^{-1})} \tag{5}$$

Net energy: this index indicates the net output energy of the farm. The negative value of this number shows that the amount of input energy is not equal to the output energy, so this indicates that energy consumption is inefficient.

$$Net\ energy = Energy\ output(MJ\ ha^{-1}) - Energy\ input\ (MJ\ ha^{-1}) \tag{6}$$

One can divide energy demand in agriculture sector by direct and indirect as well as renewable and nonrenewable sectors (Zangeneh et al., 2010). Direct energies involve the fuel consumed for different agricultural operations, electricity, and working force (labor). Indirect energy includes energy consumed in producing equipments, seeds, fertilizers, and chemicals. Renewable energy is categorized by labor and seeds. Nonrenewable energy comprises fuel energy, fertilizers and pesticides (Erdal et al., 2007).

Artificial neural network

Artificial neural network (ANN) was used to predict the GHG emission and basil yield. ANN models have been successfully used in the prediction of problems in bio-processing and chemical engineering (Mahdavian et al., 2012). In essence, ANN, a system modeled on the human brain, consists of an input layer, some hidden layers, and an output layer. In this study, the back-propagation algorithm was used for the training of all ANN models. ANN with back-propagation algorithm learns by changing the weights and these changes are stored as knowledge. Back-propagation training algorithms, gradient descent, and gradient descent with momentum are often too slow for practical problems because they require smaller rates for stable learning. Moreover, success in the algorithms depends on the user-dependent parameters learning rate and momentum constant. Faster algorithms such as conjugate gradient, quasi-Newton, and Levenberg–Marquardt (LM) use standard numerical optimization techniques. LM method is, in fact, an approximation of the Newton's method.The LM algorithm uses the second-order derivatives of the cost function so that better convergence behavior can be obtained. In the ordinary gradient descent search, only the first order derivatives are evaluated and the parameter change information contains solely the direction along which the cost is minimized, whereas the Levenberg–Marquardt technique extracts a biter parameter change vector. To obtain the best prediction by the network, several architectures were evaluated and trained using the experimental data. This algorithm uses the supervised training technique where the network weights and biases are randomly initialized at the beginning of the training phase. The error minimization process is achieved using the gradient descent rule. A fully connected multi-layer perceptron (MLP) was used in this study. Several transfer functions including sigmoid, logarithmic and linear functions together with supervised training algorithms and feed forward

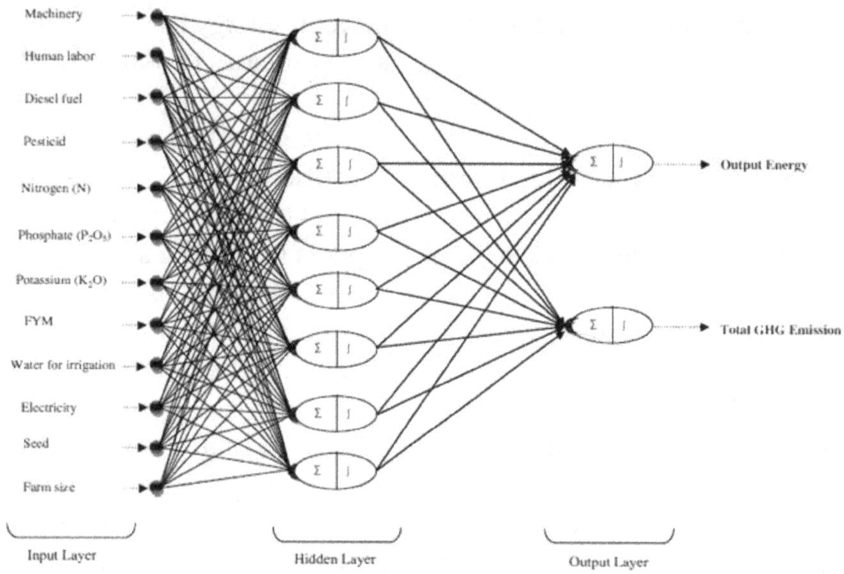

Figure1. Schematic Diagram of Selected Multilayer Feed-Forward Neural Network

back propagation approach was evaluated. To ensure that each input variable provides an equal contribution to the ANN, the inputs of the model were preprocessed and normalized, after which, 70% and 15% of 72 input patterns were devoted to training and validation data sets, respectively. The remaining part of the data wasspecified for the prediction. The learning rate of 0.2 and momentum of 0.1 was adjusted to all the tested networks. Optimum topologies were defined based on the highest R^2 and lowest MSE values. The complexity and size of the network were important, so the smaller ANNs had the priority to be selected. The schematic diagram of the selected multilayer feed-forward neural network is shown in Figure 1. The required codes were developed by MATLAB 2013 software (MathWorks Inc., Natick, Massachusetts, USA).

RESULTS AND DISCUSSION

Estimation of energy consumption

Once the questionnaires were filled out by the participants, values of input consumed energies were determined by studying greenhouses. Likewise, input and output energies were calculated using energy equation coefficients and the results were summarized in Table 3 below. Results indicated that the total consumed energy was 119,852.9 MJ/ha, whereas the output energy was 61,040 MJ/ha. Khoshnevisan et al. (2014b) in a study on strawberry productionreported that the estimated total average of energy input and output was at 35,092.4 MJ/ha and 10,405.9 MJ ha^{-1}, respectively.

As can be seen in Table 3, electricity (43.55%) and plastic (19.37%) are the most important and effective factors in energy consumption followed by chemical fertilizers (11.6%). Among the chemical fertilizers, nitrogen had the highest value (10.61%). The highest electricity consumption was reported in irrigation operations where wells were used for water pumping.

Khoshnevisan et al. (2014a) reported that from among all energy inputs in the production of tomato, natural gas with 66% share was the key input followed by electricity (27%) and chemical fertilizers (4%). This high contribution of natural gas and electricity showed the low efficiency of heating systems and electric pumps utilizedin the production process. Unakitan et al. (2010) found that nitrogen fertilizer (62.79%) and diesel fuel (24.45%) were the most energy consumers in rapeseed production.

a) Includes human labor, electricity, diesel fuel and irrigation

b) Include machinery, FYM, chemical fertilizers, plastic and chemical pesticides

c) Includes human, FYM and irrigation

d) Includes electricity, diesel fuel, machinery, plastic, chemical fertilizers and chemical pesticides

As can be seen in Table 4, one can categorize

Table 3
Energy Input and Output in Basil Production

Input/output	Consumed production (unit area^{-1})	Energy Equivalent (MJ area^{-1})	Percentage
A. Input			
1. Human labor	1100	2156	1.8
2. Seed	35	-	-
3. Irrigation	6900	7038	5.87
4. FYM	18000	5400	4.5
5. Chemical fertilizers		13894	11.6
N	212	12720	-
K_2O	34	227.8	-
P_2O_5	42	466.2	-
Fe	4	480	-
6. Chemicals			8.11
Fungicides	45	9720	-
7. Machinery	20	1254	1.05
8. Diesel fuel	113	4970.87	4.15
9. Electricity	14500	52200	43.55
10. Plastic	258	23220	19.37
Total energy input		119852.9	100
B. Output			
1. Basil yield(kg)	28000	61040	
Total energy output		61040	

the total energy consumption by direct (55.37%), indirect (44.63%) and also renewable (12.18%) and nonrenewable (87.82%) energy sources?. Electricity energy had the highest ratio (43.55%). Moreover, from among the indirect energies, plastic had the highest value (19.37%). Nonrenewable energy had the highest share, and the reason lies on the application of fossil fuels, electric pumps with lower energy consumption efficiency, and the use of chemical fertilizers in higher levels. Energy efficiency, energy productivity, specific energy, and net energy are shown in Table 4. Energy efficiency was measured as 0.45, indicating that the energy consumption for basil production in the studied region is inefficient. This ratio was reported by Pahlavan et al. (2012) as 0.25. The values of energy efficiency, specific energy, and net energy were equal to 0.121 kg/J, 4.77 MJ/kg, and -72,706.9 MJ/area, respectively. These indices were estimated by Pahlavan et al. (2012) as 0.00 kg/J, 9 MJ/kg, and -177,377 MJ/ha, respectively. Unakitan et al. (2010) reported that the shares of renewable and nonrenewable energy resources in the pro-

Table 4
Energy Indices for Basil Production inthe Studied Region

Item	Unit	Amount	Percentage%
Energy ratio	-	0.45	-
Energy productivity	Kg MJ^{-1}	0.21	-
Specific energy	MJ kg^{-1}	4.77	-
Net energy gain	MJ area^{-1}	-72706.9	-
Direct energy a	MJ area^{-1}	66364.87	55.37
Indirect energy b	MJ area^{-1}	53488	44.63
Renewable energy c	MJ area^{-1}	14594	12.18
Non- Renewable energy d	MJ area^{-1}	105258.9	87.82

a) Includes human labor, electricity, diesel fuel and irrigation
b) Include machinery, FYM, chemical fertilizers, plastic and chemical pesticides
c) Includes human, FYM and irrigation
d) Includes electricity, diesel fuel, machinery, plastic, chemical fertilizers and chemical pesticides

Table 5
GHG Emission of Inputs in Basil Productiongion

Input/output	GHG Emissions Equivalent(kg Co_{2eq} area^{-1})	Percentage%
Chemical fertilizers	290.8	3.03
Machinery	1.42	0.01
Diesel fuel	311.88	3.25
Electricity	8816	91.88
Chemical pesticides	175.5	1.83
Total GHG emissions	9595.6	100

duction of rapeseed in Turkey were 0.94% and 99.6%, respectively. The shares of direct and indirect energy were 24.69% and 75.31%, respectively. The results presenting estimation of the produced greenhouse gases are shown in Table 5.

The total gas emission from greenhouse basil production in the studied region was 9,595.6 (kg CO_{2eq} area^{-1}). Electricity had a significant role in greenhouse gas emissions (91.88%). Accordingly, diesel fuel (3.25%) and chemical fertilizers (3.03%) had a considerable share in emission of greenhouse gases. Based on the results presented in Table 3, it can be concluded that electricity has the highest energy share. This implies that most utilized tools in greenhouses work with electricity. Some researchers examined the energy consumption in tomato greenhouses and reported that emissions of greenhouse gases was 34,758 kg (CO_{2eq} area^{-1}) and that consumed electricity had the highest share (Khoshnevisan et al., 2014a).

The contents of energy pertinent to inputs

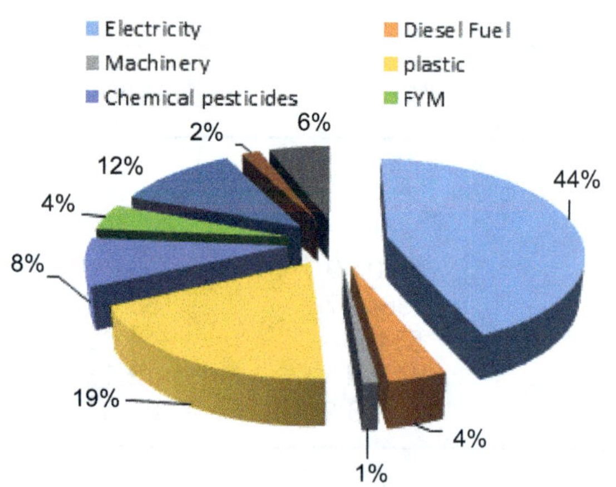

*Figure 2.*The Share of Energy Inputs of Basil in Chaharmahal-Bakhtiary

and produced products in the agriculture sector as well as the total input and output energy were calculated on basil. Figure 2 shows the share of energy inputs (manufacturing the machinery, irrigation, labor, and inputs).

According to Figure 2, the share of machinery in energy consumption is equal to 1% (equivalent to 1,254 MJ/ha), which is insignificant regarding to the total input energy followed by the labor, fuel, and organic fertilizer with 2, 4, and 4% shares, respectively.In a study on the energy required for basil production in Isfahan Province of Iran, itwas reported that the total input energy was 236,057 (GJ/ha) and that electricity (75.68%) had the highest share in energy consumption followed by plastic (9.69%) and chemical fertilizers (7.28%) (Pahlavan et al., 2012). Utilizing new pumps with higher energy efficiency in irrigation systems may be suitable to reduce the share of electric energy consumption. Additionally, excessiveuse of chemical fertilizers as the input energy may cause environmental detriments such as underwater contamination (Khan et al., 2009). In this regard, it is suggested to use organic fertilizers, compost, and plant residues.

Results presented in Table 5 indicate that electricity has the highest share in energy consumption of basil production. The ranks in this case are followed by plastic and chemical fertilizers which have a significant share in energy consumption. One can optimize the energy consumption through reducing chemical fertilizers and substituting them with combined control and precision agriculture approaches. In addition, substituting renewable energies with electricity and its optimal utilization can

Table 6
Summary ofANN Model Evaluations

Activation function	Neurons in hidden layer 1	Neurons in hidden layer 2	Training error	R (training)	R (validation)	R (test)	Epoch
Log/Tan	10	0	0.000465	0.9999	0.9999	0.9999	65
Log/Tan	15	0	0.00027	0.9999	0.9999	0.9999	87
Tan/Log	10	0	0.00021	0.9999	0.9999	0.9999	23
Tan/Log	15	0	0.000405	0.9999	0.9999	0.9999	54
Log/ Log	10	0	0.0018	0.9999	0.9999	0.9999	44
Log/Log	15	0	0.00027	0.9999	0.9987	0.9979	34
Log/Tan/Tan	10	6	0.00027	0.9999	0.9999	0.9999	76
Log/Tan/Tan	15	6	0.0153	0.9868	0.9981	0.9989	29
Log/Tan/Tan	20	6	0.0003	0.9999	0.9999	0.9999	17
Log/Log/Tan	10	6	0.000495	0.9999	0.9999	0.9999	91
Log/Log/Tan	15	6	0.000225	0.9999	0.9998	0.9999	62
Log/Log/Tan	20	6	0.000465	0.9999	0.9999	0.9999	43
Log/Tan/Tan	10	6	0.00036	0.9999	0.9999	0.9999	54
Log/Tan/Tan	**15**	**6**	**0.00023**	**0.9999**	**0.9999**	**0.9999**	**87**
Log/Tan/Tan	20	6	0.00033	0.9999	0.9976	0.9999	41

lead to a remarkable decrease in the consumption of this energy resource.When it comes to calculating the indices, energy ratio indices and energy efficiency are 0.45 and 0.21 which shows that energy efficiency was low in agriculture sector and that the net energy was -72,706.9. Furthermore, direct energy and indirect energy were 55.37% and 44.63%, suggesting that direct energy has a high share and that the highest energy consumption was related to renewable energy which can be a menace to the environment.

It was observed that electricity (92%) had the highest share in GHG generation whose reason is using inefficient electric equipment. This requires optimum use of this energy and substituting it with new energies. Followed by electricity are chemical fertilizers and biodiesel fuel (3%). Moreover, in a study by Khoshnevisan et al. (2014a), the total GHG emission was calculated as 803.4 kg CO_{2eq} ha^{-1}, 35083.5 kg CO_{2eq} ha^{-1} for OF, and GH production, respectively. Using organic fertilizers instead of chemical fertilizers can result in a reduction in GHG emissions.

ANN models for predicting the GHG emission and Basil yield

Several topologies were evaluated to obtain the maximum R^2 and minimum MSE values. The results are presented in Table 6. It can be inferred from the table that a network with 2 hidden layers (15 and 6 neurons in first and second layer, consequently) using Levenberg–Marquardt (LM) learning algorithm and tangent-sigmoid transfer function would provide an efficient response, helping to predict the output parameter. A coefficient of determination (R^2) of 0.9999 and a training error of 0.00023 were resulted from the network training. Results showed that an artificial neural network was advantageous in the prediction of basil yield and CO_2 emission. Results of a regression analysis of experimental data and network outputs are illustrated in Figures 3 and 4. The maximum value of R^2 for training, validation, and prediction stages was 0.9999.

Khoshnevisan et al. (2014a) studied artificial neural networks and the fuzzy-neural deduction system in an attempt ot to model consumed energy in producing greenhouse tomato in Isfahan Province of Iran. Comparison of artificial neural networks and the fuzzy-neural deduction system suggest that ANFIS was more accurate in modeling through making use of fuzzy rules. Other researchers employed artificial neural networks to determine the energy consumption in wheat production (Safa & Samarasinghe, 2011). They compared the artificial neural network using multivariate linear regression and

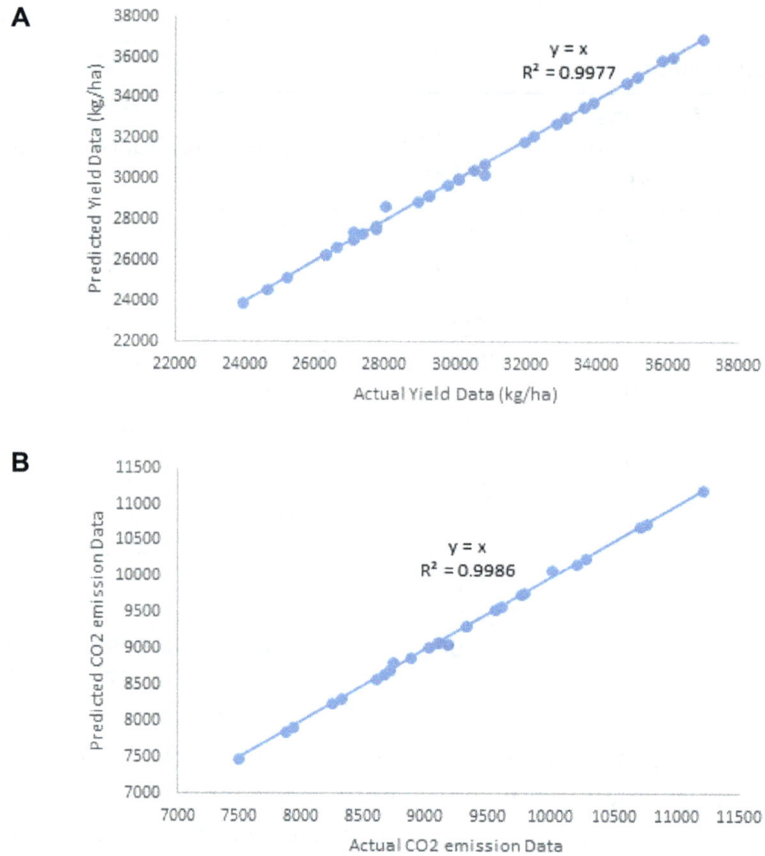

Figure 3. **(A)** Predicted Outputs Versus Measured Yield Values **(B).** Predicted Outputs versusCo2Emission Measured Values

found that these networks were better predictors of energy consumption compared to regression mode.

CONCLUSION

The relationship between input and output energy in production of greenhouse basil was studied. The total consumed energy in basil production was recorded as 119,852.9 (MJ area-1). Electricity energy had the highest share in input energies followed by plastic and chemical fertilizers. Hence, one can declare that this greenhouse is not justifiable in terms of consumption rate, efficiency, and energy efficiency. The reasons would be counted as the lack of observing the true principles of greenhouse insulation with outer space, the existence of inefficient water pumps, excessive use of fertilizers, and low-costing side of energy inputs in Iran, which require propermanagement. Carefulmanagement of inputs can reduce environmental risks. The results of model showed that the trained neural network was accurate in estimating the performance of basil and CO_2 emission (R^2=0.9999 and MSE= 0.00023). The best network was categorized by topology of 15 neurons in the first layer and 6 neurons in the second layer.

ACKNOWLEDGEMENT

Research Council of Shahrekord Universityis thank fully acknowledged for the financial support to carryingout the work (grant No: 94 GRD1M1796).

REFERENCES

1- Al-Ghandoor, A., Jaber, J., Al-Hinti, I., & Mansour, I. (2009). Residential past and future energy consumption: Potential savings and environmental impact. *Renewable and Sustainable Energy Reviews, 13*(6), 1262-1274.

2- Alam, M.S., Alam, M.R., & Islam, K. (2005). Energy flow in agriculture: Bangladesh. *American*

Journal of Environmental Sciences, 1(3), 213- 220.

3- Alluvione, F., Moretti, B., Sacco, D., & Grignani, C. (2011). EUE (energy use efficiency) of cropping systems for a sustainable agriculture. *Energy, 36*(7), 4468-4481.

4- Barut, Z. B., Ertekin, C., & Karaagac, H. A. (2011). Tillage effects on energy use for corn silage in Mediterranean Coastal of Turkey. *Energy,36*(9), 5466-5475.

5- Canakci, M., Topakci, M., Akinci, I., & Ozmerzi, A. (2005). Energy use pattern of some field crops and vegetable production: Case study for Antalya Region, Turkey. *Energy Conversion and Management, 46*(4), 655-666.

6- Chauhan, N. S., Mohapatra, P. K., & Pandey, K.P. (2006). Improving energy productivity in paddy production through benchmarking An application of data envelopment analysis. *Energy Conversion and Management,* 47(9), 1063-1085.

7- Coxworth, E., Leduc, P., & Hultgreen, G. (1995). Analysis of crop production systems for reducing carbon emissions, stashing carbon in soils and providing raw materials for bioenergy production. Agriculture Canada.

8- Dalgaard, T. (2000). Farm types-How can they be used to structure, model and generalize farm data? *In Agricultural data for life cycle assessments.* Agricultural Economics Research Institute (LEI).

9- Dyer, J. A., & Desjardins, R. L. (2006). Carbon dioxide emissions associated with the manufacturing of tractors and farm machinery in Canada. *Biosystems Engineering, 93*(1), 107-118.

10- Erdal, G., Esengun, K., Erdal, H., & Gunduz, O. (2007). Energy use and economical analysis of sugar beet production in Tokat Province of Turkey. *Energy, 32*(1), 35-41.

11- Hatirli, S. A., Ozkan, B., & Fert, C. (2005). An econometric analysis of energy input–output in Turkish agriculture. *Renewable and Sustainable Energy Reviews, 9*(6), 608-623.

12- Khan, S., Khan, M., Hanjra, M., & Mu, J. (2009). Pathways to reduce the environmental footprints of water and energy inputs in food production. *Food Policy, 34*(2), 141-149.

13- Khodi, M., & Mousavi, S. (2009). Life cycle assessment of power generation technology using GHG emissions reduction approach. In *7th National Energy Congres*, (pp. 22-23).

14- Khoshnevisan, B., Rafiee, S., Omid, M., Mousazadeh, H., & Clark, S. (2014a). Environmental impact assessment of tomato and cucumber cultivation in greenhouses using life cycle assessment and adaptive neuro-fuzzy inference system. *Journal of Cleaner Production, 73,* 183-192.

15- Khoshnevisan, B., Shariati, H. M., Rafiee, S., & Mousazadeh, H. (2014b). Comparison of energy consumption and GHG emissions of open field and greenhouse strawberry production. *Renewable and Sustainable Energy Reviews,29,* 316-324.

16- Lal, R. (2004). Carbon emission from farm operations. *Environment International, 30,* 981-990.

17- Mahdavian, A., Banakar, A., Mohammadi, A., Beigi, M., & Hosseinzadeh, B. (2012). Modelling of Shearing Energy of Canola Stem in Quasi-Static Compres-sive Loading Using Artificial Neural Network (ANN). *Middle-East Journal of Scientific Research, 11*(3), 374-381.

18- Mohammadshirazi, A., Akram, A., Rafiee, S., Avval, S.H.M., & Kalhor, E.B.(2012). An analysis of energy use and relation between energy inputs and yield in tangerine production. *Renewable and Sustainable Energy Reviews,16,* 4515-4521.

19- Mousavi-Avval, S. H., Rafiee, S., Jafari, A., & Mohammadi, A. (2011). Improving energy use efficiency of canola production using data envelopment analysis (DEA) approach. *Energy, 36*(5), 2765-2772.

20- Ozkan, B., Akcaoz, H., & Karadeniz, F. (2004). Energy requirement and economic analysis of citrus production in Turkey. *Energy Conversion and Management, 45*(11), 1821-1830.

21- Pahlavan, R., Omid, M., & Akram, A. (2012). The relationship between energy inputs and crop yield in greenhouse basil production. *Journal of Agricultural Science and Technology,14*(6), 1243-1253.

22- Pishgar-Komleh, S., Ghahderijani, M., & Sefeedpari, P. (2012). Energy consumption and CO_2emissions analysis of potato production based on different farm size levels in Iran. *Journal of Cleaner Production, 33,* 183-191.

23- Rahman, M., & Bala, B. (2010). Modelling of jute production using artificial neural networks. *Biosystems Engineering, 105*(3), 350-356.

24- Royan, M., Khojastehpour, M., Emadi, B., & Mobtaker, H. G. (2012). Investigation of energy inputs for peach production using sensitivity analysis in Iran. *Energy Conversion and Management, 64,* 441-446.

25- Safa, M., & Samarasinghe, S. (2011). Determination and modelling of energy consumption in wheat production using neural networks: A case study in Canterbury province, New Zealand. *Energy, 36*(8), 5140-5147.

26- Soltani, A., Rajabi, M., Zeinali, E., & Soltani, E. (2013). Energy inputs and greenhouse gases emissions in wheat production in Gorgan, Iran. *Energy, 50,* 54-61.

27- Tabatabaeefar, A., Emamzadeh, H., Varnamkhasti, M. G., Rahimizadeh, R., & Karimi, M. (2009). Comparison of energy of tillage systems in wheat production. *Energy, 34*(1), 41-45.

28- Unakitan, G., Hurma, H., & Yilmaz, F. (2010). An analysis of energy use efficiency of canola production in Turkey. *Energy, 35*(9), 3623-3627.

29- Zangeneh, M., Omid, M., & Akram, A. (2010). A comparative study on energy use and cost analysis of potato production under different farming technologies in Hamadan Province of Iran. *Energy, 35*(7), 2927-2933.

11

Modelling of Greenhouse Gas Emissions from Wheat Production in Irrigated and Rain-Fed Systems in Khorasan Razavi Province, Iran

Seyed Masoud Motamedolshariati [1], Hassan Sadrnia [2], Mohammad Hossein Aghkhani [3] and Mehdi Khojastehpour [2]*

Abstract

Agriculture has a key role in greenhouse gas emissions. As such, the present study aimed to evaluate the greenhouse gas emissions from wheat production in irrigated and rain-fed systems. The primary data were collected from 116 wheat farmers. The results showed that the total greenhouse gas emissions from wheat production in irrigated and rain-fed systems were 637.8 and 65.12 $kgCO_{2eq}$, respectively. The diesel fuel was the largest contributor to the total greenhouse gas emissions in irrigated systems with the share of 33%. Moreover, these inputs accounted for the highest share of greenhouse gas emissions in rain-fed system. The results of Cobb-Douglas model highlighted that the effects of inputs, including machinery, diesel fuel, electricity, and farmyard manure were positive on the yield in irrigated systems. However, the effect of chemical fertilizer and biocide inputs was negative on wheat yield. On the other hand, the effects of all inputs were positive on wheat yield in rain-fed system. The results of the sensitivity analysis showed that one kg increase in greenhouse gas emissions from chemical fertilizer and biocide would result in 0.28 and 0.15 kg loss of yield, respectively.

Keywords:
carbon dioxide, diesel fuel, environment, greenhouse gas emissions, modelling

[1] *Former M.Sc. Student, Department of Biosystems Engineering, Ferdowsi University of Mashhad, International Campus, Mashhad, Iran*

[2] *Associate Professor, Department of Biosystems Engineering, Ferdowsi University of Mashhad, Mashhad, Iran*

[3] *Professor, Department of Biosystems Engineering, Ferdowsi University of Mashhad, Mashhad, Iran*

* *Corresponding author's email: hassan.sadrnia@um.ac.ir*

INTRODUCTION

Management of the energy resources in the agricultural sector is a reall challenge, and there is considerable potential for the use of renewable energy resources. To optimize food production efficiency, research is needed to investigate the environmental impacts of agricultural production so as to achieve sustainable development. Increased level of mechanization and the use of fossil fuels have caused greenhouse gas (GHG) emissions where their management poses a serious challenge (Liang et al., 2013; Nikkhah et al., 2015a). Agriculture is one of the main sources of GHG emissions such as CO_2, CH_4 and N_2O (Van der Maas et al., 2009; Nikkhah et al., 2015b; Soltanali et al., 2015). It is necessary to evaluate environmental impacts of different sectors of agriculture.

Several research studies have been conducted on GHG emissions in crop production systems in Iran. GHG emissions from corn cultivation was investigated, and the authors reported that the highest share of GHG emissions from corn production in Karaj city of Iran belonged to machinery (74%) followed by diesel fuel (22%) (Pishgar-Komleh et al., 2011).

There have been some studies on GHG emissions from wheat production. For example, Khoshnevisan et al. (2013) showed that the total GHG emissions from wheat production were 2712 $kgCO_{2eq}$ ha^{-1} in Isfahan Province, Iran. Moreover, the greatest share of GHG emissions belonged to the electricity followed by diesel fuel. Soltani et al. (2013) showed that the average of GHG emissions from wheat production was 291 to 11137 $kgCO_{2eq}$ ha^{-1} in Goelestan Province, Iran. In yet another study, Mirhaji et al. (2013) evaluated the environmental impacts of wheat production in the Marvdasht region of Iran using Life Cycle Assessment (LCA) methodology. They claimed that the eutrophication impact category had the largest negative impacts on the environment. GHG emissions of low- and high-input wheat production systems in Western Iran were evaluated, and it was found that the highest share of GHG emissions for low and high input wheat production systems were related to N fertilizer (32%) and electricity (36%), respectively. It was also shown that one hectare of the high-input system will produce 17 times as great greenhouse effect as low-input systems (Yousefi et al., 2016).

The sustainable production of wheat in Khorasan Razavi Province of Iran requires the consideration of environmental management in the production systems. However, to the best knowledge of the authors' knowledge, no previous analytical work has been reported on the environmental impacts of wheat production in that province. Therefore, the aim of the present study was to model the GHG emissions from wheat production in irrigated and rain-fed systems in Khorasan Razavi Province.

MATERIALS AND METHODS

The study area and data collection

The study was conducted in the Sarakhs region of the Khorasan Razavi Province, Iran. The sample size was calculated using the Cochran method (Snedecor & Cochran, 1980):

$$n=(N(s\times t)^2)/((N-1)\ d^2+(s\times t)^2) \quad (1)$$

$$d=(t\times s)/\sqrt{n} \quad (2)$$

where, n = sample size, N = number of holdings in the target population, t = the confidence coefficient (1.96), s = the variance, and d = precision (Fallahi et al., 2016). Based on this calculation, data were collected from 116 farmers using a questionnaire administered face-to-face in 2012-2013. Each farmer was asked to detail activities as inputs to wheat production recorded as machinery use (hr), diesel fuel (l), chemical fertilizer (kg), and biocides (kg), and yield (kg) as the output.

GHG emissions evaluation

The GHG emissions from wheat production were determined by multiplying the input activity data by an emission factor (see Table 1) (Khojastehpour et al., 2015). The CO_2 emission from machinery contributes to the emissions from manufacturing and the use of these inputs in the farm (Firouzi et al., 2016).

GHG *emissions modeling*

The Cobbe-Douglas model was then used to find the effect of GHG emissions on the yield to produce wheat in the region (Royan et al., 2012):

$$Lny_i = a_0 + \sum_{j=1}^{n} \alpha_j \ln(x_{ij}) + e_i = 1,2,\dots,n \qquad (1)$$

where, y_i denotes the yield of the ith farmer; x_{ij} denoteseach of the inputs used in the production process (units as noted above); the constant a_j is the coefficients of inputs which are estimated from the model, and e_i is an error term. With this assumption, yield function of energy inputs, Eq. (3), can be expanded to Eq. (4) (Soltanali et al., 2016):

$$Lny_i= a_0+ \alpha_1 lnx_1+ \alpha_2 lnx_2+ \alpha_3 lnx_3+ \alpha_4 lnx_4+ \alpha_5 lnx_5+ \alpha_6 lnx_6+ e_i \qquad (4)$$

where, x_1, x_2, x_3, x_4, x_5, and x_6 are the energies of seed, human labor, machinery, diesel fueled, chemical fertilizer, and biocide, respectively. The impact of the GHG emissions on the output yield was quantified by using the standard beta. Finally, the sensitivity of yield in the region to GHG emissions was investigated using the marginal physical productivity (MPP) method, which shows the change in the output for one unit change in a given input, keeping all other factors constant (Pishgar-Komleh et al., 2013; Nikkhah et al., 2016). The MPP of the various inputs was calculated by (Rafiee et al., 2010):

$$MPP_{xj}= GM(Y)/GM(X_{ij}) \times \alpha_{ij} \qquad (5)$$

where, MPP_{ij} is the marginal physical productivity of jth input; α_j is the regression coefficient of jth input; $GM\ (Y)$ is the geometric mean of the yield, and $GM\ (X_j)$ denotes the geometric mean of the jth GHGinput on per hectare basis (Mobtaker et al., 2012).

RESULTS AND DISCUSSION

GHG *emissions results*

Table 2 shows the GHG emissions of wheat

Table 1
Greenhouse Gas Emissions Coefficients

Inputs	Unit	(Kg CO_{2eq} unit^{-1})	Reference
Machinery	MJ	0.071	(Dyer & Desjardins, 2006)
Diesel fuel	Lit	2.76	(Dyer & Desjardins, 2003)
Chemical Fertilizer			
(N)	Kg	1.3	(Lal, 2004)
(P_2O_5)	Kg	0.2	(Lal, 2004)
(K_2O)	Kg	0.2	(Lal, 2004)
Biocide			
Fungicides	Kg	3.9	(Lal, 2004)
Insecticides	Kg	5.1	(Lal, 2004)
Herbicides	Kg	6.3	(Lal, 2004)

Table 2
GHG Emissions from Wheat Production in Khorasan Razavi Province, Iran

Inputs	Production systems			
	Irrigated		Rain-fed	
	Average (kgCO_{2eq}. ha^{-1})	Standard deviation	Average (kgCO_{2eq}. ha^{-1})	Standard deviation
Chemical Fertilizers	63.79	57.00		
Nitrogen (P_2O_5)	56.34	49.92		
Phosphorus	7.46	9.18		
Electricity	75.84	69.02		
Biocide	6.14	2.92		
Machinery	154.91	103.89	14.76	3.98
Diesel fuel	207.80	56.54	50.35	19.12
Farmyard manure	129.32	103.89		
Total GHG emissions	637.80	303.32	65.12	29.83

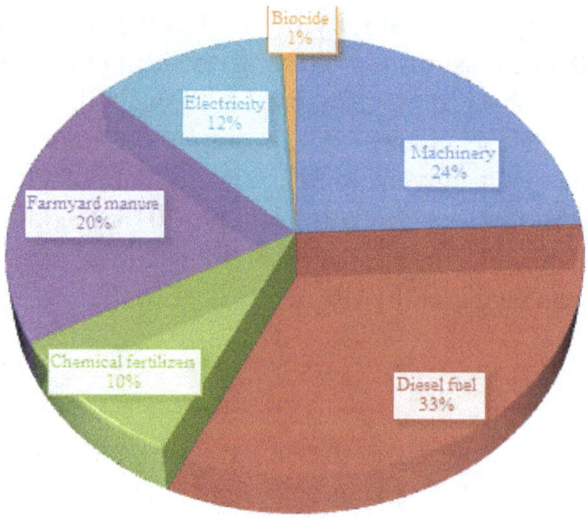

Figure 1. The share of energy inputs for wheat production in irrigated system in Khorasan Razavi Province, Iran

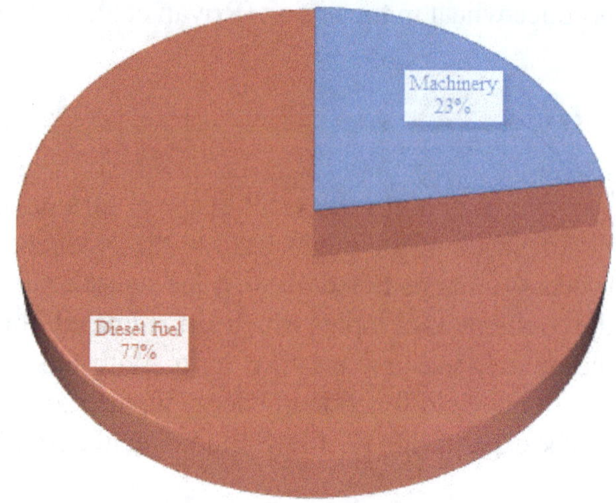

Figure 2. The share of energy inputs for wheat production in rain-fed system in Khorasan Razavi Province, Iran

production in irrigated and rain-fed systems in Khorasan Razavi Province, Iran. GHG emission of the diesel fuel was 207.80 kgCO_{2eq} ha^{-1}. The diesel fuel had the highest share (33%) in the total GHG emission from wheat production in irrigated systems in Khorasan Razavi Province (see Figure 1). The emission of machinery accounted for 24% of total emissions. The amount of GHG emissions from this input of wheat production was 154.91 kgCO_{2eq} ha^{-1} (see Table 2). Diesel fuel and chemical fertilizer had the highest GHG emissions in wheat and potato production (Soltani et al., 2013; Pishgar-Komleh et al., 2012a). GHG emissions from chemical fertilizer were 63.79 kgCO_{2eq} ha^{-1}.

GHG emission from the diesel fuel in wheat production in rain-fed system was 50.35 kg CO_{2eq} ha^{-1}. The diesel fuel had the highest share (77%) in the total GHG emission from wheat production in rain-fed system in Khorasan Razavi Province (see Figure 2). The emission of machinery accounted for 23% of total emissions. The amount of GHG emissions from this input of wheat production was 14.76 kgCO_{2eq} ha^{-1} (see Table 2).

Total GHG emission from wheat production in irrigated and rain-fed systems were obtained as 637.80 and 65.12 kgCO_{2eq} ha^{-1}, respectively (see Table 2). It means that the GHG emissions from wheat production in an irrigated system were higher than that of rain-fed system. Other researchers reported the total GHG emissions as 1195 kgCO_{2eq} ha^{-1} forcotton and 993 kgCO_{2eq} ha^{-1} for potato (Pishgar-Komleh et al., 2012a; Pishgar-Komleh et al., 2012b).

Table 3

Estimation of the Effect of GHGE Missions on Irrigated Wheat Yield in Khorasan Razavi, Iran

	Coefficient	t-ratio	p-value	MPP
Model: $Lny_i = a_0 + \alpha_1 lnx_1 + \alpha_2 lnx_2 + \alpha_3 lnx_3 + \alpha_4 lnx_4 + \alpha_5 lnx_5 + \alpha_6 lnx_6 + e_i$				
Machinery	0.041	1.26	0.141	0.0126
Diesel fuel	0.033	1.12	0.139	0.108
Chemical Fertilizers	-0.902	-2.06	0.019	-0.283
Electricity	0.932	1.54	0.033	0.157
Biocide	-0.054	-4.26	0.009	-0.146
Farmyard manure	0.012	1.14	0.133	0.151
R^2	0.77			
R^2_{Adj}	0.71			
Durbin Watson	2.01			
Return to scale	0.06			

Table 4
Estimation of the Effect of GHG Emissions on Rain-Fed Wheat Yield in Khorasan Razavi, Iran

	Coefficient	t-ratio	p-value	MPP
Model: $Lny_i = a_0 + \alpha_1 lnx_1 + \alpha_2 lnx_2 + e_i$				
Machinery	0.39	2.33	0.03	0.121
Diesel fuel	0.016	0.93	0.238	0.090
R^2	0.71			
R^2_{Adj}	0.68			
Durbin Watson	2.12			
Return to scale	0.41			

Modeling of GHG emissions

The results of the Cobb-Douglas model showed that the impacts of GHG emissions from machinery, diesel fuel, electricity, and farmyard manure on irrigated wheat yield were positive, while the impacts of GHG emissions from the chemical fertilizers and biocide were negative (see Table 3). The results of sensitivity analysis revealed that one kgincrease in GHG emissions from machinery, diesel fuel, chemical fertilizers, electricity, biocides, and farmyard manure changed the yield by 0.0126, 0.108, -0.283, 0.157, -0.146 and 0.151 kg, respectively.

Table 4 displays the estimation of the effect of GHG emissions on rain-fed wheat yield in Khorasan Razavi, Iran. The results of the Cobb-Douglas model showed that the impacts of GHG emissions of machinery and diesel fuel were positive on wheat yield. The results of sensitivity analysis indicated that one kg increase in GHG emissions from machinery and diesel fuel changed the yield by 0.121 and 0.090 kg, respectively.

ACKNOWLEDGMENT

The financial support provided by Ferdowsi University of Mashhad, Iran is duly acknowledged.

REFERENCES

1- Dyer, J. A., & Desjardins. R. L. (2003). Simulated farm fieldwork, energy consumption and related greenhouse gas emissions in Canada. *Biosystems Engineering, 85*(4), 503-513.

2- Dyer, J. A., & Desjardins, R. L. (2006). Carbon dioxide emissions associated with the manufacturing of tractors and farm machinery in Canada. *Biosystems Engineering, 93*(1), 107-118.

3- Fallahi, H., Abbaspour-Fard, M. H., Azhari, A., Khojastehpour, M., & Nikkhah, A. (2016). Ergonomic assessment of drivers in MF285 and MF399 tractors during clutching using algometer. *Information Processing in Agriculture,3* (1), 54-60.

4- Firouzi, S., Nikkhah, A., Khojastehpour, M., & Holden, N., (2016). Energy use efficiency, GHG emissions, and carbon efficiency of paddy rice production in Iran. *Energy Equipment and Systems, 4*(2), 169-176.

5- Khojastehpour, M., Nikkhah, A., & Hashemabadi, D. (2015). A comparative study of energy use and greenhouse gas emissions of canola production. *International Journal of Agricultural Management and Development, 5* (1), 51-58.

6- Khoshnevisan, B., Rafiee, S., Omid, M., Yousefi, M., & Movahedi, M. (2013). Modeling of energy consumption and GHG (greenhouse gas) emissions in wheat production in Esfahan province of Iran using artificial neural networks. *Energy, 52,* 333-338.

7- Lal, R. (2004). Carbon emission from farm operations. *Environment International, 30*(7), 981-990.

8- Liang, S., Xu, M., & Zhang, T. (2013). Life cycle assessment of biodiesel production in China. *Bioresource Technology, 129*, 72-77.

9- Mirhaji, H., Khojastehpour, M., & Abaspour-Fard, M. H. (2013). Environmental Effects of wheat production in the Marvdasht region. *Journal of Natural Environment, 66* (2), 223-232 (In Persian).

10- Mobtaker, H. G., Akram, A., & Keyhani, A. (2012). Energy use and sensitivity analysis of energy inputs for alfalfa production in Iran. *Energy for Sustainable Development, 16*(1), 84-89.

11- Nikkhah, A., Emadi, B., & Firouzi, S. (2015a). Greenhouse gas emissions footprint of production in Guilan province of Iran. *Sustainable Energy Technologies and Assessments, 12*, 10–14.

12- Nikkhah, A., Khojastehpour, M., Emadi, B., Taheri- Rad, A., & Khorramdel, S. (2015b). Environmental impacts of peanut production system using life cycle assessment methodology. *Journal of Cleaner Production, 92,* 84-90.

13- Nikkhah, A., Emadi, B., Soltanali, H., Firouzi, S., Rosentrater, K.A. & Allahyari, M.S., (2016). Integration of life cycle assessment and Cobb- Douglas modeling for the environmental as sessment of kiwifruit in Iran. *Journal of Cleaner Production, 137,* 843-849.

14- Pishgar-Komleh, S. H., Omid, M., & Heidari, M. D., (2013). On the study of energy use and GHG (greenhouse gas) emissions in greenhouse cucumber production in Yazd Province. *Energy,59*, 63-71.

15- Pishgar-Komleh, S.H., Keyhani, A., Rafiee, S., & Sefeedpary. P. (2011). Energy use and economic analysis of corn silage production under three cultivated area levels in Tehran province of Iran. *Energy, 36,* 3335-3341.

16- Pishgar-Komleh, S.H., Sefeedpari, P., & Ghahderijani, M. (2012b). Exploring energy consumption and CO [sub 2] emission of cotton production in Iran. *Journal of Renewable and Sustainable Energy, 4*, 33115-33114.

17- Pishgar-Komleh, SH., Ghahderijani, M., & Sefeedpari, P. (2012a). Energy consumption and CO_2 emissions analysis of potato production based on different farm size levels in Iran. *Journal of Cleaner Production, 33,* 183-191.

18- Rafiee, S., Mousavi-Avval, S.H., & Mohammadi, A. (2010). Modeling and sensitivity analysis of energy inputs for apple production in Iran. *Energy, 35*, 3301-3306.

19- Royan, M., Khojastehpour, M., Emadi, B., & Mobtaker, H. G. (2012). Investigation of energy inputs for peach production using sensitivity analysis in Iran. *Energy Conversion and Management, 64*, 441-446.

20- Snedecor, G.W., & Cochran, W.G. (1980). Statistical methods. Iowa State University Press, USA.

21- Soltanali, H., Emadi, B., Rohani, A., Khojastehpour, M., & Nikkhah, A. (2015). Life cycle assessment modeling of milk production in Iran. Information *Processing in Agriculture, 2*(2), 101-108.

22- Soltanali, H., Emadi, B., Rohani, A., Khojastehpour, M., & Nikkhah, A. (2016). Optimization of energy consumption in milk production units through integration of DEA approach and sensitivity analysis. *Iranian Journal of Applied Animal Science, 6*(1), 15-23.

23- Soltani, A., Rajabi, M.H., Zeinali, E., Soltani, E. (2013). Energy inputs and greenhouse gases emissions in wheat production in Gorgan, Iran. *Energy, 50*, 54-61.

24- Van der Maas, C.W.M., Coenen, P.W.H.G., Zijlema, P.J., Brandes, L.J., Baas, K., Van den Berghe, G., Van den Born, G.J., Guis, B., Geilenkirchen, G., Te Molder, R., Nijdam, D.S., Olivier, J.G.J., Peek, C.J., Van Schijndel, M.W., & vander Sluis, S.M. (2009). *Greenhouse Gas Emissions in the Netherlands 1990–2007.* National Inventory Report Netherlands Environmental Assessment Agency (PBL), Bilthoven.

25- Yousefi, M., Mahdavi Damghani, A., Khoramivafa, M. 2016. Comparison greenhouse gas (GHG) emissions and global warming potential (GWP) effect of energy use, in different wheat agroecosystems in Iran. *Environmental Science and Pollution Research, 23*(8), 7390-7397.

12

Factors Affecting Commercialization of Agricultural Innovation in Kermanshah Science and Technology Park, Iran

Nasibeh Pourfateh [1], Nader Naderi [2] and Farahnaz Rostami [3]*

Keywords:
agricultural innovation; commercialization; science and technology parks; Structural Equation Modeling

Abstract

Science and technology parks have been a major driver of both the commercialization of agricultural innovation and the financial success of many farm and agribusiness firms. Therefore, the main purpose of this study was to explain the factors affecting commercialization of agricultural innovation in Kermanshah Science and Technology Park. The experts of this center were sampled by the census method (N=110). Structural equation modeling was used to analyze the data. Research findings indicated that factors affecting commercialization of agricultural innovation in Kermanshah Science and Technology Park included support of small and medium enterprise firms, relationship of parks with universities and research centers, and consequence of commercialization for agricultural sectors and research centers.

[1] *M.Sc., Razi University, Kermanshah, Iran*
[2] *Assistant Professor, Razi University, Kermanshah, Iran*
[3] *Assistant Professor, Razi University, Kermanshah, Iran*
* *Corresponding author's email: fr304@yahoo.com*

INTRODUCTION

Providing a context of science leads to economic and technological growth for society and also economic value for organizations. The commercialization of innovation is so important that many research institutes have recognized the commercialization of their innovations using collaboration in research projects and consultation. Studies have shown that science and technology parks are still at the beginning of their activities and they have not been fully matured in structural and institutional (Arasteh & Jahed, 2010). The weak relationship of knowledge production centers such as universities and research centers with industry is the one of the barriers to the commercialization of science. Because of the importance of these problems, this study is intended to explain the factors affecting commercialization of innovations which output the agricultural faculties and agricultural research centers and its consequences to help managers and policy makers to leave old pattern and it is the most important feature for organization development (Arasteh & Jahed, 2010). Jimenez and Cegarra (2008) relate it to the fact that organizations that have the capacity for innovation will be able to respond to environmental challenges faster and better than non-innovative organizations and this in turn increase performances. The commercialization process of innovation is closely related to transfer of innovation and on the other hand it innovation transfer process from research centers to industry. Commercialization starts with the development of an idea, takes shape with the production of goods and the development-based services, and completes with the sale of goods and services to the end users. Commercialization is also considered as an innovation transfer from a person to another or from a group to another group. Aghajani and Talebnejad (2011) proposed a framework with four dimensions of personnel management, infrastructures, service, and output to evaluate growth centers in Iran. Infrastructure includes location, physical features required for growth center, management includes managers who can coach and founder of the new company and familiar with business, service includes office and business support services and administration, and outputs are products and services generated by the technological growth center. They concluded that the role of the growth center of the process and acts is more than its effect on the gained results by companies. The mentioned factors are divided into seven categories including structural factors, political factors, funding, high-risk, intellectual property systems, human factors, and cognitive and attitudinal factors. Arasteh and Jahed (2010) noted six roles for science and technology park in commercialization: generating and remaining competitive, supporting small companies, improving mutual relations with the government, academia, and society, technology entrepreneurship, creating opportunity for knowledge workers and employers, and working on risk reduction projects in science and technology. Sanni et al. (2010) stated that critical success factors for parks are physical facilities, low-cost incubator space, enough funding, strong regional universities offering graduate programs in management and engineering. Vila and Pages (2008) believe that critical factors influencing the success of parks are open standards and official activities, strong connection with the university and full-time professional management. Law (2005) has provided a framework for the assessment of technology incubators in the science and technology park, namely the sharing of resources, resource integration, consulting services, geographic proximity, and budget subsidies. Fukugawa (2006) showed that the science park in Japan based on new technology has more willingness to participate in collaborative research with institutions. Sohn and Moon (2004) provided a model to identify structures influencing the commercialization of the technology and assess the success of technology commercialization. They divided these factors into four categories: methods of technology transfer, technology transfer, technology receptors, and environmental conditions. Siegel and Westhead (2003) have stated that the main barriers to effective technology transfer from university to industry are cultural gap between industry and academia, lack of flexibility, bureaucracy, poor reward systems and other

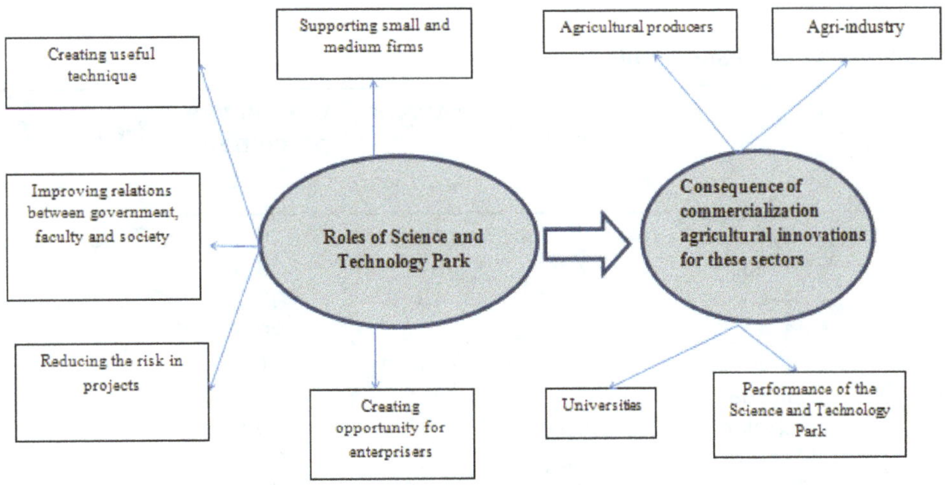

Figure 1. Conceptual Framework of the Research

management non-effective technology transfer offices. Shinn and Lamy (2006) believe that the way for commercialization of knowledge is that the boundaries between universities, research centers, and enterprises should be removed or wiped out. Arasteh and Jahed (2010) divided the general role of science and technology parks in commercialization into six stages. We present them with a little change here in five stages. These steps include: supporting small and medium firms, improving the relationship between government, society and universities, technology entrepreneurship, reducing risk science and technology projects, and creating opportunities for knowledge workers (Figure 1). So the main purpose of this study was to explain the factors affecting commercialization of agricultural innovation in Kermanshah Science and Technology Park.

MATERIALS AND METHODS

The study was designed as a descriptive-correlation survey to determine components of the agricultural innovation and commercialization model and achievements of commercialization in the Center of Agricultural Science and Technology Park in Kermanshah Province by using a confirmatory factor analysis. The method tests specific hypotheses about the structure of factor loadings and their internal consistency (Statsoft, 2009; Pallant, 2010). A total of 110 expert members of Science and Technology Park in Kermanshah Province were selected by the census method as a statistical population because of its small size. Structural equation modeling and multivariate analysis of variance were used to analyze the data. For the purpose of this study, a four-part questionnaire was developed. Part 1 was the cover letter explaining the objective of the study for respondents. Part 2 included questions about personal characteristics such as age, level of education, organizational position, career history, and field of study. Part 3 explained factors affecting commercialization of agricultural innovation, including supporting small and medium-size firms or SMEs (Y1), improving relations among government, universities and society (Y2), technology creation (Y3), risk reduction of Science and Technology Projects (Y4), and an opportunity for researchers and employers (Y5). Part 4 included results and consequence of commercialization in this sectors: agriculture (Z1), Science and Technology Park (Z2), university (Z3) and industry (Z4). In these two parts, the respondents were asked to quantify the items on a five-point Lickert scale. The mean scores were calculated and responses to importance were ranked. A panel of experts validated the instrument for face and content validity. Mean Cronbach's alpha for part 3 (α=0.84) and part 4 (α=0.88) was found to be 0.86, which made the Lickert scale sufficiently reliable (Bagozzi & Yi, 1988).

RESULTS AND DISCUSSION

The results of personal features showed that professional experts had an average age of 30 years, and also around 4.4 years' experience. The study of organizational position of respondents indicated that 57 percent of staff experts

Table 1
Personal Characteristics of Respondents

Item		Frequency	Percent	Cumulative percentage	Mean	SD	Mode
Age (years)	≤26	12	10.9	10.9	30.8	5.25	28
	27-33	74	67.3	78.2			
	34-40	17	15.5	93.7			
	41-47	6	5.9	99.6			
	≥48	1	.4	100			
Level of education	Diploma	10	9.1	9.1			Bachelor
	Bachelor	66	60	69.1			Expert
	Master	32	29.1	98.2			
	Ph.D.	2	1.8	100			
Organizational position	Directing Manager	53	48.2	48.2			
	Expert	57	51.5	100	4.4	3.47	5
Years of experience	1-3 years	52	47.3	47.3			
	4-9 years	48	43.6	90.9			
	10-16 years	8	7.3	98.2			
	More than 17	2	1.8	100			Industry
Field of Study	Agronomy	25	22.72	22.72			
	Agricultural	18	16.36	39.08			
	Extension	14	12.72	51.80			
	Animal Science						
	Agricultural	10	9.09	60.89			
	Machinery	5	4.54	65.43			
	Water Engineering						
	Soil Science	3	2.72	68.15			
	Plant Pathology	10	9.9	78.05			
	Industry and other fields	30	27.27	100			

and are the rest of the post was directing managers. Educational status of the respondents indicated that 60% of participants had a bachelor's degree, 85 percent had studied in the field of agriculture (Agriculture Extension, Animal Science, Machinery, Water engineering, Soil science, and Plant pathology), and 15 percent were in the industry and the economic fields (Table 1).

Importance and priority of agricultural innovation, commercialization process

As can be seen in Table 2, in cooperation of the Park with the private sector, has the most important, in support of small and medium enterprises and also attract credit from various institutions (private and public) has been the least important. Overall, the average rating for the support of small and medium enterprises is equal to 4.11 from 5. Also in the park relationships with universities and research centers, the results suggest that effective communication between park management and local executive authorities has the highest importance and communication between the park with the research, education and extension service has the lowest importance. In general, the average rating of factors associated with universities and research centers is equal to 4.01 out of 5. From the studied experts' point of view, the availability of scientific and high quality skills has the highest importance and the creation of a platform for cooperation between inventors and investors has the lowest importance. Generally, average ratings of the importance of technological entrepreneurship, commercialization of agricultural innovations is equal to 3.97 out of 5. In risk reduction stage of science and technology projects, specialized training and awareness courses, short-term and long-term feasibility of entrepreneurship and startup companies, has the highest importance and availability of evaluating the regularity of measure effectiveness of implementing projects has the lowest importance. In addition, the average rating for reducing the risk factors related to science and technology projects is equal to 3.84 out of 5. In the field of opportunity for knowledge

Table 2
The Prioritization of Factors Affecting the Commercialization of Agricultural Innovation and Its Items

Item	Mean	SD	Rank
Small and medium enterprise support			
Providing appropriate physical facilities for each project, including laboratory equipment	4.10	1.05	8
Attracting different credit agencies (private and public)	4.00	1.05	9
Specialized educational services provided by the park	4.18	0.87	4
Availability of widespread electronic communication related industries based on agricultural production	3.95	0.77	2
Science and Technology Park in collaboration with the private sector	4.25	0.69	1
Organizing seminars, expert meetings and educational courses develop the science and technology by the park	4.03	1.00	7
Availability of clear mechanisms and structures for the management of intellectual property through patents and protect its research findings	4.04	0.83	3
Granting funds in accordance with the needs of new companies at every stage of their development	4.18	0.9	5
Availability of financial and legal incentives and support company	4.22	0.97	6
Parks relationships with universities and research centers			
Effective communication between park management and local executive authorities, provinces, universities and research centers	4.12	0.88	1
Proximity of the park to the universities and research centers	3.88	0.92	5
Communication and information exchange network and services between companies and universities in park	3.98	0.95	6
Relationship between the park with the Research, Education and Extension Service	4.01	1.04	7
Optimal use of existing capacities of universities, research institutes, manufacturing and service organizations.	4.08	0.93	2
Park relationship associated with academic professionals and faculty members	4.01	0.92	3
Joint research projects between universities and IT companies	4.00	0.95	4
Tech Entrepreneurs			
Consulting services - commercial and business education and marketing of science and technology park	4.04	0.83	2
Assistance in the utilization of capital and preparation for commercialization and marketing of new ideas, commercialization of R&D and innovative activities	4.01	0.87	4
Provision of field research projects related to technologies, processes, methods and new ideas	3.94	0.82	3
The availability of scientific expertise and high quality business skills	3.96	0.77	1
Creating cooperation between inventors and investors	4.03	1.01	7
Availability of positive interactions between firms	3.88	0.96	6
The use of services, resources and facilities shared by academic institutions and industry in the production technology	3.91	0.88	5
Science and technology risk reduction projects			
Having plans to support companies in the first phase of new product sales (such as support for the exhibition)	3.99	0.98	3
Having a team of market professionals	3.95	0.98	4
Availability of a combination of the functional and risk teams in organizing new products	3.84	0.89	2
Ability to predict long-term regular basis to make investment decisions	3.80	0.99	5
The infrastructure necessary to develop technologies such as workshops	3.80	1.00	6
Availability of assessment system for the effectiveness of implementing projects	3.55	1.05	7
Specialized training and briefings and short term and long term courses in entrepreneurship and emerging enterprises	3.96	0.89	1
Opportunity for knowledge workers and employers			
Availability of marketing plans groups	3.80	1.02	3
Attracting and retaining staff, researchers and managers do high-quality Agricultural Sciences	3.75	0.98	2
The use of information technology to diagnose Park Mall, process, creation of new methods for identifying customer needs and meeting customer requirements, client needs and service	3.94	0.95	1
Being close to the areas and towns, industrial parks and major transportation centers such as international airports	3.37	1.05	4

Scale: (1=very low, 2=low, 3=average, 4= high, 5= very high)

workers and their employers, Park usage of information technology to identify markets, creating new methods and process, and identifying customer needs and satisfying customer needs and service has the highest importance and also park proximity to industrial area and major transportation centers such as international airports has the lowest importance. Overall, the average rank of factors affecting opportunity for knowledge workers and employers is equal to 3.79 out of 5.

In addition to explaining each of the components of the five-fold in the commercialization of agricultural innovation factor, we have used analytical verification factor. For this purpose, the structural loadings of each indicator on the commercialization of innovation in Agricultural Science and Technology Park are estimated by t-value to evaluate their significance. At this stage, first-order factor and structural equation model were used to assess the role of indicators used for the formation of the Science and Technology Park in the commercialization of agricultural innovation analysis.

Regarding Table 3, it is indicated that all indicators of structural performance in commercialization of Science and Technology Innovation Park have t-value of greater than 1.96. Therefore, the hypothesis of being 0 which means it is not significant is rejected and significant relation based on the conformation analytical factor is confirmed. The results show that all selective indicators for assessing structural role in the commercialization of innovations in Agricultural Science and Technology Park are enough prices and also their validity and reliability are confirmed. According to the parameters given in Table 4, it can be seen that the Science and Technology Park Performance measurement model structures in agricultural innovation to commercialization have acceptable fitting index indicators for each construct is valid and acceptable. External model in AMOS software showed, evaluated fitted model for commercialization of agricultural innovation with variable affected (Figure 2).

Table 3
Factor Loading Model

Structure	Indicator	Evaluated Parameter	t-value	SE	AVE	CR	α
Science and technology park performance in agriculture product commercialization	Y5	0.701	---	---	0.51	0.81	0.94
	Y4	0.800	7.534	0.144			
	Y3	0.780	7.369	0.129			
	Y2	0.704	6.712	0.144			
	Y1	0.824	7.795	0.122			

Table 4
Fitting Index Model

Fitting index	Acceptability index	Reported value
Chi-square with degrees of freedom	≤3	1.92
NFI, Soft indicators fitness	0.90 ≤	0.96
Adaptive fitness index, CFI	0.90 ≤	0.98
Indicators of Compliance, GFI	0.90 ≤	0.96
The mean squared residue, RMR	≤ 0.05	0.01
The square root of the estimated variance of the error of approximation, RMSEA	≤ 0.08	0.09

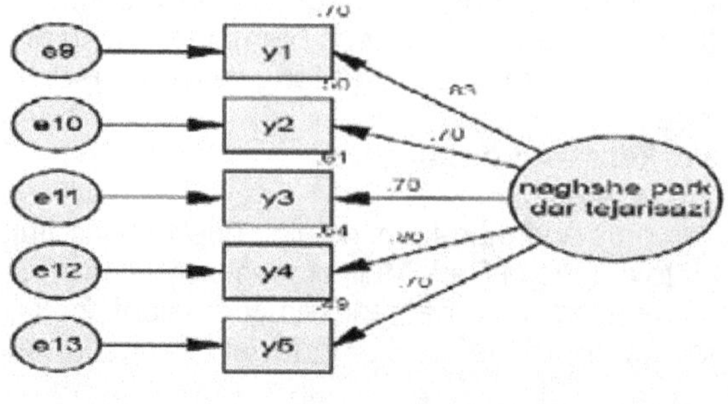

Figure 2. Evaluated Fitted Model for Commercialization of Agricultural Innovation

Importance and priority of outcomes and consequences of commercialization of agricultural innovations

As shown in Table 5, the results and outcomes of commercialization in agriculture, "the promotion of entrepreneurship and business development innovation in agriculture" is the most important factor, and the "protection of natural

Table 5
Prioritize Agricultural Innovation, Commercialization Achievements and Outcomes

α	Items	Mean	SD	Rank
	Agriculture			
	Promotion of entrepreneurship and business development of innovative technologies in agriculture	4.24	0.76	1
	Promotion of science and technology, agriculture, reducing gaps in knowledge and technology sector	4.50	0.84	2
	Shaping and implementation of innovation networks in agriculture through improved interaction and collaboration among science and technology universities, manufacturers, markets and other systems	4.01	0.75	3
	Development of the market and increase in the value-added chain of production and wealth creation in agriculture	4.18	0.79	4
	Enhancement of the quality of agricultural production	4.07	0.82	5
	Increasing rural incomes and farmers	4.09	0.86	6
	Longer life cycle in agricultural technology (using supplementary technology)	3.95	0.85	7
	Increasing prosperity in the rural small businesses affiliated with minor processing industries	4.07	0.91	8
	Increased innovation in agricultural production and expanding consumer markets	4.08	0.92	9
	GDP growth in local area	4.00	0.94	10
	Increasing concentration of capital in the manufacturing sector by reducing imported technologies	4.08	0.97	11
	Protection of natural resources through greater compliance with environmental technology	3.81	1.00	12
	Science and Technology Park			
	Promoting a culture of innovation and constructive competition among companies and institutions based in the science park	4.02	0.85	1
	Partnership and collaboration with professionals as a resource for teaching and applied research and economic development	3.98	0.87	2
	Achieving technology-driven economic development	3.92	0.86	3
	Minimizing duplication of research and development through frequent contact and exchange of ideas among researchers, institutions	3.98	0.89	4
0.91	Facilitating the process of technology transfer to industry	3.85	0.87	5
	Economic development zone	4.03	0.92	6
	More cooperation between the universities and research institutes and industry park	3.95	0.94	7
	Promoting economic development and increasing the value of investments in the field of technology	3.90	0.98	8
	Compatible favorable environment to attract academics and scientists, companies and entrepreneurs who want to start new investment	3.92	1.01	9
	Creating innovations in science and technology	3.80	1.12	10
	Investments in science, research, technology development and training opportunities	4.11	1.95	11
	Universities and research centers			
	Development of infrastructure and capacity for research and technology in universities and research centers	4.03	0.80	1
	Development and evolution of organizational structure and management of universities and research centers in cooperation with another resource	3.94	0.84	2
	Promoting investment and intellectual property at universities and research centers with respect to the ideas and technology	4.01	0.86	3
	Evolution and synergies towards achieving the university's mission of university entrepreneurs	3.96	0.85	4
	Increasing the efficiency and effectiveness of academic research findings in the field of agriculture through improved results	3.93	0.88	5
	Improving the social status of universities and research centers	3.91	0.93	6
	Increasing agriculture students transfer and employment using with the development of technology-based businesses	3.98	0.98	7
	Directing and making more real of universities, research programs and research centers	3.79	0.99	8
	Earning funds for universities and research centers (reducing dependence on public funding)	3.86	1.12	9
	Increasing the social value of urban agriculture disciplines	4.34	0.79	10
	Industry			
	Industry support and enhance the ability to export engineering services abroad	3.98	0.95	1
	Enhancing export competitiveness in domestic and gain higher share in world trade	3.97	0.99	2
	Raising the level of design and construction of the domestic products	3.88	0.99	3
	Promotion of new technologies and production using advanced technologies	3.87	0.99	4
	Creation of new jobs and new field of activity for professional work forces	4.00	1.04	5
	The development of related industries, both upstream and downstream	3.79	0.99	6
	Creating confidence in the applied research and development	3.90	1.16	7

Table 6
Factor Loading of Indicator and Structures and Commercialization Outcomes Measurement Mododel

Structure	Indicator	Evaluated Parameter	t-value	SE	AVE	CR	α
Achievements and outcomes	Z4 (Industry)	0.870	---	---			
	Z3 (University)	0.733	8.968	0.76			
	Z2 (Science and Technology Park)	0.915	12.272	0.66	0.55	0.82	0.94
	Z1 (Agriculture)	0.748	9.235	0.69			

resources through greater compliance with environmental technology" is the least important factor. Overall, the average rank of the achievements and outcomes of commercialization of agriculture is equal to 4.06 out of 5. Also in the part of the Science and Technology Park, according to the results of "upgrade/agriculture, culture of innovation and constructive competition among the companies in the park and on the institution of science and learning" has the most importance and "investments in science, research, technology development and training opportunities" has the lowest importance. Overall, the average rank of commercialization outcomes of the Science and Technology Park is equal to 3.95 out of 5. In terms of achievement in the commercialization of universities and research centers, developing infrastructure and capacity of universities and research and technology as a result of the acquisition of new resources and joint ventures has the highest importance and increasing social value fields and urban agriculture has the lowest importance. Overall, the average rating of the importance of outcomes related to commercialization of universities and research centers is equal to 3.98 out of 5. Achievements in the commercialization of the industry, as well as "industry support and enhance the ability to export engineering services to the highest important factor and to create confidence in the applied research and development" is the least important factor. Overall, the average rank of the commercialization for industry achievement is equal to 3.91 out of 5.

In order to assess the achievements and impact of the used indicators on shaping the agricultural innovation, commercialization, the first order analytical factor was used, whose results are shown in Table 6.

Regarding Table 6, it can be stated that all indicators of structural achievements and outcomes of commercialization of agricultural innovations have t-value of greater than 1.96. Thus, the assumption of being zero of indicator, meaning absence of significant structural role in the formation indicator is rejected and significant relationship on the basis of confirmatory factor analysis is confirmed. The results show that all selective indicators for assessing structural outcomes and consequences of the commercialization of agricultural innovations are prices enough and also validity and reliability are confirmed.

According to the indicators of fitness as seen in Table 7, it can be said that achievements measurement and impact model of the constructs in commercialization of agricultural innovations to fit the relationship between structure and indicators for each construct is valid and acceptable. So it can be said that the research findings are based on a model with the used indicator for assessing the factor structure have acceptable fitting.

Table 7
Fitting Index Model

Fitting index	Acceptability index	Reported value
Chi-square with degrees of freedom	≤3	3.259
NFI, Soft indicators fitness	0.90 ≤	0.97
Adaptive fitness index, CFI	0.90 ≤	0.98
Indicators of Compliance, GFI	0.90 ≤	0.97
The mean squared residue, RMR	≤ 0.05	0.01
The square root of the estimated variance of the error of approximation, RMSEA	≤ 0.08	0.14

Results showed that in evaluating the performance of Science and Technology Park in commercialization of innovations in agriculture in support of small and medium sized enterprises, Science and Technology Park in collaboration with the private sector is the most important and absorbing account of the various organs (private and public) is the least important factors. It seems that a park partnership with the private sector can produce better and easier-to-market innovations. As a result, they can expand the market for innovations generated in the park. These are similar to results reported by Davoodi and Kalantari (2011) and Kalantari (2010). About park relationships with universities and research centers, the utmost importance is effective communication between park management and local executive authorities, while the relationship between the park and the Research, Education and Extension system is the least important. It can be concluded that the parks are facing the financial difficulties due to limited demand for attendance at the park. Limited budget in research impacts park performance because these centers depend on government budgets. So the park is going to be able to use the resources available in the relationship with the executive authorities of the city and in this way, more features can be provided for companies in the park. These are similar to Salami et al. (2010) and Siegel and Westhead (2003). Based on research findings, the experts in the field of science and technology parks in the commercialization of agricultural innovations in the field of entrepreneurship believe that the availability of scientific expertise with high quality and skill have the highest importance and creating cooperation between inventors and investors have the lowest importance. Thus, it can be argued that park in the field of business consulting for startup companies puts special advisers, to strengthen Company's competitiveness on the track as well as the transfer of skills and business experience which enhances knowledge enterprises on market principles, so one of the benefits of the presence of the knowledge-based enterprises in the science and technology parks is the use of synergy for business skills. These findings are in agreement with Gordfaramarzi (2011), Arasteh and Jahed (2010) and Aghajani and Talebnejad (2011). Based on research findings about the reduction of project risks, specialized training and briefings and short term and long term courses in entrepreneurship and emerging enterprises are the most important factors and the least important is the assess system to evaluate the effectiveness of implementing projects. Risk of investing in agriculture sector is higher than in industry and services sectors. Also, the market has very high risks in agricultural production because of the low investment in agricultural innovation support mechanisms and the fact that the market for agricultural technology is also very limited. Therefore, entrepreneurship education programs to young companies can reduce project risk and cost of the loss of a large percentage of providing human, financial and material, and increasing the risk-taking by entrepreneurs, researchers and companies. Davoodi and Kalantari (2011) and Arasteh and Jahed (2010) reported similar results. Based on research findings, it was shown that the performance in commercialization of innovations in agricultural sector by science and technology park, in the field of opportunity for knowledge workers and employers, use of information technology to diagnose market, creating new processes and procedures and identifying customers' requirements and meeting customer needs and service are the most important and being close to the towns and industrial areas and major transportation centers such as international airports are the least important. It was concluded that information technology is a critical tool for organizations to achieve competitive advantage and organizational innovation. Thus, science and technology parks through information technology as the communication infrastructure within the enterprise can make a proper way to supply their products to customers. Using the powerful tools of information technology can help the marketing of products, and thus arises the concept of Internet Marketing. Internet marketing is the process of creating and maintaining relationships with clients in the areas of Internet to facilitate the exchange of ideas, goods and services in such a way that aims to satisfy two sides. These results are aligned with the results reported by Manian and Saremi (2009) and Luftman et al. (2005). Based on the results observed in the outcome and consequences of agricultural innovation, commercialization, the promotion of entrepreneurship and innovative business development in the agricultural sector is the most important data, and the protection of natural resources through greater technology compliance is considered the least important. Thus, it can be concluded that the commercialization of agricultural innovations can increase its role in economic growth, job creation and the economic development of the society. On the other hand, it makes a massive employment of unemployed students in agricultural fields as well as private companies are presented in the agricultural sector that will allow the creation of new services. These findings are consistent with the findings of Nasaj Hoseini and Meraji (2005). Experts in the field of the achievements and outcomes of commercialization of agricultural innovations for the Science and Technology Park, promoting a culture of innovation and constructive competition among companies and institutions based on the science and knowledge are the most important drive, but investments in science, research, technology development and training opportunities are least important. Science and technology parks and incubators have comprehensive information about producers and consumers of science due to the nature of the tasks they are doing and at the same time have taken part in production processes, and weaknesses in any of the processes will lead to the reduced production and conveyance of knowledge, credibility and endangerment of their survival. So, they are trying to improve their competitive ability and their enterprises' and institutions' competitive ability to help their survival and growth. Improving competitiveness means improving quality, quantity, customer satisfaction, knowledge, intellectual and social capital, and these factors may increase the rate of commercialization of research findings. These are consistent with Arasteh and Jahed (2010). The results showed that in terms of achievements and outcomes of university research centers, innovation and the commercialization, businesses and joint ventures are the most important resources and increased social value of urban agriculture field drive for urban communities is the least important. This is because universities and research centers can pick up new ideas by contacting the park to enhance the information in relation to the market and the latest creates a specialized activity, as well as joint ventures between the park and the university. Based on research findings, studied the outcome and consequences of innovation to industrial commercialization, raising the potential for export of engineering services is the most important and confidence towards applied research and development is the least important. This is because of the fact that parks help technological units to increase industrial production in the country and this makes the industrial companies able to export their products to overseas, resulting in economic growth of the country that is by itself associated with an increase in the welfare of the people. Economic development can be followed over time. The results are consistent with Gordfaramrzi (2011). Confirmatory factor analysis results show that all the variables used to measure the performance in commercialization of agricultural innovations in Science and Technology Park were selected properly and their validity is confirmed. Therefore, the measurement model of measuring performance in Science and Technology Park for commercialization of agricultural innovations with respect to all variables is appropriate, acceptable and significant on the basis of the theory of conformity assessment. The results of Shinn and Lamy (2006), Mohammadi and Bigdelli (2012) and Manian and Saremi (2009) that considered the relationship between the park and universities take Park performance in supporting of the firm's effectiveness. Cocks (1985), Kalantari (2010) and Aghajani and TalebNejad (2011) mention supporting infrastructure such as location and business skills training as the functions of park. According to Malekzadeh and Kazemi (2010), Soleimani (2011), Lofsten and Lindalof (2001), Law (2005) and Salami et al. (2010), critical success factors of science and technology parks in the commercialization process and expedite the process of transforming ideas into product development and innovation include spatial factors, including proximity to the university and research centers, proximity and easy access to convenient transportation facilities, proximity to markets and suppliers, critical factors include consulting services, marketing, training business by park resources and specialized equipment in parks, public services, management factors including directors dominating the debate on the Science and Technology Park, the local effective communication between park managers and executives of provinces, universities and research center and social and cultural factors, including the adoption of intellectual property, the spirit of entrepreneurship and innovation in society, to make a healthy competition. Arasteh and Jahed (2010), Rezaei et al. (2009), Fukugawa (2006), and Minshal (1983) also confirm these results. Confirmatory factor analysis results show that all variables used to measure the achievements and outcomes of commercialization of agricultural innovations were properly selected and their validity is confirmed. Therefore, the measurement model for measuring achievements and outcomes of commercialization of agricultural innovations with respect to all variables is appropriate and acceptable on the basis of the theory of conformity assessment. The results of Gordfaramarzi (2011) showed that the achievements of the park for the industrial sector include increasing technology in the country, creating confidence and development, raising the level of design and construction.

CONCLUSIONS

According to the results, there is a need to focus on improvement of the relationship between agricultural faculties and science and technology parks. Therefore, it is recommended to locate Science and Technology Park near the university or research centers. In addition, because of financial shortage, it is necessary to meet the physical facilities managers who rank low should do innovative solutions like other incubators to provide flexible space, the allocation of shared space and facilities and rental rates up allowed to renew the lease with regard to optimize the exploitation of their resources. Incubators can communicate with institutions such as universities, and research institutes and the private sector provide access to some manufacturing facilities such as workshops, laboratories and research, and development for entrepreneurs. This relationship requires a mutual benefit for both parties. According to research findings, since access to the skills and expertise is a priority to improve quality of business, the establishment of a network of information and transfer of experience and a network of business sponsors can increase the success rate of knowledge-based enterprise.

ACKNOWLEDGEMENT

The authors are grateful to the staff and experts of the Science and Technology Park in Kermanshah County for their sincere cooperation with researchers.

REFERENCES

1- Aghajani, H., & Talebnejad, A. (2011). Comparison Assessment of selected growth center performance in Iran. *Development and Entrepreneurship, 4*(2), 13.

2- Arasteh, H., Jahed, H. (2010). Science and technology parks and research center role in innovation commercialization. *Industry and University magazine, 3*(9-10), 35-43.

3- Bagozzi, R. P., & Yi, Y. (1988). On the evaluation of structural equation models. *Journal of the Academy of Marketing Science, 16*(1), 74-94.

4- Cocks, R.N. (1985). Lessons from 30 Years of Science Parks in the USA. *Science Parks and Innovation Centres: Their Economic and Social Impact, 4*, 32-35.

5- Davoodi, H. & Kalantari, Kh. (2011). Studying of problems in development of science and technology Tehran University Park. *Policy, Science and Technique Magazine, 4*(2), 18-22.

6- Fukugawa, N. (2006). Science parks in Japan and their value-added contributions to new technology-based firms. *International Journal of Industrial Organization, 24*(2), 381-400.

7- Gordfaramarzi, H. (2011). Commercialization as affecting factor on knowledge-based firms and national economy development. *Park Policy,Science and Technique Magazine, 8*, 29-33.

8- Jimenez, J. D., & Cegarra, N. J. (2008). Fostering innovation: the role of market orientation and organizational learning. *European Journal of Innovation Management, 11*(3), 389-412.

9- Kalantari, Kh. (2010). *Structural equation modeling in social-economic researches* (1st ed.), Tehran: Farhang Saba.

10- Law, John. (2005). *Practising Nature and Culture:an Essay for Ted Benton, in Sandra Moog and Rob Shields (eds), Nature, Social Relations and Human Needs: Essays in Honour of Ted Benton*. London: Palgrave.

11- Lofsten, H., & Lindalof, P. (2001). Science parks in Sweden Industrial renewal and development. *R&D Management, 31*(3), 18-23.

12- Luftman, J., Kempaiah, R., & Nash, E. (2005). Key issues for it executives. *MIS Quarterly Executive, 5*(2), 81-99.

13- Malekzadeh, Gh. & Kazemi, M. (2010). Studying of influencing factors on small company performance in the beginning of growth stage. *Excavations of Business Administration, 2*(3), 24-31.

14- Manian, A., & Saremi, M. (2009). A conceptual model to assessment readiness of organizations for strategic IT cooperation and business. *Management of Information Technology, 1*(1), 44-57.

15- Minshall, C. W. (1983). An overview of trends in science and high technology parks. *Economic and Policy Analysis, 37*, 9-25.

16- Mohammadi, A., & Bigdelli, N. (2012). Studying of support tools for knowledge-based companies in science and Technology Park in the Middle East. *Approach Magazine, 50*(1), 67-71.

17- Nasaj Hoseini, S., & Meraji, A. (2005). *Agriculture growth center for improvement agriculture status and job opportunity*, paper presented at: the second conference of parks and growth center of science and technology in Iran, Tehran International Exhibition.

18- Pallant, J.(2010). SPSS survival manual (4th ed). Berkshire, UK: Mc Graw-Hill Rezaei, B., & Karimizadeh, M. (2009). A literature review of the effect of science and technology parks on firm performance: A new model of value creation. 5(30), 11999–12007.

19- Salami, S., Behgozin, S., & Shafiei, M. (2010). Assessment and introduction of affecting factor on science and technology park success from experts point of view. *Park Policy, Science and Technique Magazine, 8* (29), 84-93.

20- Sanni, M., Egbetokun, A.A., & Siyanbola, w.o., (2010). A model for the Design and Development of a Science and Technology Park in Developing Countries. *International Journal and Enterprise Development, 8* (1), 62-81.

21- Senowbar, N., Salmani, B., & Tajvidi, M. (2011). The effect of innovation on innovation capacity of knowledge-based companies. *Policy, Science and Technique Magazine, 4* (2), 91-107.

22- Shinn, T., & Lamy, E. (2006). Paths of commercial knowledge: Forms and consequences of university–enterprise synergy in scientistsponsored firms. *Research Policy, 35*(10), 1465-1476.

23- Siegel, D., & Westhead, P. (2003). Science parks and the performance of new technology based firms: A review of recent UK evidence and an agenda for future research. *Small Business Economic, 20*(2), 57-71.

24- Sohn, S.Y., & Moon, T.H. (2004). Decision tree based on data envelopment analysis for effective technology commercialization. *Expert Systems with Applications, 26* (2), 279–284.

25- Soleimani, M. (2011). *Report of the Science and Technology Parks, weekly news (431)*, President Deputy Strategic Planning And Control. (In Persian), Tehran, Science and Technology Parks press.

26- Statsoft (2009). Electronic statistics textbook. Retrieved on November 19, 2009, from http://www.Statosft.com/textbook/structuralequation- modeling.

27- Vila, P.C., & Pages, J.L. (2008). Science and Technology Parks: Creating new environments favorable to innovation. Paradigmes. Retrived from:www.gencat.cat/diue/doc/doc_ 25819940 x_3.pdf.

13

Advantages and Disadvantages of Participatory Plant Breeding (PPB) in Iran: A Study Based on Breeders' Perceptions

Amir Naeimi [1], Mostafa Karbasioun [2] and Fatemeh Abbasi [3]*

Keywords:
agricultural extension, participatory approach, participatory plant breeding, breeders' perceptions

Abstract

The purpose of this study was to investigate the advantages and disadvantages of Participatory Plant Breeding (PPB). Seventy-six breeders were randomly selected from among 95 breeders who were working at the Seed and Plant Improvement Institute, Karaj, Iran. Data were collected by a questionnaire as the main data collection instrument. The face and content validity of the questionnaire was approved by a panel of agricultural extension and plant breeding experts. The reliability of the questionnaire was calculated and it turned out to be high, ranging from 0.83-0.92. The descriptive findings of the study suggested that the breeders had generally positive perception toward using PPB. The correlation analysis results showed that there is a positive and significant ($p>0.01$) correlation among the levels of familiarity with PPB, advantages, and breeders' perception toward the use of PPB. In addition, there was a negative and significant ($p>0.01$) correlation between breeders' perception of the disadvantages faced by PPB and their perception of using participatory plant breeding. Finally, stepwise regression analysis indicated that the advantages and disadvantages of PPB explained about 58% of the variations in breeders' perception toward using PPB.

[1] *Assistant Professor, Faculty of Agriculture, Department of Extension, Communication and Rural Development, University of Zanjan, Iran,*
[2] *Assistant Professor, Faculty of Agriculture, Department of Agricultural Development, Shahrekord University, Shahrekord, Iran*
[3] *Ph.D. Student of Agricultural Exatension and Education, Bu'AliSina Universtiy, Iran*
* *Corresponding author's email: a.naeimi@znu.ac.ir*

INTRODUCTION

One of the major concerns of agricultural extension systems is to encourage the farmers to participates extension programs and use new technologies in agriculture (Fami, 2006). So far, efforts have been made to develop agriculture through the use of variation technologies. The goal of all these attempts has been to increase quantity of production that was started along with the green revolution in 1950s and has been going on up to now. Yet, such efforts have led to a number of crises: soil erosion, pollution of water resources, overuse of chemical materials, as well as destruction of natural wildlife and insects are expressed as only small amount of concerns voiced by environmentalists, producers, and policy makers (Gang et al., 2010).

In addition, one of the current challenges in management of agricultural extension is farmer's participation in extension programs, dissemination of technology, and new innovation. Therefore, in order to achieve sustainability in agriculture, political efforts should be followed more seriously and also practical methods must be designed based on the local knowledge and skills which will result in improvement and co-operational solutions in sustainable agriculture (Fami, 2006). Effective political decision could bring different organizations together in order to investigate the reality and unpredictable situations. This will provide new guidelines in partnership agriculture (Rolling & Pretty, 1997). One of these guidelines that has focused on partnership with farmers is the PTD (Participatory Technology Development) model. This model is built upon devoting attention to partnership with farmers in extending new technologies. Such programs will increase farmers' self-confidence and creativity (David, 1999). Due to high needs of fertilizers and other chemicals in new variety developed by conventional breeding programs, many small farmers who are not able to afford fertilizers cannot adopt new varieties. These farmers normally work on small farms in hard working conditions and do not benefit from new varieties because of their high associated cost. These problems led to the development of programs called (PPB) in 1980 (Ceccarelli, 2008).

A PPB program started in Syria in 1996 by ICARDA and was successfully performed and extended to Tunisia, Morocco, Jordan, Egypt, Yemen, Eritrea, and Ethiopia, and finally, to Iran in 2006 (Bhargav & Meena, 2014). PPB is a complement to conventional breeding approaches. Participatory plant breeding is perceived by several scientists as a way to overcome the limitations of conventional breeding by offering farmers the possibility to decide which varieties suit better their needs and conditions without exposing their household to any risk during the selection progress (Bhargav & Meena, 2014). Moreover, PPB program is a dynamic process in which the roles of patterns and the extent and also the manner of their collaboration change over the time (Ceccarelli, 2012).

Despite the differences between crops producers and breeders, in all breeding programs, it is possible to identify three main stages:

- Generating genetic variability;
- Selection of the best genetic material within the genetic variability created in the first stage and;
- Testing of breeding lines.

One of the main goals of international agricultural research centers that work on participatory research is to develop a database of farmers' indigenous knowledge and its requirements and reasons. Accordingly, systematic records of farmers' knowledge and their use in breeding programs could be regarded as an important outcome of Participatory Plant Breeding (PPB) (Ceccarelli, 2009). As such, effective knowledge and technology transfer to the farming communities, using a combination of scientific and practical expertise, is of high importance (Mayer, 2013). According to researches, more than 75% of poor people in the world are livelihood farmers who live in developing countries. These farmers produce more than 90% of consumed seeds in developing countries (Ceccarelli et al., 2000). Therefore, in these countries, the implementation of PPB programs would increase the efficiency and effectiveness of plant breeding programs. As a result, the major part of breeding programs in PPB is adequately shifted from research stations to farmers' fields (Ceccarelli & Grando, 2008). Likewise, PPB encourages two kinds of participation:

Functional participation:

• Plant breeders can direct their research according to the needs of the specific groups of farmers (women, men, rich and poor). Hence, physical and economic resource bases of different people necessitate tailored research approaches.

• Farmers can assure plant breeders that they are assessing tradeoffs among traits correctly.

• On-farm research assures that varieties will produce well under "real life" conditions. On farm research can be managed by the researcher, farmer, or both.

• PPB ensures greater success of innovation adoption by the farmers.

Empowering participation:

• Increasing farmer knowledge and skills so that farmers can better participate in collaborative breeding efforts and be more successful in their personal efforts (Bhargav & Meena, 2014).

The main goal of involving farmers, then, is to improve client orientation, and highly client-oriented breeding (COB) describes this purpose, whereas PPB describes an activity (Mayer, 2013). Results of this project in Iran in 2007 showed that PPB is an effective strategy for farmers' empowerment to be able to solve their own problems and to decrease their production costs (Heidari et al., 2007). In a study by Maghirang et al. (2014), some farmers learned how to keep their production records, sales and selections, as well as the seeds that they produce, sell, and distribute among other farmers. A number of varieties were developed and released through PPB in Guangxi, south west of China during 2000 to 2011 (Yiching & Jingsing, 2011). Bhargav and Meena (2014) expressed more opportunities for cost sharing in research and less expensive means of diffusing varieties. Reaching user needs effectively and higher degree of farmers' satisfaction are some examples of the benefits of PPB.

Although PPB has many potential advantages, it also has several potential shortcomings such as high overall cost of breeding programs, high costs of participating farmers, and additional training needed for scientists which impose extra expenses as well (Morris & Bellon, 2004).

Aref (2010) reported the lack of proper planning for rural development as a deterrent factor in sustainable agriculture. Some other factors he underlined could be noted as: the lack of funding resources, the shortage of human resources development, lack of necessary funds to provide governmental supports for farmers, lack of local agricultural organization and farmers' limited, and insufficient knowledge. Lahmar (2010) mentioned the living – physical conditions, political, institutional and technical, cultural, and social ambiances as the most important barriers of advancing sustainable agriculture programs like PPB. In addition, Surangsari and Nuta (2005) mentioned the most important reasons for rejecting PPB methods by farmers as: lack of tangible positive results during the short term, lack of enough knowledge, and the shortage of farmers' encouragement toward participatory approaches. Taking all these into account, PPB was developed as an alternative and complementary breeding approach to conventional plant breeding. Conventional plant breeding is generally carried out by trained breeders in standard laboratory and controlled environment which is of course under favorable farming conditions. While PPB oppositely breeders, farmers and other consumers or end users such as rural farm associations or cooperatives involves in plant breeding activities. This approach enables plant breeders to better understand local farming conditions, farmers' traditional ways for managing plant diversity, as well as their specific needs and preferences. Although, PPB caries many advantages, it has also some disadvantages as stated above, which should be considered. Thus, the main goal of this study was to determine advantages and disadvantages of applying Participatory Plant Breeding (PPB).

MATERIALS AND METHODS

This study used a descriptive, correlation design. The statistical population comprised all plant breeders who worked at the Seed and Plant Improvement Institute (N=95). The sample size was determined by Krejcie and Morgan table (n=76). Breeders were studied through a simple randomized sampling method.

The major data collection instrument used in the present study was a questionnaire that consisted of four parts: Part one (18 items) was designed to identify the perception of breeders toward PB. These statements were measured on a five-point Likert scale with responses ranging from 1="strongly disagree" to 2="disagree", 3= “no opinion", 4="agree", and 5="strongly agree". In section 2 (12 items) was aimed to determine the advantages and part 3 (8 items) to measure disadvantages of PB. In these parts (parts two and three), all 20 items were measured on a ten-point scale that ranged from 0 = "low importance" to 10= "high importance". This scale was divided to three-point Likert scale (0-3.99 = "low importance", 4.00-6.99= "medium importance", 7.00-10= "high importance") during the analysis. Finally, demographic information of breeders was collected through the fourth part.

The instrument was assessed and confirmed for the content and face validity by a panel of agricultural extension and plant breeding experts. The instrument was pilot tested using a sample of breeders (N=30) at the research center in Kermanshah that was excluded from the population study. Next, to determine the reliability index of the instrument, Cornbach’s alpha was calculated for Likert- type items of the questionnaire and it turned out to be somewhere in between 0.83-0.92.

The data were collected via applying the survey instrument. Twenty-three questionnaires (30.26%) out of 76 questionnaires were returned during four weeks. A follow-up reminder was used for non-respondents four weeks after initial survey (Miller & Smith, 1983). As a result of this second effort, an additional 46 questionnaires (60.52%) were received. Finally, 69 questionnaires (90.78%) were returned. Descriptive statistics such as frequencies, percentages, means, and standard deviation were used. Additionally, inferential statistics such as correlation coefficient was applied to determine the relationships among the variables. In addition, the stepwise linear regression was used to estimate the variance explanation of dependent variable by diverse independent variables. SPSS version 16 was used for data analysis.

RESULTS

Demographic characteristics of breeders

The average age of breeders was 42. About a quarter (22.4%) of them was 35 or younger and nearly half of them (55.2%) were between 36 and 45. The majority (59.2%) of respondents were male. Sixty-one percent of breeders had MSc degree and the remaining 39% had PhD degree. Most of them had studied breeding and biotechnology (69.7%) and others (30.3%) had studied related majors such as genetic, plant physiology, agronomy, and so on. The mean of respondent’s job experience was more than 13 years (13.84). Concerning employment situation of breeders, a little more than one half (53.9%) had official or permanent job and others were contractual. A large number of breeders (86.8%) had participated in more than three training courses related to PB, and only 13.2% had participated in less than three training courses. Most of the breeders (59.2%) had a brief understanding of PPB, and only 18.4% of them were in high level of familiarity with PPB.

Perception of breeders toward using PPB

In general, the breeders tended to present positive view (between somewhat and agree level) toward PPB usage (Mean = 3.17; SD = 0.49) (see Table 1). In the questionnaire, breeders were asked to indicate their perception toward using PPB with 18 statements. Means and standard deviations for these 18 perception statements are represented in Table 1. A number of five out of 18 statements had a mean value close to four level of agreement. The highest means were related to the statement of “attention to the local and indigenous knowledge” (Mean=3.98, SD=0.8); “acceptance of new variety by farmers” (Mean=3.85, SD=0.93) and “strong relationship between farmers and breeders” (Mean=3.84 and SD=0.83). This allowed breeders to better understand the local farming conditions, the farmers’ traditional ways of managing plant diversity, as well as their specific needs and preferences. Furthermore, PPB can empower groups, especially women or less well-off farmers that are traditionally left out of the development process. Compared to the Participatory Varietal Selection

Table 1
Mean, Standard Deviation and Breeders' Perception preferences towards Using PB (n=76)

Items	Mean	SD	Rank
Attention to local and indigenous knowledge	3.98	0.80	1
Acceptance of new variety by farmers	3.85	0.93	2
Strong relationship between farmers and breeders	3.84	0.83	3
Higher genetic diversity and more options for farmers	3.60	1.04	4
Using participatory breeding method toward sustainable agriculture	3.53	0.99	5
Higher acceptance of new released varieties	3.32	1.03	6
Improving farmers' life quality	3.30	1.03	7
Using local varieties in participatory breeding	3.27	1.04	8
Using drought tolerant plants in participatory breeding can improve water efficiency	3.19	1.11	9
Scientific value of participatory breeding is very notable in dry and marginal lands	3.17	1.05	10
Decrease in technical and economic efficiency could diminish farmers' attendance	3.10	1.01	11
The poorer the farmers, the lower their role in participatory breeding	3.07	1.06	12
Farmers have a major role in participatory breeding	3.05	1.00	13
Participatory breeding is comparable with other breeding programs in terms of scientific value	2.92	1.05	14
Participatory method is not suitable for the country at present time**	2.59	1.09	15
In all participatory programs farmer should be part of the program	2.52	1.31	16
No success will be achieved in participatory breeding without farmers co-operation	2.47	1.10	17
Participatory breeding is suitable for the region in which diversity will decrease pests' and diseases' attacks	2.35	0.77	18
Total (Sum of Means*)	57.12	18.2	

* Mean computed on a scale 1 = strongly disagree to 5 = strongly agree and could range from a low scale of 18 to high scale of 90 with the theoretical midpoint of 54.
** Reversed type: 5 = strongly disagree to 1 = strongly agree.

(PVS) where farmers are only involved in variety selection among a line's pre-determined group being field tested, PPB is found to have a higher empowerment effect. Eight statements had a mean value between 3.00 and 3.50, indicating that breeders had ''no opinion" about using PPB. Moreover, three statements (14, 15 and 16) had a mean value close to three, which is 'no opinion' (see Table 1).

Advantages of using participatory breeding

Results indicated that, overall, breeders pinpointed the "medium importance" level of advantages of using PPB (Mean= 6.68, SD= 2.51). More specifically, breeders were asked to describe the advantages of using PPB through 12 statements. The means and standard deviations of 12 advantage statements are reported in Table 2. As it is perceived, four of statements had a mean value between 7 and 10, indicating "high importance". The highest mean was related to these three statement: "Ability to use in farms (on-farm)" (Mean= 8.15, SD= 2.16); "higher farmers' acceptance because of farmers cooperation" (Mean=7.53, SD=2.48) and "ability to show new varieties in different microclimates" (Mean=7.19, SD=2.26). There are four main activities where farmers can contribute: The farmers can share their knowledge and experiences, they can contribute in genetic materials; they are able to conduct trials; and finally, select and evaluate germplasm. A wide range of impacts were achieved through the use of participatory approaches. Participatory Plant Breeding (PPB) generally involves a greater and more complex degree of farmers' involvement due to their engagement in decision-making process in earlier and more fundamental stages of the variety development chain. PPB can have a large empowerment effect on participating farmers, when they are intimately involved in the programs from the early stages of the research design. In case of on-farm varietal selection, ideas from a variety of members of the community, not just selected members to participate in the breeding program, can be more easily incorporated upon identifying community needs and preferences. Other statements had the mean score between 4 and 6.99 indicating "Medium importance" of

Table 2
Mean, Standard Deviation and Rank of Advantages of PB (n=76)

Advantages	Mean*	SD	Rank
Using PB as on-farm	8.15	2.16	1
Higher acceptance among farmers because of farmers co-operation	7.53	2.48	2
Ability to show new varieties in different microclimates	7.19	2.26	3
Making farmers more capable to do farming by establishing connections with researchers	7.13	2.35	4
Applying a more demand driven method	6.93	2.68	5
Increasing usefulness and effectiveness of breeding	6.73	2.70	6
Selection of the varieties with appropriate production potential in different climatic conditions	6.72	2.44	7
Integrating local and new science	6.67	2.80	8
Participation of farmers in all plat breeding steps	6.14	2.65	9
Systematic record of farmers' knowledge	6.01	2.40	10
Protecting local germplasm	5.81	2.67	11
Paying more attention to small farmers	5.21	2.53	12
Total	6.68	2.51	

* Mean computed on a scale 0 = Low Importance to 10 = High Importance and could be categorized to a three-point scale: (0-3.99 = "low importance", 4.00-6.99= "Medium importance", 7.00-10= "High importance").

these PB advantages. The lowest mean value referred to the statement "paying more attention to small farmers" (Mean= 5.21, SD= 2.3) (Table 2).

Participatory breeding disadvantages

Breeders were asked to describe the disadvantages of the use of PPB using eight statements. Means and standard deviations of the statements are reported in Table 3. Looking at the table, the findings showed that the average of disadvantages items had a "high importance" concerning the use of PPB (Mean= 7.08, SD= 2.60). Half of the disadvantages had a mean value higher than seven, indicating the "high importance". The highest mean referred to "lack of enough relationship and coordination between different parts" (Mean= 8.35, SD= 2.21). The last four statements had the mean score between 4 and 6.99, indicating the "medium importance". The lowest mean value was related to the statement "no optimum conditions for testing the method in the field" (Mean= 6.30, SD= 2.57) (Table 3).

Determining the correlation between research variables and breeders' perception toward the use of PB

Pearson correlation coefficient was calculated to describe the relationships between research variables and breeders' perception toward PB usage (Table 4). There was a positive significant relationship between "levels of familiarity with PB", "advantages" and breeders' perception toward PPB usage. Moderate negative relationship ($r = -0.345$, $p<0.01$) was found between "disadvantages" and breeders' per-

Table 3
Mean, Standard Deviation and Rank of Disadvantages to use PB (n=76)

Disadvantages	Mean*	SD	Rank
Lack of enough relationship and coordination between different parts	8.35	2.21	1
Lack of team-working and participative behaviour among the farmers and researchers	7.56	2.60	2
Minor role of agricultural extension	7.25	2.58	3
Little farmers' knowledge about this method	7.22	2.49	4
Lack of suitable facilities for this method	6.68	2.74	5
Lack of proper environment for farmers' participation	6.65	2.58	6
Lack of appropriate laws and regulations	6.60	3.03	7
Lack of optimum situation for testing the method in the field	6.30	2.57	8
Total	7.08	2.60	

* Mean computed on a scale 0 = Low Importance to 10 = High Importance and could be categorized to a three-point scale: (0-3.99 = "low importance", 4.00-6.99= "Medium importance", 7.00-10= "High importance").

Table 4
Correlation between research variables and breeders' perception (n=76)

Variable	r	p-value	DCIC [1]
Age	0.17	0.137	-
Work experience	0.17	0.141	-
Number of passed training courses	0.10	0.352	-
Education level*	-0.16	0.155	-
Level of Familiarity with PB	0.34**	0.002	Moderate
PB Advantages	0.74**	0.001	Very high
PB Disadvantages	-0.34**	0.002	Moderate

*For this variable spearman (rho) correlation coefficient (rs) was used.
**P<0˙01

Table 4
Stepwise regression of breeders' perception with regard to independent varibles (n = 76)

Variable	B	Beta	t	p-value
Constant	27.94	-	8.29	≤0.01
PB Advantages (x_1)	0.28	0.70	8.94	≤0.01
PB disadvantages (x_2)	0.17	-0.19	2.45	≤0.01

$R = 0.761$ $R^2 = 0.58$ $R_{Ad} = 0.568$ $F = 50.32$ $P \leq 0.001$ $DW = 2.02$
$Y = Constante + b_1(x_1) + b_2(x_2)$
$Y = 27.94 + 0.28\, x_1 + 0.17\, x_2$

ception; however, no relationship was revealed between "age", "work experience", "number of passed training courses", "education level" and breeders' perception toward the use of PB. Overall, similar to the findings of Davis (1971), most measured relationship rates were between moderate and very strong levels (see Table 4).

Stepwise multiple regression of breeders' Perception and independent variables

Durbin-Watson's statistical value was calculated, and it was equal to 2.02, which indicated the possibility of using regression for this research (Moemeni & Ghayumi, 2007). Stepwise multiple regression analysis, as such, was used to determine the regressions' equation. Variables such as: "familiarity level with PB", "PB advantages" and "PB disadvantages" which had significant correlation with dependent variable were selected for the regression analysis. After three steps, only "PB advantages" and "PB disadvantages" remained in the equation. These variables explained 58% of the variance in breeders' perception toward the use of PB (see Table 5).

DISCUSSION

As it was illustrated earlier, varieties created by conventional breeding are beneficial for the farmers who have fertile lands and easy access to agricultural inputs and water. Therefore, small and poor farmers are not able to adopt these modern and efficient varieties, and as a result, their lifestyle is not expected to improve. Working with these small and farmers in breeding new varieties could help them to alleviate their problems (Ceccarelli et al., 2007; Murphy et al., 2005). Accordingly, the major goal of this study was to determine the advantages and disadvantages of Participatory Breeding system in Iran.

The results of this study indicated that most of the experts accentuate "attention to local knowledge" as the key of success in participatory breeding system. Farmers have improved and bred crops over last decades and therefore have valuable experience in that respect, which, unfortunately, has always been neglected. Consequently, collecting the data of farmer's local and indigenous knowledge and conducting research based on them is strongly recommended.

The findings of the research also demonstrated that on-farm trials and farmers 'acceptance are

[1] *Davis Correlation Intensity coefficient*

assumed as most important advantages of the participatory breeding. These results are in agreement with the study of Mustafa et al. (2005) who reported that farmers in critical situations and high stress environment do not accept varieties produced by conventional plant breeding, and according to his research, surprisingly, the farmers' acceptance rate was zero percent.

Additionally, this research showed that "lack of necessary and sufficient coordination between different divisions" and "the lack of existence of morality of participatory work among agriculturists and researchers" are the most crucial problems of participatory breeding. These results are in line with the study of Ceccarelli et al. (2000) who reported that the lack of cooperation between agricultural extension and education institutes is the main bottleneck of participatory programs. In order to achieve a successful participatory breeding program, cooperation of all parts in agriculture would be necessary. Furthermore, the cooperation of researchers and agriculturists and also agricultural extension and education organization and executive parts would be essential. In order to help farmers and breeders and to apply successful implementation of this method, cooperation of agricultural extension specialists looks vital.

Correlation analyses of the study suggested that there was a positive significant relationship between the levels of familiarity with PB, PB advantages, and breeders' perception toward PPB usage, and also, a negative relationship was found between PB disadvantages and breeders' perception. On the other hand, as participatory breeding compatibility increases, breeders' familiarity and addressing the aforementioned issues could enhance the acceptance rate of this method. Accordingly, it is proposed that more workshops, seminars and congresses are held on PPB in order to get more people acquainted with this method and its advantages.

At the end of the questionnaire, respondents' perceptions was asked about the use of participatory breeding and more than half of the breeders had a positive and favorable attitude towards utilizing participatory breeding. It is, then recommended that an official place should be dedicated for the implementation of these beneficial plans by the Ministry of Agriculture.

Ultimately, some more suggestions could be listed as below:

• Increasing the cooperation among breeders, farmers and extension agents at national and international level for better achievements of PPB project.

• Using farmer-researcher oriented systems for increasing the effectiveness of PPB projects is advised.

• Creating a powerful database of indigenous knowledge for documentation and access to valuable experience of farmers is recommended.

ACKNOWLEDGEMENT

The researchers would like to express their sincere gratitude for the helps and supports they received from the managers and experts of Seed and Plant Improvement Institute, Karaj, Iran.

REFERENCES

1- Aref, F. (2010). Barriers of agricultural development in Iran: A case study of Fars Province. *Journal of American Science, 6* (11), 155-158.

2- Bhargav, D.K., & Meena, H. P. (2014). Participatory Plant Breeding: Farmers as Breeders. *Popular Kheti, 2*(1), 7-14.

3- Ceccarelli, S. (2008). *Importance of participation in research regional and international breeding programs*. The first national breeding grain workshop, Kermanshah, 13-14 October, Iran.

4- Ceccarelli, S. (2009). Evolution, plant breeding and biodiversity. *Journal of Agriculture and Environment for International Development. 103*, 131-145.

5- Ceccarelli, S. (2012). *Plant breeding with farmers.*A Technical manual. International Center for Agricultural Research in the Dry Areas (ICARDA), Aleppo, Syria, Pp. 6-12.

6- Ceccarelli, S., & Grando, S. (2008). *Decentralized plant breeding with participatory method.*Arman Ardalan (translator). Tehran: Institute of Sustainable Development of Environment, P.128

7- Ceccarelli, S., Grando, S., & Baum, M. (2007). Participatory plant breeding in water-limited environments. *Experimental Agriculture, 43,* 411-435.

8- Ceccarelli, S., Grando, S., Tutwiler, R., Baha, J., Martini, A. M., Salahieh, H., Goodchild, A.,& Michael, M. (2000). A methodological study on participatory barley breeding I. Selection Phase. *Euphytica, 111*, 91-104.

9- Davis, J. A. (1971). *Elementary survey analysis*. Englewood, Prentice Hall. NJ. Fami, SH. H. (2006). *Principles of agricultural extension and education*. Tehran: Payam Noor University Press. P. 373

10- David, R. (1999). *Participatory Action Research and social change*. New York: Ithaca, Cornell University. p. 273

11- Gangn, C. N., Cheng, SH., Huang, L., & Sambou A. (2010). Sustainable agriculture in Benin:Strategies for applying the Chinese circular agriculture model. *Journal of Sustainable Development, 3*(1), 25-32.

12- Heidari, H., Impiglia, A., Darayi, L., & Mirzaei,F. (2007). Farmer field school deliver results in Iran. *Integrated Pest Management, Pesticides News, 76*, 43-54.

13- Lahmar, R. (2010). Adoption of conservation agriculture in Europe: Lessons of The KASSA Project. *Land Use Policy, 27*, 4–10.

14- Mayer, R. (2013). Technology options for feeding 10 billion people Plant breeding and innovative agriculture. Institute for Technology Assessment and Systems Analysis (ITAS). Retrieved from http://www.europal.europa.eu/ stoa/cms/studies.

15- Maghirang, R. G., Rodulfo, G. S., Madrid, I., Ferry, E., Cruz, C., Vilbar, L., & Misterio, J. S. (2014). *Participatory Breeding on Organic Vegetables*. Proceedings of the 4th ISOFAR Scientific Conference. 'Building Organic Bridges', at the Organic World Congress, 13-15 Oct., Istanbul, Turkey. P. 912.

16- Miller, L.E., & Smith, K.L. (1983). Handling non response issues. Retrieved from the www.joe.org/joe/1996 February/ rb2.html

17- Moemeni, M., & Ghayomi, A. (2007). Statistical analysis with using SPSS. Tehran: Ketab-e-No.

18- Morris, M. L., & Bellon, M. R. (2004). Participatory plant breeding research: opportunities and challenges for the international crop improvement system. *Euphytica, 136,* 21–35.

19- Murphy, K., Lammer, D., Lyon, S., Carter, B., & Jones, S.S. (2005). Breeding for organic and low-input farming systems: An evolution ary-participatory breeding method for inbred cereal grains. *Renewable Agriculture and Food Systems Journal Renew, 20*, 45-55.

20- Mustafa, Y., Ceccarelli, S., & Grando, S. (2005). Benefit-cost analysis of participatory breeding program in Syria. International Center for Agricultural Research in the Dry Areas (ICARDA).P.O. Box 5466, Aleppo, Syria.

21- Rolling, N., & Pretty, J. N. (1997). *Extension role in sustainable agricultural development.* In: Improving Agricultural Extension, A reference manual (Eds. Swanson *et al.*), FAO, Rome. p. 512.

22- Surangsri, W. C., & Nunta, B. (2005). *Factors Affecting the adoption and non-adoption of sloping land conservation farming practices by small-scale farmers in Thailand.* In Kheoruenromne, I.; Riddell, J. A.; Soitong, K. (Eds.). Proceedings of SSWM 2004 International Conference on Innovative Practices for Sustainable Sloping Lands and Watershed Management, 5-9 September 2004. Bangkok, Thailand: Department of Agricultural Extension.Pp.257-267.

23- Yiching, S., & Jingsong, L. (2011). *The role of biodiversity, traditional knowledge and participatory plant breeding in climate change adaptation in Karst mountain areas in SW China.* A country case study on the impacts of climate change and local people's adaptation. International Institute for Environment and Development, Retrieved from the: http://www.eldis. org/go/ home&id=60588&-type=Document#.WNdUsGfVDDc.

14

Production Planning Optimization Using Genetic Algorithm and Particle Swarm Optimization (Case Study: Soofi Tea Factory)

Mansour Soofi [1*] and Maryam Mohseni [2]

Keywords:
production planning, Genetic algorithm, Particle Swarm Optimization Algorithm, Securities and Exchange Organization

Abstract

Production planning includes complex topics of production and operation management that according to expansion of decision-making methods, have been considerably developed. Nowadays, managers use innovative approaches to solving problems of production planning. Given that the production plan is a type of prediction, models should be such that the slightest deviation from their reality. In order to minimize deviations from the values stated in the tea industry, two Particle Swarm optimization algorithm and genetic algorithm were used to solve the model. The data were obtained through interviews with Securities and Exchange Organization and those in financial units, industrial, commercial, and production. The results indicated the superiority of birds swarm optimization algorithm in the tea industry.

[1] Department of Industrial Management, Rasht Branch, Islamic Azad University, Rasht, Iran
[2] PhD in Industrial Management, Tehran University
** Corresponding author's email: msoufi@aol.com*

INTRODUCTION

One of the most important tasks of production management is production planning. Good planning can make the company competitive in a strategic position. Production planning, one of the principal duties of managers, aims to meet the demand for the product on the market in the shortest time and lowest cost. Operations management tries to achieve the organization's goals with the aim of minimizing the cost of production, without compromising other goals such as product quality (Vonderembse & White, 2014). Today, managers tend to have flexible organizations consistent with the accelerating changes in market demand and to reduce production costs, improve the quality of products or services, promote innovation and risk-taking is so important (Pavão & Ravagnani, 2016). Of course, to achieve these goals, managers must make decisions in areas such as production, warehousing, and levels in the workforce. To this end, managers who designed appropriate mathematical models, according to good demands such as reducing production costs and increasing the quality of products, can find appropriate decision variable set by finding an optimal solution to this model (Braune et al., 2000).

In production planning, it can be recognized how much production is needed in each course of planning to meet the demand for the goods. Planning methods that are used to solve production issues have changed due to improvements in operation research techniques. Classical and traditional optimization methods, long used in solving optimization problems, have been unsuccessful in solving complex problems, because they cannot solve problems with complex variables and objective functions. The managers are currently facing various goals for the future where these goals are not compatible with each other (Fatemi Qomi, 2004). Accordingly, nowadays, managers use heuristic methods to solve production planning problems where these methods aim to look for very justified answers. These include repetitive algorithms that any iteration includes the search for better answers that is better than the previous answer (Zha et al., 2016). It should be noted that the majority of companies still rely on the knowledge and experience of their experts for production planning and modeling, as well optimization of production planning. We should also know that optimization of any factory depends on the characteristics, needs, and constraints of the production system. In other words, optimization goals in each factory may vary from one factory to another and can include a variety of maximization or minimization (Naima et al., 2015).

The tea industry, like other industries, is faced with more objectives and constraints and its process is composed of many steps and a variety of stations. For this reason, to use the maximum capacity of machines and manpower, accurate production planning is needed. To have appropriate production program to meet fluctuations in demand in the tea industry, it is necessary to provide a snapshot of conditions for tea production in our model. This research trying to find a suitable mathematical model for Soofi Tea Factory to optimize production planning, determining the type and amount of product mix and production plan tailored to meet seasonal fluctuations in demand in the factory. For this purpose, a suitable model, due to limitations faced by factories, is introduced for tea industry using the production planning techniques and then using ideal planning and advanced techniques of operations research (Genetic Algorithm and Particle Swarm Optimization), the designed model is solved; and by these techniques, the optimal solution, that is the optimal amount of output per workstation and normal and overtime hours that they work per month within the industrial company Soofi Tea is obtained, so that all Tea manufacturing companies by slightly change can use the model to plan their production and maximize efficiency.

The model is presented in the form of multi-product, multi-stage, and multi-period production planning system which contains two target functions: the first function is to reduce the cost of company that in this study are non-linear. As mentioned earlier, it is better to use modeling to optimize production planning to overcome the

[1] *Recently, a fourth dimension has been added:– the institutional dimension of development*

modern change. Of course, finding the optimal point in the case when faced with a complex model is not easy. In fact, classical and traditional optimization methods have long been used to solve optimization problems; however, they have failed in solving complex problems. Because they have not the ability to obtain the global optimal point in optimization problems with high variables and complex target functions and to solve this problem, give us only one local optimum point. Most classical optimization methods use gradient or higher-order derivatives of target function and start from an initial point, gradually the answer found has been improved in ascending or descending order. In the gradient methods, local minimum risk of exposure is high (Modares, 2013). One of the ways to overcome this problem is to use intelligent population optimization methods such as genetic algorithms and Particle Swarm Optimization method that are designed to find the global optimal solution to complex problems. Production planning is to determine production targets for a period of time in the future that is called the planning horizon as well as optimal use of resources to meet the specified requirements (Wang et al., 2015). The purpose of production planning is to plan and arrange the order sequence based on strategic objectives and the actual conditions of the manufacturing plant. Production planning attempts to improve the material flow and efficient use of machinery in its production and includes different management goals such as reducing work in process, lead times and costs, as well as improving responsiveness to fluctuations in demand (Zolghadri et al., 2008). In some manufacturing units, the production process is such that the final product must pass several stages of production to complete. The sequence of production steps leads the input provide to each stage of previous output.

Goal programming was first created in 1960 by Charnes and Cooper and is employed by Lee in production planning in 1973. Lee pointed out that goal programming models allow disparate variables to be collected in a single objective function. The main problem of production planning is fluctuating demand over time. Fluctuations in product demand to be answered by adopting different strategies, such as changing the speed of production, changes in labor, overtime or negligence, side contract and change the warehouse level. So far several methods is used to solve such problems including the decision search programming and linear programming method, but these methods are not widely used in industry. The main reason for this is that these models greatly simplifying the fact and leaves no room to impose manager tastes or method to solve problems. In addition, solving production planning problem by any of the above methods involve actual items production costs, staffing and storage that access to real values of these costs is very difficult. Because most of these cost are not linear and are in grade two and above and they are not simple estimate. Since the objective function of each of the above methods has cost structure, perhaps the use of excessive approximation cause that the answer obtained is not be reasonable, although it may be optimized for used model. Each comprehensive production planning issue is trying to determine the optimal production staffing levels and warehouse, so that at the least cost to meet demand over the planning horizon. The main problem in determining these values is to contrast of the three targets (production, human resources, and warehouse) together. That means access to each of Purposes undermines other purposes.

Therefore, the above objectives cannot be gathered in a single objective function, such as would be used in linear programming. On the other hand, often for any of these purposes, certain amount to be levied by managers which makes full access is not possible yet with goals. Therefore the problem in terms of the above models is impossible. This problem can be solved by using goal programming models. In addition, it must be said that the three causes of the goal programming method to solve problems of production planning is recommended, 1) Non-retractable objectives, 2) Fluctuating demand and, 3) Lack of access to the real cost items. In goal programming, unwanted deviations from target values set by the decision making can be minimized in order to reach an acceptable solution. Unwanted distortions measured by positive and negative deviation variables that are defined for every goal. They show the supra and infra achieve to any goal (Liu, et al., 2016). Genetic Algorithm that highly influenced with natural phenomena had been provided by Charles Darwin in the mid-19th century (Nunez-Letamendia, 2015). Genetic algorithms first introduced by John Holland in 1962, then Goldberg expanded it in 1986 (Yand & Chiang, 1996). Due to the high potential in solving complex optimization problems much attention has been paid (Mitchell, 1996). In genetic algorithms, each chromosome represents a point in the search space and a possible solution to the problem. Chromosomes itself (solutions) are formed of a fixed number of genes (variable). Set of chromosomes make up a colony. With the impact of genetic operators on each population, the new population is formed of the same number of chromosomes. Steps of genetic algorithm can be stated as follows:

1. The initial population is composed of people who are usually randomly selected.
2. Current population estimates are based on the relative fitness.
3. If the final criterion that is considered to be meets, the ultimate solution is selected.
4. The new population is created based genetic operators.
5. Actions are repeated from step 2 again and until the end criterion is met, the repetition is continued (Asgarpour Khansary & Hallaji Sani, 2014).

Genetic Algorithm uses two different genetic operator named cutting (link) and mutation that the main operator in genetic algorithm is cutting action. This operator, by combining the two chromosomes as their parents creates two new chromosomes as children which are somewhat similar in terms of specifications to parents and inherit the properties of both. Through this inheritance is that the gradual evolution arises ultimately. The population will be modified over time. It should be noted that interbreeding usually does not apply on all chromosome pairs selected for crossover. Usually the probability of crossover is considered for each chromosome pair within 0.6 to 0.95. This number is called crossover rate or crossover probability. But the mutation operator create a sudden or random change on some of genes or chromosomes that cause firstly, the likelihood of algorithm engaging in a locally optimal would be less and secondly, new areas is search and new chromosomes are added to the cycle. The possibility of mutation action on each chromosome is called mutation rate mutation or mutations probability. Usually this number is considered very small (0.001) (Kuo et al., 2012).

To select members of the next generation from among the members of the current generation and offspring from operators, several selection methods have been proposed. The main idea in the selection methods is that good people prefer on bad people where the best and worst of people defined by the fitness function f.

In the present study, tournament method is used to select. In this way a few chromosomes are randomly selected and the best of them will go to the next generation (Depending on the size of the tournament, which is usually between 2 to 7). This work will continue to evolve the next generation. The number of chromosomes that are randomly selected called tournament size. Usually two people are randomly selected from the population. Then a random number r is chosen between 0 and 1. If $r < k$ (where k is a parameter, for example, 0.75), a more fitness person, and otherwise less fit person to be elected as a parent. The two then returned to the initial population and again place in the selection process (Liu et al., 2016). Usually, different stop criteria are considered to stopping the algorithm. For example, the end time of the algorithm, or the number of algorithm iteration are examples of the algorithm stop criteria.

Particle Swarm Optimization method (PSO) is one of the most recent population-based search methods (Giftson & Rajan, 2015). In PSO, each member of a community called a particle. These particles all are floating in multidimensional space and moving based on the exchange of information with other members. Change the position of each particle is based on his experience and knowledge and its neighbors. Thus, the particle in the search space is floating and moving. And while movement of particles, each of them during the way movement store a memory of their best value and its position. From now, Xi (t) is used to show the position of the particle Pi at time t.

In PSO, the best value and position of each particle is stored in the variable ip best and best position of the particles in all the repetitions stored on ig best (Kennedy, 1997). The new move of particles takes place based on the resultant p best i (The past experience of each particle) and g best (best past experience of all the particles) in the previous moving direction of particles (Mansour et al., 2008).

MATERIALS AND METHODS

The mathematical model designed for this study is called goal programming model. The model consists of 276 decision variables; two objectives functions that include minimizing the cost and maximizing sales; six categories of restrictions including restrictions on cars production capacity (indicating that each machine cannot produce more than its capacity), balancing limits (to create effective communication between the lines that in this limits the amount of each product at each station is balance with the required amount of it to product in the next step), demand restrictions (storage of one product in one specified month equal to the product storage in the previous month as well as production of that product in that month minus the projected demand for that product per month), restrictions of overtime hours and normal hours of work (number of hours that people work in the area that the company has set), restrictions of the number of people who work extra hours (the number of people who work extra hours cannot exceed the number of people in the factory) and restrictions of the amount of inventory (the amount of inventory that must be kept in stock cannot exceed the limit that has been set for it). To build capacity constraints and the balance, the production process should be considered and then presented the respective constraints for each station. The production process is shown in Figure 1.

In such systems of production, a balance between the various stages of the production sequence is very important (Jewski & Ritzman, 2001).

$$\min z = \sum_{k=1}^{k}\sum_{i=1}^{m}\sum_{j=1}^{n}\left[C_{ijk}X_{ijk}\right]$$

St:

$$\sum_{i=1}^{m}\sum_{k=1}^{k}\left[a_{ijkl}x_{ijk} \le b_{jl}\right] \quad (l=1, 2, \ldots, l)\ (j=1, 2, \ldots, n)$$

$$\sum_{k=1}^{k} X_{ijk} = \sum_{k=1}^{k} X_{i,j+1,k} \quad (i=1, 2, \ldots, m)\ (j=1, 2, \ldots, n-1)$$

$$\sum_{k=1}^{k} X_{imk} = Di \qquad i=1, 2, ..m$$

All variables≥0

where:

X_{ijk}: Product *i* which by the method *k* make in *j* stage

a_{ijkl}: The amount of product consumption *i* by method *k* in the process *j* in material balance stage 1

b_{jl}: amount of raw material 1 in the process *j*

k: Production methods.

The overall goal programming model is as follows.

$$Min\sum_{k=1}^{q}\sum_{i=1}^{m}P_k(d^-_i + d^+_i)$$

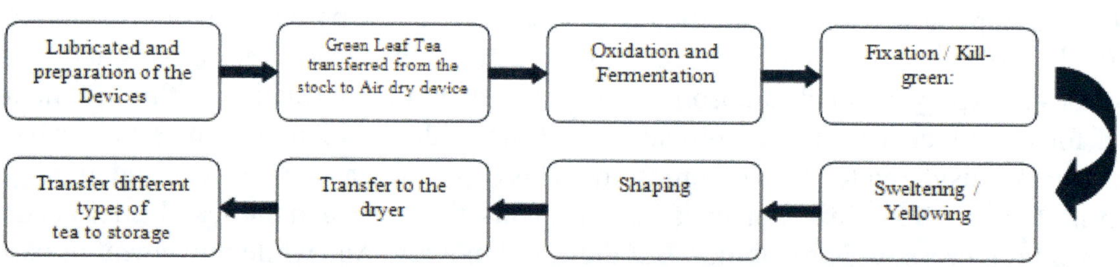

Figure 1. The production process in Soofi Tea Factory.

St:

$$\sum_{j-1}^{n} C_{ij}X_j + d^-_i + d^+_i = b_i \quad (J=1,2,\ldots,n)(I=1,2,\ldots,m)$$

$$\sum_{j-1}^{n} a_{rj}x_j \le b'_r \quad (r=0,1,2,\ldots,s)\ (j=1,2,\ldots,n)$$

cXj , di+ , di- ≥ 0

X_j: Model decision variables that can adopt any non-negative number

D_i^+, d_i^-: Show Positive and negative deviations from the ideal variables

B_i: Express the right number or level of desire of its goal i

P_n: specifies the kth priorities (k=1,2, ..., N)of goal.

A_{rj}: Provides technical coefficients of the model

C_{ij}: Coefficients of the decision variables

b_r: The right numbers of functional limitations

To implement a fitness-proportionate selection, several sampling methods have been proposed such as roulette wheel. Fitness or the selection probability of each chromosome is calculated as follows:

$$P_k = \frac{f_k}{\sum_{j=1}^{pop-size} f_j}$$

where fj, is the value of chromosomes matching function.

The particles movement rate i is defined as follows:

$$\vec{v}_i(t) = \vec{x}_i(t) - \vec{x}_i(t-1)$$

The following steps are to be taken when employing the Particle Swarm Optimization method: (Karimi et al., 2010)

1. Choose the initial population (population of particles are randomly selected in the search space).

2. Calculate the value of particles using its present position (for every particles, its value is obtain according to the criterion function).

3. Compare the present value of the particles and best experience and the following conditional Green Leaf Tea ((In fact, the value of each particle compared with pi besti and if it's current value is pi pest, we replace the value of the current position by p best I and the position related it). If f (pi)≥ p best I , then:

$$f(p_i)=p\ best_i \quad (A$$

$$p\ best_i = \dot{X}_I(t\dot{X}) \quad (B$$

4. Compare the current value of each particle with the best previous experience on all Population particles and Green Leaf Tea of below: (In fact, the value of each particle compared with g besti and if the particles value is more than g besti, we replace the value of the position by g besti and the position related it). If f (pi) ≥ gi best, then:

$$f(p_i)=p\ best_i \quad (A$$

$$g\ best_I = \dot{X}i\ (t\dot{X}) \quad (B$$

5. Change the speed of the particles according to the following equation:

$$_i=\vec{v}i(t-1)+\rho 1\ (\vec{x}pbest_i-\vec{x}_i(t))+\rho 2\ (\vec{x}gbest_i-\vec{x}_i(t))\dot{V}$$

where ρ_1 and ρ_2 are positive random numbers.

6. Move the particle to its new position

$$X_i(t) = \vec{x}_i(t-1)+\vec{v}_i(t)$$

7. Go to step 2 and repeat algorithm until convergence.

Decision variables

f_1: Values greater than the cost of products specified in October

f_2: Values greater than the cost of products specified in November

f_3: Values greater than the cost of products specified in December

f_4: Values greater than the cost of products specified in January

f_5: Values greater than the cost of products specified in February

f_6: Values greater than the cost of products specified in March

h_1: Values less than the sale of products specified in October

h_2: Values less than the sale of products specified in November

h_3: Values less than the sale of products specified in December

h_4: Values less than the sale of products specified in January

h_5: Values less than the sale of products specified in February

h_6: Values less than the sale of products specified in March

PC_{1ijt}: Oxidation and fermentation Tea that is generated in the band i with difficulty j in period t.

$PC2_{ijt}$: Product that is generated in the Fixation i with difficulty j in period t.

Ip_{tt}: The Green Leaf Tea that is stored in the warehouse of raw materials in band during the period t.

RM_{jt}: The oxidation Tea produced with difficulty j in period t.

RS_t: Oxidation Tea amount stored in the period t

Iot: The Natural flavors of tea is stored in the intermediate product warehouse in period t.

CM_{ijt}: The amount of tea which are produced by the mill i with the genus j in period t.

ZL_t: The additives for production in the period t

Il_t: The amount of tea stored in the period t in the warehouse number one.

IM_t: The amount of tea stored in the period t in the warehouse numbers two.

IS_t: The amount of tea stored in the period t in the warehouse number three.

S_{ijt}: The amount of tea are ready to sell from the final station with a label i with genus j in period t.

N_t: Normal hours of work per worker in period t

L_t: Number of employees who work extra hours during t

M_t: The number of hours of overtime work per worker in period t

IS_{ijt}: The final balance of the tea with label i and genus j in period t.

Model parameters

The model parameters include projected demand of different products according to the type of material and the period of maximum-band capacity and the efficiency of various devices and the beginning inventory of products with label i for genus j in period t. Goals amount of sales and the cost of the salary of regular working hours, overtime, the cost of one square meter warehousing product, the number of workers who work in overtime; minimum number of normal working hours and the maximum and minimum number of overtime hours per worker and the selling price and cost of production per Kg of product.

$C.cap1_{ijt}$: The maximum capacity of the circling to dry i to produce a production with difficulty j in period t.

$M.cap2_{ijt}$: The maximum capacity of each Oxidation and Fermentation device i to produce a production with difficulty j in period t.

$I.cap_t$: The maximum capacity of Green Leaf Tea warehouse to store raw materials crushed in the period t.

$RM.cap_{jt}$: The maximum capacity of the Oxidation and Fermentation batching machine to produce a production with difficulty j in period t.

$IO.cap_t$: The maximum capacity of the central warehouse to store the Shaping Tea in band in the period t.

$RS.cap_t$: Maximum capacity of the form in period t.

$K.cap_t$: The maximum capacity of the roasting i to produce a production with degree j in period t.

$CM.cap_{ijtt}$: The maximum capacity of Oxidation and Fermentation device to produce a production with degree j in period t.

$IL.cap_t$: The maximum capacity of the device number one to shaping tea in period t.

$IM.cap_t$: The maximum capacity of the device number two to shaping tea in period t.

$IS.cap_t$: The maximum capacity of the device number three to shaping tea in period t.

$S.cap_{ijt}$: The maximum capacity of the selling station of products with devices i with genus j in period t.

$N_{t\,min}$: Normal minimum number of hours each employee works in period t

$N_{t\,max}$: Normal maximum number of hours each employee works in period t

$L_{t\,max}$: Maximum employees who work in period t

$M_{t\,max}$: Maximum hours that each worker has overtime in the period t

$M_{t\,min}$: Minimum hours that each worker has overtime in the period t

IS_{ijt-1}: The amount of tea inventory with label i and genus j at the beginning of the first month.

D_{ijt}: The demand for teas with label i and genus j in period t.

C_{ijt}: The cost of one Kg tea with label i and genus j in period t.

P_{ijt}: Sale price of Kg products with label i and genus j in period t.

C_{RL}: The cost of one normal work hours per worker in period t

C_{RO}: The cost of one hour of overtime per worker in period t.

C_{IO}: The cost of inventory per square meter products in the period t.

gt: The goal set for maximizing sales revenue.

%X: The efficiency of Oxidation and fermentation unit

%Y: Efficiency of shaping devices

%A: Tea percent

% (1-A): % of additives

The mathematical model

$$Z1\,t = \sum_{i=1}^{m}\sum_{j=1}^{n} S\,ijt * C\,ijt + \sum_{t=1}^{t}(L\,tmax * N\,t * C\,rl) + \sum_{t=1}^{t}(M\,t * L\,t * C\,ro) + \sum_{i=1}^{m}\sum_{j=1}^{n}(IS\,ijt + S\,ijt - D\,ijt) * C\,io \le C\,t \quad (1)$$

$$Z2t = \sum_{i=1}^{m}\sum_{j=1}^{n}(P\,ijt * S\,ijt) \ge g\,t \quad (2)$$

$$N\,t\,min \le N\,t \le N\,t\,max \quad (3)$$
$$L\,t \le L\,tmax \quad (4)$$
$$S\,ijt + IS\,ijt\text{-}1 - Isijt \ge D\,ijt \quad (5)$$
$$M\,tmin \le M\,t \le M\,t\,max \quad (6)$$
$$PC\,1ijt \le C.cap\,1ijt \quad (7)$$
$$PC\,2ijt \le C.cap\,2ijt \quad (8)$$
$$IP\,t \le I.cap\,t \quad (9)$$
$$RM\,jt \le RM.cap\,jt \quad (10)$$
$$RS\,t \le RS.cap\,t \quad (11)$$
$$K\,ijt \le K.cap\,ijt \quad (12)$$
$$IO\,t \le IO.cap\,t \quad (13)$$
$$CM\,ijt \le CM.cap\,ijt \quad (14)$$
$$IL\,t \le IL.cap\,t \quad (15)$$
$$IM\,t \le IM.cap\,t \quad (16)$$
$$IS\,t \le IS.cap\,t \quad (17)$$
$$S \le S.cap\,ijt \quad (18)$$

$$\sum_{i=1}^{m}\sum_{j=1}^{n} PC\,1ijt + \sum_{i=1}^{h}\sum_{j=1}^{n} PC\,2ijt = IP\,t \quad (19)$$

$$\sum_{j=1}^{n} RM\,jt = IP\,t \quad (20)$$

$$\%X\sum_{j=1}^{n} RM\,jt = RS\,t \quad (21)$$

$$RSt = \sum_{i=1}^{m}\sum_{j=1}^{n} K\,ijt \quad (22)$$

$$\%A[IO\,t] + (1 - \%A)(ZLt) = \sum_{i=1}^{m}\sum_{j=1}^{n} CM\,ijt \quad (23)$$

$$\%y\sum_{i=1}^{m}\sum_{j=1}^{n} K\,ijt = IO\,t \quad (24)$$

$$\sum_{i=1}^{m-2}\sum_{j=1}^{n} CM\,ijt = IL\,t \quad (25)$$

$$\sum_{i=m-1}^{m-1}\sum_{j=1}^{n} CM\,ijt = IM\,t \quad (26)$$

$$\sum_{i=1}^{m}\sum_{j=1}^{n} CM\,ijt = IS\,t \quad (27)$$

$$IL\,t + IM\,t + IS\,t = \sum_{i=1}^{m}\sum_{j=1}^{n} S\,ijt \quad (28)$$

As mentioned, the study aimed to satisfied two objectives: Reducing costs and increasing sales that two equations (1) and (2) in the above model reflects the two objectives functions. Thus, the (1) relate to 4 group of cost includes cost of production, normal wage labor costs, the cost of overtime wage and factory storage cost and Ct represents the goal cost. Equation (2) is also refer to the second goal, that is the proceeds from the sale of tea products and gt refers to goal sale. Equation (3) ensures that normal hours of labor in the factory does not exceed from less than the minimum working hours and the maximum working hours.

Equation (5) ensures that labor overtime hours in the factory does not exceed from less than the minimum overtime hours and the maximum overtime hours. As equation (6) suggests, the amount of a product inventory in the previous month in addition to the amount of product in that month minus the amount of inventory in that month must be greater than anticipated demand for the product on the same month. Equation (7) indicates the capacity constraints of Oxidation and fermentation device and equation (8) indicates the form capacity constraints and ensures that the product produced by the mold for a given period and certain hardness cannot exceed the maximum capacity of Fixation / Kill-green devices. Equation (9) refers to raw material storage capacity constraints.

Equation (19) means that the products of different machines in the first station are equal to the input to the warehouse of Green Leaf Teas. Equation (10) states that the prepared oxidation Tea should not exceed the maximum amount of Fixation / Kill-green devices to produce production. Equation (20) means that the input Green Leaf Tea to oxidation Tea construction equipment is equal to the amount of material in the warehouse for raw materials. According to equation (11) the amount of material stored in each period cannot exceed the maximum capacity of vibrating table. Equation (21) means the amount of input material in the silo of raw materials is equal to manufactured products from raw mill minus waste moisture produced by the grinding of raw materials. The main part takes place when drying that according to (12) the product is not more than the maximum capacity of the Fixation / Kill-green devices.

Equation (22) indicates that total powdered material stored in the silo of raw materials is equal to imported materials to oxidation Tea machine building. This means that all materials go from silo of raw materials into the devices. Equation (13) ensures that the additive that is stored in the intermediate products stock in each period does not exceed the maximum storage capacity. Equation (24) states that all materials that are inserted into the machine is not converted into oxidation Tea, but depends to station efficiency (% y); So, the material leaves the device to the next round is not equal to the material inserted into the machine. Equation (13) refers to the limited capacity of oxidation tea shaping for the production. According to equation (23), a percentage of additives are added to the tea that is removed from the device to achieve oxidation tea in the next step. Equations (15), (16) and (17) refers o the limited capacity of three Oxidation Tea silos. And according to the equation (25), (26) and (27), all products will be imported to the warehouse where they are stored. Equation (19) refers to labels of produced oxidation Tea and limited capacity to a certain degree production. Equation (28) means that all products go to sales station.

RESULTS

After making the appropriate model, decision maker can act on any kind of experience and applied it. As described earlier, the present study with the help of Genetic Algorithm and Particle Swarm simulation algorithm sought to solve the proposed mathematical model. In the first step, the basic parameters of the algorithm should be determined. In Particle Swarm simulation algorithm, each particle is a potential answer. As such, each particle has 276 decision variables. In addition, in genetic algorithms, every chromosome is a potential answer that has 276 decision variables. Accordingly, the search space in each of these two algorithms will have 276 dimensions. In the studied factory, as mentioned the value of all objective function to manage all the same and rate the weighted factor for both functions is considered one. In the optimization model in genetic algorithm,

Table 1
Parameters of Genetic Algorithm

Parameters	Value
Mutation rate	0.1
Crossover rate	0.8
Initial population size	40
Number of generations	100

Table 2
PSO algorithm parameters

Parameters	Value
Number of particles	40
Number of generations	100
Acceleration coefficient g best (c2)	2
Acceleration coefficient p best (c1)	2

the extended sampling space is considered and selection method is tournament. Basic parameters of GA and PSO are shown in Tables 1 and 2.

Moreover, r1 and r2 random numbers in the period [0, 1] and , the inertia weight decreases linearly during the iterations of the algorithm. In the beginning, is considered equal to 1 and finally can reach 0.5 with trial and error to achieve better results (Soleimani & Kannan, 2014). Next, by collecting the necessary data from the Soofi Tea Factory and after importing the inputs to the model by the programming software called MATLAB, the obtained model using GA and PSO were solved and the optimum solutions were obtained. The answers of the model include the amount of output per workstation, the number of overtime hours and the normal number of hours that employees work per month and the number of employees who work extra hours per month, end inventory for each product per month, and the deviation rate of the objective functions from the goals assigned to them.

Data collection

It is noteworthy that before entering data and parameters into the model in the present study, the data were obtained through interviews with SEO and those in financial units, industrial, commercial and production. Storage cost per Kg of tea is 1000. The factory has 350 workers at its manufacturing plant in normal time work. The cost of each worker's salary for normal working hours is 2000 and the cost of wages per worker per hour of overtime is 3500.The collected information is shown in Tables 3 to 6.

Table 3
Data on Goals (IRR)

Goal	October	November	December	January	February	March
Cost	2210000000	2200000000	1900000000	1890000000	19650000000	2160000000
Sale	4210000000	4162000000	3580000000	3570000000	3710000000	4130000000

Table 4
Data on the Genus Product and Related Information (IRR)

Teas		cost of per Kg (to IRR)	selling price per Kg (to IRR)	Max inventory of end period (to kg)	Tea stock at the beginning of the first month planning (to kg)
Black	I	25000	50000	0	0
	II	30000	56500	3000	3000
Green	I	26000	51500	0	0
	II	315000	58000	3000	2000
Tea bag	I	25500	51000	0	0
	II	32500	57500	3500	3000

Table 5
Data on Projected Demand for the Next 6 Months (kg)

Tea Type		October	November	December	January	February	March
Black	I	25000	24000	20000	20000	22000	25000
	II	40000	38000	30000	30000	30000	40000
Green	I	4000	3000	3000	3000	4000	3000
	II	3000	2000	2000	2000	3000	3000
Tea bag	I	6000	5000	5000	5000	4000	3000
	II	3000	5000	6000	6000	6000	7000

Table 6
OtherData

	Specifications
Loading Place	Total: 2 Power loading of Tea in months/ kg: 90000
Green leaf tea stock	Total: 3 Capacity in kg: 20/180/160
Oxidation tea production	Monthly production capacity in kg: 2000 For each ton of entries: 600 kg
Additives Stock	Monthly capacity in kg: 90000 Scrap percentage: 40%
Fixation / Kill-green devices	Total capacity of mold in months: 264000
Number of shaping devices	Total: 12 Power Monthly: 178000
Green Leaf tea storage	Capacity in Kg/ Month: 528000
Tea	Monthly production power/hours: 800 Monthly production power/ kg: 70000 Monthly production power/hours:200
Oxidation tea	Monthly production power/ kg: 40000 -Additives for oxidation Tea capacity per month: 900000 kg ; Combined ratio: 95%
Products storage	-Capacity of oxidation Tea corrective materials/month: 300000Combined ratio: 5%

The answerers obtained from the model

To solve the problem in two ways Genetic Algorithm and Particle Swarm algorithm, the optimum amount of product on each work-station according to number of machine, genus, as well as tag of product and production rate in different months/Kg can be obtained. Tables 7 and 8 show the values obtained from the algorithm:

The regular working hours overtime and the

Table 7
Optimum Amount of Product in Each Station Using GA (kg)

Production station		Type of genus produce	Production October	Production November	Production December	Production/ January	Production February	Production March
Loading Place	Lot By Lot	Black	25000	24000	20000	20000	22000	25000
		Green	4000	3000	3000	3000	4000	3000
		Tea bag	6000	5000	5000	5000	4000	6000
	Quantity	Black	37000	38000	30010	33480	30000	36500
		Green	3500	2000	2000	1000	1500	3000
		Tea bag	2500	5000	6000	3500	7000	3000
Fixation Tea	Fixation IM1	All three	26000	25670	22010	21990	22830	25500
	Fixation IL2	All three	26000	25670	22010	21990	22830	25500
	Fixation IH3	All three	26000	25670	22010	21990	22830	25500
Shaping	i=1	All three	26010	25680	22020	21990	22830	25500
	i=2	All three	26010	25680	22020	21990	22830	25500
	i=3	All three	26010	25680	22020	21990	22830	25500
Stock intermediate products	i=1	All three	71580	70530	58970	58930	61580	70000
Amendments	i=1	All three	200000	200000	200000	200000	200000	200000
Dryer	i=1	All three	238590	235080	196560	196440	205260	233340
Raw materials Silo	i=1	All three	238600	235090	196560	196430	205260	233330
Raw Mill	i=1	All three	477180	470190	393120	392850	410520	466680
Raw material storage	i=1	All three	477190	470180	393110	392850	410530	466670
Circling 1	i=1	All three	178590	175080	136560	136440	145260	173340
Circling 2	i=2	All three	178590	175080	136560	136440	145260	173340
Additive	i=3	All three	120000	120000	120000	120000	120000	120000

Table 8
Optimum Amount of Product in Each Station Using PSO (kg)

Production station		Type of genus produce	Production October	Production November	Production December	Production/ January	Production February	Production March
Loading Place	Lot By Lot	Black	25000	24000	20000	20000	22000	25000
		Green	4000	3000	3000	3000	4000	3000
		Tea bag	600	5000	5000	5000	4000	6000
	Quantity	Black	37030	38000	30020	33450	30000	36500
		Green	3500	2000	2000	1500	1700	3100
		Tea bag	2500	5020	6080	3400	7000	3000
Fixation Tea	Fixation IM1	All three	26010	25675	22030	22115	22900	25530
	Fixation IL2	All three	26010	25675	22030	22115	22900	25530
	Fixation IH3	All three	26010	25675	22030	22115	22900	25530
Shaping	i=1	All three	26020	25685	22040	22125	22910	25530
	i=2	All three	26020	25685	22040	22125	22910	25530
	i=3	All three	26020	25685	22040	22125	22910	25530
Stock intermediate products	i=1	All three	71610	70580	59070	59240	61820	70094
Amendments	i=1	All three	200000	200000	200000	200000	200000	200000
Dryer	i=1	All three	238690	235270	196900	197800	206070	233650
Raw materials Silo	i=1	All three	238700	235280	196900	197800	206070	233660
Raw Mill	i=1	All three	477380	470540	393800	395600	412150	467300
Raw material storage	i=1	All three	477390	470550	393790	395600	412150	467310
Circling 1	i=1	All three	178695	175275	136895	137800	146070	173655
Circling 2	i=2	All three	178695	175275	136895	137800	146070	173655

Table 9
Number of Hours Worked and the Number of People Who Work Overtime Based on GA

Month	Regular hours	Overtime hours	number of people who work overtime
October	160	50	30
November	150	50	40
December	140	50	30
January	130	50	40
February	130	50	50
March	150	50	30

Table 10
Number of Hours Worked and the Number of People Who Work Overtime Based on PSO

Month	Regular hours	Overtime hours	number of people who work overtime
October	160	50	40
November	150	50	50
December	140	50	40
January	130	50	30
February	130	50	55
March	150	50	40

number of people who work overtime per month in studied plants are expressed in terms of variables and values obtained from using genetic algorithm and Particle Swarm algorithm shown in Tables 9 and 10.

The amount of each product must be in storage per month using Genetic Algorithm and Particle Swarm also shown in tables 11 and 12.

Tables 13 and 14 shows the deviation from the objective functions, each of which has been implemented five times by each of the algorithms. In different performances of the two algorithms, the amount of output per workstation was identical, while the number

Table 11
The Ending Inventory Based on GA (kg)

Month	Ending inventory					
	Black		Green		Tea bag	
	LBL i=1	Q i=2	LBL i=1	Q i=2	LBL i=1	Q i=2
October	0	0	0	2500	0	2500
November	0	0	0	2500	0	2500
December	0	20	0	2500	0	2500
January	0	3500	0	1500	0	0
February	0	3500	0	0	0	0
March	0	0	0	0	0	0

Table 12
The Ending Inventory Based on PSO (kg)

Month	Ending inventory					
	Black		Green		Tea bag	
	LBL i=1	Q i=2	LBL i=1	Q i=2	LBL i=1	Q i=2
October	0	30	0	2500	0	2500
November	0	30	0	2500	0	2520
December	0	50	0	2500	0	2600
January	0	3500	0	2000	0	0
February	0	3500	0	700	0	0
March	0	0	0	800	0	0

Table 13
The Deviations from GA with the Population Size 40

Performance	1	2	3	4	5
f1	1540000	6085000	34055000	6733000	4343000
f2	1266000	43698000	40000	27150000	37340000
f3	67526000	25256000	55888000	26525000	9921000
f4	5246000	17728000	5809000	22231000	28446000
f5	54050000	8435000	290000	33369000	12890000
f6	2540000	26258000	2540000	891000	11991000
h1	10750000	10750000	10750000	10750000	10750000
h2	2000000	2000000	2000000	2000000	2000000
h3	13441000	13441000	13441000	13441000	13441000
h4	9559000	9559000	9559000	9559000	9559000
h5	15500000	15500000	15500000	15500000	15500000
h6	10750000	10750000	750000	750000	750000
F	194168000	177180000	160622000	186918000	186931000

of overtime hours, the number of regular working hours and the number of designated people were fixed. Because the changes of overtime hours, regular hours, and the number of people work overtime per month affect only the costs objective function and will not affect the sales objective function.

The model was aimed at reducing the total distortions. According to the results that have been obtained by PSO, whole unwanted deviations from target values set by the experts participating in PSO is less than when the model is solved using GA. And from this, it can be concluded that the results obtained from of PSO are better than the ones that can be achieved by GA.

Table 13
The Deviations from PSO with the Population Size 40

Performance	1	2	3	4	5
f1	2440000	4194000	2440000	20940000	3443000
f2	690000	5895000	690000	42320000	56690000
f3	7021000	7021000	54601000	7021000	7021000
f4	16846000	16846000	25267000	16846000	16846000
f5	6590000	6590000	6590000	6590000	6590000
f6	49075000	5690000	5690000	8840000	5691000
h1	9055000	9055000	9055000	9055000	9055000
h2	850000	850000	850000	850000	850000
h3	8276000	8276000	8276000	8276000	8276000
h4	50000	50000	50000	50000	50000
h5	3900000	3900000	3900000	3900000	3900000
h6	4950000	4950000	4950000	4950000	4950000
F	109743000	73317000	122359000	129638000	123362000

CONCLUSION

According to the results and outputs of the model by the Genetic Algorithm and Particle Swarm optimization algorithm, it can be contended that the production rate by PSO is higher than that of GA. As we know, organizations are trying to produce more to gain further profit. In the answer obtained from Genetic Algorithm and Particle Swarm Optimization Algorithm (see Table 7), the total products that achieved in sales within the six months that the organization is planning using genetic algorithms is 433 990 (the sum of all values in station sales) and using Particle Swarm Optimization Algorithm is 434 800 (Table 8). This amount associated with an increase of 810 Kg of tea than to use genetic algorithms. Also compared to tables 7 and 8, it can be seen that the production in other production stations is more obtained by PSO algorithm. This means that the production stations capacity more used. As such, given that in any organization, management aims to increase production in view of the costs, we can argue that the answer obtained from birds swarm optimization algorithm is better than the one obtained from the genetic algorithm.

Moreover, given that the objective function in our model is to reduce the sum of cost and sales revenue deviations, in comparison to Table 13 and 14, it can be argued that the total target deviation (cost and sales revenue) in the values specified in Genetic Algorithm is higher in amount compared to Particle Swarm algorithm. According to what was explained it can be concluded that particle Swarm Optimization algorithm performance is better than genetic algorithm, and by using it, more satisfactory results can be achieved.

Given the results of the model using GA, the company is able to produce 433990Kg in 6 months for which production planning has taken place and launch it to the consumer market for sales. This amount is compared to the company's present production that is 367 788 Kg, which has increased to 18% over six months. It can, then, be contended that the obtained results from GA are satisfactory. On the other hand, according to the results yielded by the model, the company is able to produce 434800Kg in 6 months for which production planning has taken place and launch it to the consumer market for sales using PSO. This amount can be compared to the company's present production, which is 367 788 Kg and shows an increase by 18.2 percent. Values of, f, h, or deviation from the ideals means that the solutions obtained to what extent are different from the ideals set for an object. Values of suggest to what degree the obtained answers are different from the ideal of cost. For example, to solve model using PSO, the amount of 1 is corresponds to obtain 2440000. This number means that the cost obtained for the month of October in 2440000 is more than the ideal cost. The values h also indicated the extent to which the obtained answers are different from the ideal of sales. For example, to solve the model using PSO, the amount of h1 corresponds to 9055000. This number means that the sales revenue obtained for the month October in 9055000 is less than the ideal sales. The values F also represents the total difference in cost goal objective function and ideal sales from objective function. For example, to solve the model using PSO, the amount of F corresponds to obtained109750000. This number means that the total objective function is different from the cost ideals and sales ideals from the objective function is 109750000.

ACKNOWLEDGEMENT

We are thankful to Dr. Alireza Amirteimoori professor in Applied Mathematics and Operations Research, for his scientific comments on this study.

REFRENCES

1- Asgarpour Khansary, M., Hallaji Sani, A. (2014). Using genetic algorithm (GA) and particle swarm optimization (PSO) methods for determination of interaction parameters in multicomponent systems of liquid-liquid equilibria. *Fluid Phase Equilibria, 365*, 141–145.

2- Braune, R., Wanger, S., Affenzeller, M. (2000). Applying Genetic Algorithms to the optimization of production planning in a real world manufacturing environment. *Institute of Systems Theory and Simulation Johannes KeplerUniversity, 69.*

3- Dilworth, J. (1989). *Production and Operation Management, Manufacturing and Non anufacturing.* Fourth Edition. London: Rondan House Inc.

4- Fatemi Qomi, M.T. (2004). *Planning and inventory control.* Amir Kabir Publications.

5- Giftson Samuel, G., & Christober Asir Rajan, C. (2015). Hybrid: Particle Swarm Optimization-Genetic Algorithm and Particle Swarm Optimization-Shuffled Frog Leaping Algorithm for long-term generator maintenance scheduling. *International Journal of Electrical Power and Energy Systems, 65*, 432–442.

6- Jewski, L.J.K., & Ritzman L.P. (2001). *Operation Management.* (Sixth Edit). Newjersy: Prentic Hall.

7- Karimi, H., Youssef, H., Omar, A. (2010). Adaptive fuzzy APSO based inverse tracking controller with an application DC motor. *Expert System with Application, 36,* 3454–3458.

8- Kennedy J. (1997). The Particle Swarm: Social Adaptation of knowledge. In *Proceeding of 1997 International Conference on Evolutionary Computing.* IEEE press.

9- Kuo, R.J., Syu, Y.J., Chen, Z.Y., Tien, F.C. (2012). Integration of particle swarm optimization and genetic algorithm for dynamic clustering. *Information Sciences, 195*, 124–140.

10- Liu, X., An, H., Wang, L., Lia, X. (2016). An integrated approach to optimize moving average rules in the EUA futures market based on particle swarm optimization and genetic algorithms. *Applied Energy, 185* (2017): 1778-1787.

11- Mansour, R., Bettayeb, M., Djamah, T., & Djnnoun, S. (2008). Fitting fractional System identification using Particle swarm optimization. *Applied Mathematics and Computation, 510–520*, 510–520.

12- Mitchell M. (1996). *An introduction to Genetic Algorithms.* MIT Press, Cambrige, MA. Modares, H. (2013). *Controller design of noise cancellation by active method using algorithm based on particles colony.*

13- Naima, K., Fadela, B., Imene, C., & Abdelkader, C. (2015). USE of Genitic Algorithm and Particle Swarm Optimisation Methods for the Optimal Control of the Reactive Power in Western Algerian Power System. *Energy Procedia, 74*, 265–272.

14- Nunez-Letamendia, L. (2015). Fitting the control parameters of a genetic algorithm an application to technical trading systems design. European *Journal of Operational Research, 179*, 179 (847-868). 2007.

15- Pavão, L., Ravagnani, M. (2016). Heat Exchanger Network Synthesis using a Genetic Algorithms-Particle Swarm Optimization Hybrid Method and Parallel Processing. Computer Aided Chemical Engineering (Vol. 38). Elsevier Masson SAS.

16- Soleimani, H., Kannan, G. (2014). A hybrid particle swarm optimization and genetic algorithm for closed-loop supply chain network design in large-scale networks. *Applied Mathematical Modelling, 39*(14), 3990–4012.

17- Vonderembse, M. A., White, G. P. (2014). *Core Concepts of Operations Management.* West Publishing Company.

18- Wang, Y., Ma, X., Xu, M., Liu, Y., Wang, Y. (2015). Two-echelon logistics distribution region partitioning problem based on a hybrid particle swarm optimization-genetic algorithm. *Expert Systems with Applications, 42*(12), 5019–5031.

19- Yand, F., & Chiang, H.D. (1996). A Parallel genetic algorithm for generation expansion planning. IEEE Trans. *On Power System, 11*(3), 478–486.
20- Zha, Y., Liang, N., Wu, M., Bian, Y. (2016). Efficiency evaluation of banks in China: A dynamic two-stage slacks-based measure approach. *Omega, 60,* 60-72.
21- Zolghadri, M., Oliver, C., Bourriers., J. (2008). Close Optimal Production and Procurement Policy for a X-network of Added Value Using Lexicographic Linear Goal Programming. *Computers and Industrial Engineering, 54*, 821–839.

15

Regression Analysis of Innovation Management in Processing and Complementary Industries of Livestock Products

Ahmadreza Ommani

Abstract

The purpose of research was analysis of innovation management in processing and complementary industries of livestock products. The method of research was correlative descriptive. The statistical population of this study was all managers in processing and complementary industries of livestock products of Khouzestan Province (N=486). The sampling method was stratified random sampling. The sample size was a (n=125). A questionnaire was developed to gather information regarding innovation management in processing and complementary industries of livestock products. Data collected were analyzed using the Statistical Package for the Social Sciences (SPSS). According to the results, the correlation between innovation management with participation in special course, accountability, risk taking, tend to be creative, competitiveness, attitude to innovation management and obtained credit was significant. Based on regression analysis participation in special course, accountability, risk taking, tend to be creative, competitiveness and attitude to innovation management may well explain for 42.9% changes (R^2=0.429) in level of innovation management. According to path analysis, competitiveness and attitude to innovation management were the most important variables influencing the innovation management.

Keywords:
Innovation Management, Regression analysis, Processing and Complementary Industries

[1] *Associate Professor, Department of Agricultural Management, Shoushtar Branch, Islamic Azad University, Shoushtar, Iran.*
* *Corresponding author's email: a.omani@iau-shoushtar.ac.ir*

INTRODUCTION

Agricultural innovation management is widely considered as a key factor in achieving food security through innovative agricultural systems and by raising production and incomes, especially of smallholder farms. Also, innovation allowing companies and firms or enterprises to stay competitive in ever changing world markets. For all of the talk about the importance of innovation, innovation management and creativity in business, the topics are hardly generally well understood (Riederer *et al.*, 2005).

Innovation is not new and it comes in many forms. Also, as many authors argue, it is important for organizations to innovate (Eveleens, 2010, Evitt, 2007, Francis and Bessant, 2005, Hamel and Prahalad, 1994). The innovation process is defined as the development and selection of ideas for innovation and the transformation of these ideas into the innovation (Eveleens, 2010). Andrew *et al.* (2007) argue that the management of an innovation project is essentially like any other business projects, though it comes with more risk and uncertainty.

According to Ommani (2015), the innovation management level in processing and complementary industries of Khuzestan province is not desirable. Therefore, practitioners should be required to provide increasing levels of innovation management. Based on the results of this study variables such as income, competitiveness, accountability, risk taking, tend to be creative, level of education and attitude to innovation management played a critical role in improving innovation management. Improve psychological characteristics such as risk taking, competitiveness, accountability for innovation through workshops, specialized training and scientific visits, will play an important role in the management of innovation.

Agricultural innovation is vital to promoting agricultural and rural development and poverty reduction. Innovation in the agriculture sector is critical to achieving the necessary growth in production in an environmentally sustainable way (Sunding and Zilberman, 2001).

Also, according Ommani (2011), 76% of farmers of Kouzestan Province had moderate to very low knowledge regarding innovation management. In this study, there was a significant relationship between the farmer's knowledge regarding innovation management with accessing to communications channels, level of education, income, crop yield, size of farm, social participation, and level of participation in extension classes. Level of education, income, crop yield, size of farm, social participation, level of participation in extension classes may well explain for 53% (R^2 =0.534) changes in farmer's knowledge regarding innovation management.

Innovations management uses the systems and business to make the organization more innovative. The aim of innovations management is to maintain and improve the competitive position of the business by usage of innovation. Figure 1

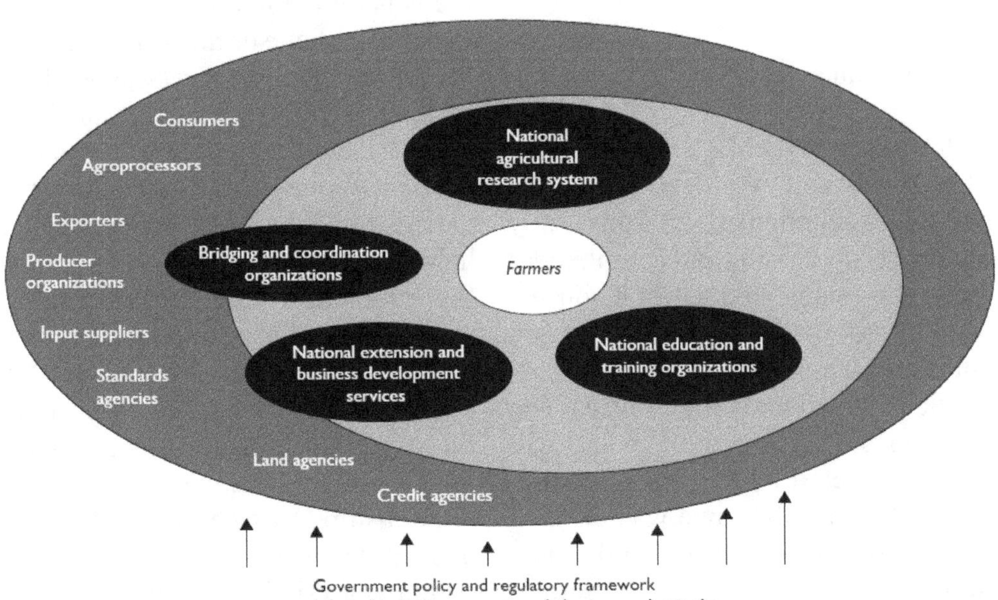

Figure 1: An agricultural innovation system Source (World Bank, 2012).

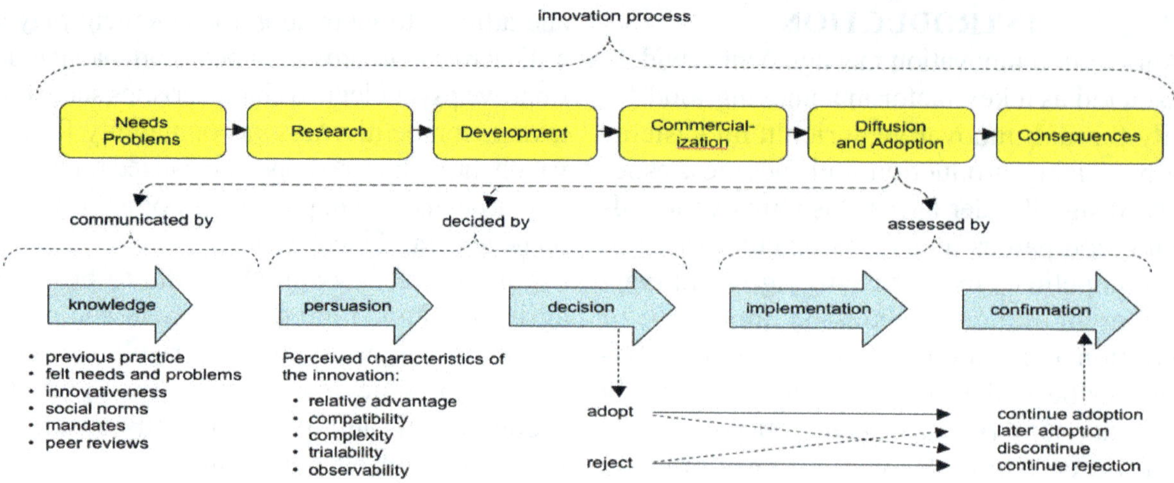

Figure 2: Agricultural innovation process.

presents a conceptual framework for an agricultural innovation system. The figure shows the main actors (typical agricultural knowledge and technology providers and users, as well as the bridging or intermediary institutions that facilitate their interaction); the potential interactions between actors; and the agricultural policies and informal institutions, attitudes, and practices that either support or hinder the process of innovation (World Bank, 2012).

Based on Figure 2 the process of innovation as part of innovation management including:

-Assessment of problem and determining need problem

- Research

-Development

-Commercialization

-Diffusion

-Adoption

-Analyzing consequences

MATERIALS AND METHODS

The method of research was descriptive-correlative. The statistical population of this study was all managers in processing and complementary industries of livestock products of Khouzestan Province (N=486). Using by stratified random sampling, a random sample was selected for participation in the study (n=125). The researchers used survey research method in which data was gathered by the questionnaire. A first step Khouzestan Province was divided into three areas, North, Central and South. Then, in proportion to the population of each section, a sample was selected.

A questionnaire was developed to gather data regarding innovation management in processing and complementary industries of livestock products. To determine the validity of the questionnaire a panel of experts was used and agricultural experts' comments were considered. The questionnaire was pilot tested and reliability was estimated by calculating Cronbach's alpha (Cronbach's alpha=0.89). Reliability was Data collected were analyzed using the Statistical Package for the Social Sciences (SPSS19). In this study, Regression model was used for data analysis. Regression analysis is a statistical procedure for estimating the relationships among variables. It includes many techniques for modeling and analyzing several variables, when the focus is on the relationship between dependent variable and one or more independent variables. More specifically, regression analysis helps one understand how the typical value of the dependent variable (or 'criterion variable') changes when any one of the independent variables is varied, while the other independent variables are held fixed. Regression analysis is widely used for prediction and forecasting. Regression analysis is also used to understand which among the independent variables are related to the dependent variable, and to explore the forms of these relationships. In restricted circumstances, regression analysis can be used to infer causal relationships between the independent and dependent variables.

Path analysis is used to describe the directed

dependencies among a set of variables. This includes models equivalent to any form of multiple regression analysis, as well as more general families of models in the multivariate analysis of variance and covariance.

For assessment the level of innovation management in processing and complementary industries of livestock products were used seven subsystems. The items of each subsystem in Likert scale were analyzed.

RESULTS

Descriptive result

The ages of the respondents ranged from 24 to 67 years. The mean age was 42 (SD=7.87, n=125). The majority (38.4%, n=48) of respondent were 41-50 years old. Most of the respondents in the study were male (88%) and only 15 persons (12%) were female. Three person of managers had a doctoral degree and 64% (n = 80) of respondents were a bachelor's degree holders.

Correlation study

Based on the results status of all subsystems of innovation management were moderate. For assessment of correlation between dependent and independent variables was used spearman correlation coefficient. Based on results, correlation between innovation management and participation in special classes, accountability, risk taking, tendency to be creative, competitiveness, attitude to innovation management and obtained credit was significant (Table 1).

Regression analysis

Based on regression analysis, participation in special courses, accountability, risk taking, tend to be creative, competitiveness and attitude to innovation management may well explain for 42.9% changes (R^2=0.429) in level of innovation management. Based on Table 2, we can see that the predictor variables of participation in special course, risk taking, competitiveness and attitude are significant because their p-values are <0.01. However, the p-value for tendency to be creative (0.731) and accountability (0.967) are greater than the common alpha level of 0.05, which indicates that it is not statistically significant.

In consideration to Variance Inflation Factor (VIF), we can argue about co-linearity statistics. If VIF is less than 10, co-linearity will not be significant. According to results, it is considered

Table 1: correlation between dependent and independent variables.

Predictor Variables	Response Variable	r_s	p-value
Income		0.154	0.086
Age		0.017	0.853
Level of Education		0.121	0.181
Experience		0.118	0.190
Participation in special courses		0.287	0.001
Tendency to be creative	Innovation management	0.373	0.000
Risk taking		0.420	0.000
Accountability		0.197	0.000
Competitiveness		0.480	0.000
Attitude		0.439	0.000
Credit		0.198	0.000

Table 2: Regression analysis between dependent and independent variables.

Independent Variables	B	SE B	Beta	t	p-value
Participation in special courses	8.493	2.406	0.270	3.531	0.001
Tendency to be creative	0.940	2.728	0.031	0.345	0.731
Risk taking	8.249	2.554	0.256	3.230	0.002
Accountability	0.126	3.044	0.003	0.042	0.967
Competitiveness	10.74	2.837	0.325	3.804	0.000
Attitude	0.733	0.232	0.243	3.164	0.002
Constant	-41.06	14.04	----	-2.925	0.004

amount of co-linearity is less than 10 for predictor variable in the last stage of regression analysis. Considering to quantity of beta (ß) can be arbitrated ratio and proportion predictor variables in explanation of dependent variable. Quantities of beta (the fourth column of Table 2) show that per unit of variation in participation in special course, risk taking, competitiveness and attitude can be varied standard deviation of dependent variable.

Path analysis

In addition to being thought of as a form of multiple regression focusing on causality, path analysis can be viewed as a special case of structural equation modeling (SEM) – one in which only single indicators are employed for each of the variables in the causal model. That is, path analysis is SEM with a structural model, but no measurement model. Other terms used to refer to path analysis include causal modeling, analysis of covariance structures, and latent variable models (Figure 3). A path coefficient indicates the direct effect of a variable assumed to be a cause on another variable assumed to be an effect. Path coefficients are standardized because they are estimated from correlations (a path regression coefficient is unstandardized). Path coefficients are written with two subscripts (Table 3).

Effects of independent variables on dependent variable:

1) Participation in special courses on innovation management:

Direct Effect= P_{71}=0.270

2) Tendency to be creative on innovation management:

Direct Effect =P_{72}=0.031

Indirect Effect = P_{62}×P_{76}=0.112×0.325= 0.036

Total effect-0.031+0.036=0.067

3) Risk taking on innovation management:

Table 3: Effects of independent variables on dependent variable.

Independent Variables	Direct effect	Indirect effect	Total Effects
Participation in special courses	0.270	---	0.270
Tendency to be creative	0.031	0.036	0.067
Risk taking	0.256	0.044	0.300
Accountability	0.003	0.098	0.101
Competitiveness	0.325	---	0.325
Attitude	0.243	0.093	0.336

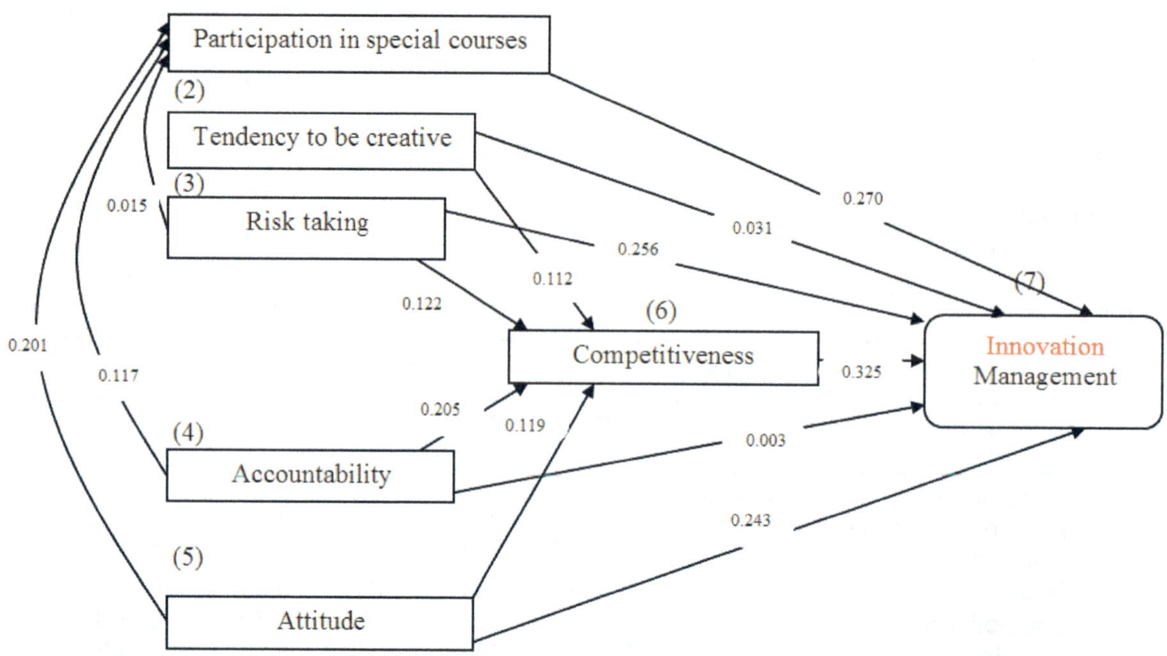

Figure 3: Path analysis of innovation management.

Direct Effect= P_{73}=0.256

Indirect Effect= ($P_{63}\times P_{76}$)+ ($P_{13}\times P_{71}$) = 0.122 × 0.325+0.015×0.270=0.04+0.004=0.044

Total effect-0.256+0.044=0.3

4) Accountability on innovation management:

Direct Effect= P_{73}=0.003

Indirect Effect= ($P_{64}\times P_{76}$)+ ($P_{14}\times P_{71}$) = 0.205× 0.325+0.117×0.270=0.098

Total effect-0.003+0.098=0.101

5) Attitude on innovation management:

Direct Effect= P_{73}=0.243

Indirect Effect= ($P_{65}\times P_{76}$) + ($P_{15}\times P_{71}$) = 0.119× 0.325+0.201×0.270=0.093

Total effect-0.243+0.093=0.336

DISCUSSION AND CONCLUSION

In today's competitive world, innovation is considered as a key factor to success and survival for organizations. According to the results, the correlation between innovation management with participation in special courses, accountability, risk taking, tend to be creative, competitiveness, attitude to innovation management and innovation management was significant. Based on regression analysis participation in special classes, accountability, risk taking, tend to be creative, competitiveness and attitude to innovation management may well explain for 42.9% changes (R^2=0.429) in level of innovation management. The some of this finding was supported by Bylin *et al.* (2004), Coash *et al.* (2003), Fulton *et al.* (2003) Kwamena (2008), Quinn (1999), Reeve and Black (1998), Riederer *et al.* (2005), Ryan and Oestreich (1991) observed incidences where a risk-averse attitude became the organizational norm due to an atmosphere of perceived fear, resulting in decreased innovation.

Improve psychological characteristics such as risk taking, competitiveness, accountability for innovation through workshops, specialized training and scientific visits, will play an important role in the management of innovation. In the current study, analysis of the results demonstrates a positive and significant relationship between the dimensions of psychological characteristics and innovation management. This finding is in line with previous studies (e.g. Abbas and Raja, 2015; Jafri, 2012; Rego *et al.* 2012).

ACKNOWLEDGEMENT

This research was supported by the Shoushtar Branch Islamic Azad University of the Islamic Republic of Iran through a grant to Ahmad Reza Ommani. The author wants to express their gratitude for this support.

REFERENCES

1- Abbas, M., & Raja, U. (2015). Impact of psychological capital on innovative performance and job stress. *Canadian Journal of Administrative Sciences,* Article first published online: 17 APR 2015 DOI: 10.1002/cjas.1314.

2- Andrews, J.P., Sirkin, H.L., Haanaes, K., & Micheal, D. C. (2007). *Innovation 2007 ABCG Senior Management Survey on Innovation*, Boston consulting group Inc. Boston, USA.

3- Bylin, C., Misra, R., Murch, M., & Rigterink, W. (2004). *Sustainable agriculture: development of an on farm assessment tools. Center for sustainable systems,* Report No. CSS04-03, University of Michigan, Ann Arbor, Michigan.

4- Coash, A., Hughes, A., Bullock, A., & Potton, M. (2003). *The relationship between training and business performance*, ESRC Centers for Business Research, University of Cambridge, Cambridge, England.

5- Eveleens, C. (2010). Innovation Management: A Literature Review of Innovation Process Models and Their Implications. Nijmegen, NL: Lector at Innovative Publieke Sector. http://www.scribd.com /doc/ 72251473/Innovation-Management-Literature-Review#scribd

6- Evitt, F. (2007). *Modelling the innovation process: A multi-case comparison.* Unpublished thesis Unitec Institute of Technology New Zealand. Retrieved from http://unitec.researchbank.ac.nz/ bitstream /handle/10652/1257/fulltext.pdf?...1

7- Francis, D., & Bessant., J. (2005). Targeting innovation and implications for capability development, *Technovation*, 25(3), 171-183.

8- Fulton, A., Fulton, D., Tabart, T., Ball, P., Champion, S., Weatherley, J., & Heinjus, D. (2003). *Agricultural extension, learning and change.* Report prepared for the Rural Research and Development Corporation. RIRDC Publication, (03/032).

9- Hamel, G., & Prahalad, C. (1994). *Competing for the Future.* Boston: Harvard Business School Press.

10- Jafri, H. (2012). Psychological capital and innovative behavior: An empirical study on apparel fashion industry. *The Journal Contemporary Management Research,* 6(1), 42–52.

11- Kwamena, A. (2008). *Innovation and experience economy.* Regional Studies Association Regions: The Dilemmas of Integration and Competition, 27th -29th May 2008. Prague, Czech Republic. Available on the: http://vbn.aau.dk/files/16527264/conference _ pa per.pdf

12- Ommani, A. R. (2011). Assessment of farmers' knowledge regarding innovation management in farming cooperatives in Shoushtar Township, Iran. *Journal of American Science*,7(1):321-324

13- Ommani, A. R. (2015). Analyze of predictive model of innovation management in processing and complementary industries of livestock products. *International Journal of Agricultural Management and Development*, 5(1), 27-32.

14- Quinn, G. F. M. (1999). *Genetic and production improvement in the Australian beef cattle industry.* Armidale: University of New England Press.

15- Reeve, I. J., & Black, A. W. (1998). *Improving farmers' management practices - through learning and group participation*. RIRDC: Canberra.

16- Rego, A., Sousa, F., Marques, C., & Cunha, M. (2012). Authentic leadership promoting employees' psychological capital and creativity. *Journal of Business Research,* 65(3),429–437.

17- Riederer, J. P., Baier, M., & Graefe, G. (2005). Innovation management–An overview and some best practices. *C-Lab report*, 4(3), 1-58.

18- Ryan, K. D., & Oestreich, D. K. (1991). *Driving fear out of the workplace.* San Francisco: Jossey Bass.

19- Sunding, D., & Zilberman, D. (2001). The agricultural innovation process: research and technology adoption in a changing agricultural sector. *Handbook of Agricultural Economics*, 1, 207-261.

20- World Bank. (2012). Agricultural innovation systems an investment sourcebook. The World Bank 1818 H Street NW Washington DC 20433. Retrieved from https://openknowledge.worldbank.org/bitstream/handle/ 10986/ 2247/ 672070PUB0EPI0067844B09 780821386842.pdf?sequence=1

16

A Comparative Study of Energy Use and Greenhouse Gas Emissions of Canola Production

Mehdi Khojastehpour [1], Amin Nikkhah [2] and Davood Hashemabadi [3]*

Abstract

In this research, the energy flow and production energy indices of canola cultivation in Trakya province of Turkey, Golestan and Mazandaran provinces of Iran were compared. Diesel fuel and chemical fertilizer inputs were the highest consumer of energy in the production of canola in these three regions. The results indicated that despite the higher energy use of machinery in Trakya province of Turkey, the energy use of diesel fuel in this province is less than the energy consumed in the two northern provinces of Iran, which could be due to available old machines for the production of canola in Mazandaran and Golestan provinces. Total greenhouse gas emissions of canola production for these regions were computed 562.85, 652.86 and 887.30 kgCO_2eq ha^{-1}, respectively. The inputs of chemical fertilizer and diesel fuel in canola production produced the highest percentage of gas emissions in these three areas. Energy consumption for potential feedstock production for one kg production of biodiesel fuel in these provinces was calculated as 14.76, 20.66 and 37.77 MJ, respectively. The amounts of greenhouse gas emissions for potential feedstock production for one kg production of biodiesel were obtained 0.45, 0.76 and 1.17 kgCO_2eq for Trakya, Golestan and Mazandaran provinces, respectively.

Keywords:
Carbon dioxide emission, Energy indices, Feedstock for biodiesel, Iran, Turkey

[1] *Department of Biosystems Engineering, Ferdowsi University of Mashhad, Mashhad, Iran.*
[2] *Young Researchers and Elite Club, Rasht Branch, Islamic Azad University, Rasht, Iran.*
[3] *Department of Horticultural Science, Rasht Branch, Islamic Azad University, Rasht, Iran.*
* *Corresponding author's email: mkhpour@um.ac.ir*

INTRODUCTION

Canola is an important industrial oilseed crop (Rempel *et al.*, 2014). It is now cultivated in many agricultural regions worldwide (Kant *et al.*, 2014). The canola growing area increased from 119,000 ha in 2004 to 170,000 ha in 2012 in Iran (Food and Agriculture Organization, 2012). The highest share of canola growing area in Iran belongs to Golestan (27.4%) followed by Mazandaran province (17.3%) (Ministry of Jihad-e-Agriculture of Iran, 2012). Moreover, the Turkish canola growing area increased from 1,700 ha in 2004 to 10,000 ha in 2007 and to 30,000 ha in 2012 (FAO, 2012). Canola is a potential feedstock for biodiesel fuel production (Ahmad *et al.*, 2011; Ghobadian, 2012). Sims and Sayigh (2003) reported that the all forms of bioenergy such as biodiesel when substituted for fossil fuels, will directly reduce carbon dioxide emissions. From an economic point of view for producing energy from edible oil seeds, there are five major items that should be considered: increasing crop yield per unit area, assessment of environmental benefits, reduce the consumption of inputs, increasing energy efficiency and reduce the lost opportunity cost of crop land (Nikkhah *et al.*, 2015; Safieddin-Ardebili *et al.*, 2011; Sims and Sayigh, 2003).

Due to the nature of the energy outputs of various crops, the comparison of energy efficiency, energy productivity, specific energy and net energy cannot give us complete information. Asgharipour *et al.* (2012) performed a research on the energy of sugar beet production in Iran. For sugar beet, energy efficiency, productivity energy, specific energy and net energy were 13.4, 0.8 kgMJ^{-1}, 1.3 MJkg^{-1}, 521,413.7 MJha^{-1}, respectively. In another study that carried out by Salehi *et al.* (2014), these indices for button mushroom production were reported 0.028, 0.017 kgMJ^{-1}, 59.5 MJkg^{-1}, -875 MJha^{-1} respectively. As a consequence, best results are obtained when the comparison of production of one crop in various regions performed.

Ozkan *et al.* (2004) performed a study on the energy use in the Turkish agricultural sector for the period of 1975–2000; they concluded that total energy input increased from 17.4 GJha^{-1} in 1975 to 47.4 GJha^{-1} in the year 2000. Moreover, total output energy increased from 38.8 to 55.8 GJha^{-1} in the same period. The energy efficiency was found to be 2.23 in 1975 and 1.18 in 2000. In a similar study that carried out by Beheshti-Tabar *et al.* (2010), the energy performance of Iranian agricultural systems during 1990-2006 years was evaluated. They reported that total energy input increased from 32.40 GJ ha^{-1} in 1990 to 37.20 GJ ha^{-1} in 2006. Total output energy increased from 30.85 to 43.68 GJ ha^{-1} at the same period. Irrigation with 40.0% and fertilizer with 28.4% had the highest share in energy use. Several studies have been performed on energy consumption of crop production in Iran and Turkey after 2006 year, showing that diesel fuel and chemical fertilizers have the largest share in energy consumption (AghaAlikhani *et al.*, 2013; Banaeian *et al.*, 2011; Çetin and Vardar, 2008; Erdal *et al.*, 2007; Heidari and Omid 2011; Mohammadi and Omid 2010; Mohammadi *et al.*, 2008; Samavatean *et al.*, 2011).

Several researches have been conducted on energy and GHG emissions in different agricultural crops production. Yilmaz *et al.* (2005) analyzed the energy consumption of cotton production in Turkey and found the total energy use 49.73 Gjha^{-1} and the diesel fuel input was the highest GHG emissions input and followed by chemical fertilizers GHG emissions. In a similar study, carried out by Pishgar-Komleh *et al.* (2012) energy use and GHG emissions of cotton production was performed and concluded that chemical fertilizers and diesel fuel were the most influential factors in energy consumption. The inputs of machinery and diesel fuel in cotton production produced the highest percentage of carbon dioxide emission in Alborz province of Iran. In a another study conducted by Ramedani *et al.* (2011) total energy consumption of soybean production in Golestan province of Iran was obtained 18026 MJha^{-1} and in between the different energy sources, the highest share of energy consumption belonged to diesel fuel (67%) followed by chemical fertilizers (14%). In a study Liu *et al.* (2013) selected five scenarios related to cassava (based fuel ethanol) planting modes to evaluate the en-

Table 1: Greenhouse gas emission coefficients

Inputs	Unit	GHG emissions coefficients ($kgCO_2eq\ unit^{-1}$)	Reference
Agricultural machinery	MJ	0.071	(Dyer and Desjardins, 2006)
Diesel fuel	lit	2.76	(Dyer and Desjardins, 2003)
Chemical Fertilizer			
N	kg	1.3	(Lal, 2004)
P_2O_5	kg	0.2	(Lal, 2004)
K_2O	kg	0.2	(Lal, 2004)
Biocide			
Fungicides	kg	3.9	(Lal, 2004)
Insecticides	kg	5.1	(Lal, 2004)
Herbicides	kg	6.3	(Lal, 2004)
Electricity	kWh	0.608	(Lal, 2004)

ergy balance and GHG emissions. They showed that, although all the five cassava scenarios have positive net energy values and GHG emissions savings in comparing with the conventional gasoline, the planting modes have significant effect on their energy and carbon dioxide emission. Rajaefar *et al.* (2014) analyzed the energy use and GHG emission of biodiesel production from soybean in Iran. The biodiesel production system consisted of six stages: soybean production, soybean transportation, soybean crushing, biodiesel conversion, and transportation. They reported that agricultural soybean production stage ranked the first in energy consumption among the five main stages where it used 50.56% of total fossil energy consumption in the biodiesel production.

Considering there is no published document in terms of energy use and GHG emission from potential feedstock (canola) production of biodiesel fuel in Iran and Turkey, this study was aimed to compare the energy consumption and GHG emissions of canola production in Trakya province of Turkey, Golestan and Mazandaran provinces of Iran.

MATERIALS AND METHODS

Site description and data collection

This study was conducted in Mazandaran and Golestan provinces of Iran and Trakya province of Turkey. Mazandaran and Golestan provinces are located in north of Iran. The data for energy inputs and output and energy indices were collected from the results of other studies (Mousavi-Avval *et al.*, 2011; Taheri-Garavand *et al.*, 2010; Unakitan *et al.*, 2010).

Greenhouse gas emission

The greenhouse gas emissions were calculated for Trakya province of Turkey, Golestan and Mazandaran provinces. In these regions, the general inputs were seed, human labor, machinery, diesel fuel, chemical fertilizer electricity and biocides and the output was canola yield. The amount of GHG emissions calculated by multiplying the input application rate (seed, human labor, machinery, diesel fuel, chemical fertilizer, electricity and biocide) by its corresponding emission coefficient are shown in Table 1. The energy use of irrigation water was converted to the amount of electricity to gain the total CO_2 emissions in irrigation water input by multiplying the electricity consumption by GHG coefficient (Khoshnevisan *et al.*, 2013). GHG emissions of canola production were determined per each hectare of land, each ton of canola grain and per 1000 MJ of total energy output.

Energy and GHG emission from potential

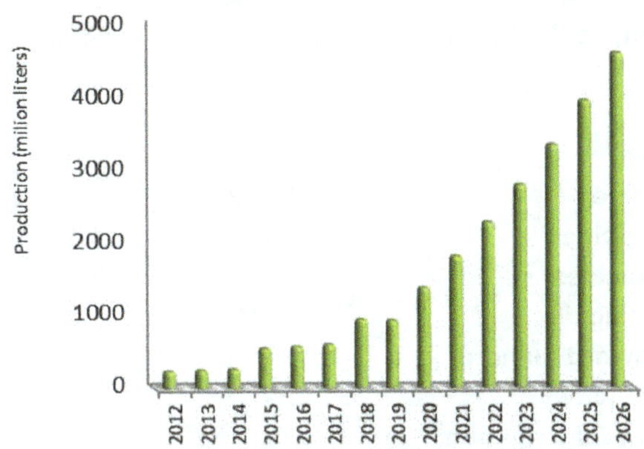

Figure 1: The outlook of biodiesel fuel share in Iran's transportation (Ghobadian, 2012)

feedstock production of biodiesel

The outlook of biodiesel fuel share in Iran's transportation is shown in Figure 1. Biodiesel production in Iran will increase during future years. There is a growing interest for biodiesel production in this region. Moreover, there is a high potential for biodiesel production from edible oil seeds in Iran. Canola is an edible oil seed and potential feedstock for biodiesel fuel production. Also, Turkey plans to widely expand its use of renewable energy, especially use of edible oil for biodiesel production (Acaroğlu and Aydoğan, 2012; Aytav and Kocar, 2013). In this study, canola weight conversion to biodiesel coefficients according to the study by Safieddin-Ardebili *et al.* (2011) was used for estimation. The energy input in cropping systems can be classed into two main categories (Salehi *et al.*, 2014): a) direct energy: energy inputs including: human labor, diesel fuel and electricity b) indirect energy: energy inputs including: biocide, chemical fertilizer, seed and agriculture machinery (Kuswardhani *et al.*, 2013; Soltani *et al.*, 2013).

RESULTS AND DISCUSSION

Energy inputs-output in canola production

The amounts of inputs-output energy for canola production in Trakya province of Turkey, Golestan and Mazandaran provinces of Iran are shown in Table 2. The highest share of energy consumption belongs to chemical fertilizers followed by diesel fuel in the production of canola in these three regions. Human labor used for canola production was as 21.93, 154.79 and 36.14 h ha-1, respectively. The inputs of biocide energy use of canola production in Trakya province is less than the energy consumed in the two northern provinces of Iran. Moreover, the inputs of human labor energy consumption in Trakya province is less than the energy consumed in the two northern provinces of Iran and the energy use of machinery in Trakya province is higher than the energy consumed in the two northern provinces of Iran. It can be concluded that the mechanization degree of canola production in Trakya province of Turkey is high.

As seen in Table 2, despite the higher energy use of machinery in Trakya province, the energy use of diesel fuel in this province is less than the energy consumed in the two northern provinces of Iran, which could be due to available old machines for the production of canola in Mazandaran and Golestan provinces.

Total energy inputs for canola production in Trakya province of Turkey, Golestan and Mazandaran provinces were obtained 18297.61, 17786.36 and 28705.31 kg CO_2eq ha^{-1}, respec-

Table 2: Energy inputs and output for canola production in Iran and turkey

	Trakya province (Turkey)		Golestan province (Iran)		Mazandaran province (Iran)	
	Energy (MJha^{-1})	Percentage	Energy (MJha^{-1})	Percentage	Energy (MJha^{-1})	Percentage
Seed	129.06	0.71	29.61	0.17	155.37	0.54
Human Labor	42.99	0.23	154.79	0.87	70.84	0.25
Agricultural machinery	1071.03	5.85	958.27	5.54	930.55	3.24
Diesel fuel	4473.83	24.45	4850.65	27.27	8604.17	29.97
Chemical Fertilizer	11823.37	64.62	8648.36	48.62	18809.84	65.53
N	11489.76	62.79	8368.88		17572.50	61.22
P_2O_5	333.61	1.82	630.07		1237.34	4.31
K_2O	-	-	149.35		-	-
Biocide	757.32	4.14	537.56	3.02	134.54	0.47
Fungicides	216.00	1.18	190.41			
Herbicides	491.28	2.68	303.55			
Insecticides	50.04	0.27	43.59			-
Farmyard manure	-	-	487.16			-
Electricity	-	-	2326.57	13.08	-	
Water for irrigation	-	-	253.56	1.43	-	
Total energy input	18297.61		18786.63		28705.31	
Total energy output	85556.96		53798.46		41230	

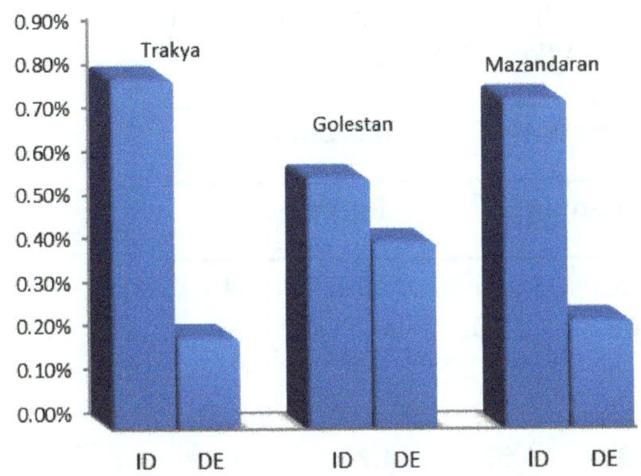

Figure 2: The percentage of direct energy and indirect energy for canola production in Iran and Turkey [1]

tively. Despite the use of electricity in Trakya province, the total energy inputs of canola production in Trakya region is less than the energy consumption in the Mazandaran and Golestan provinces. In addition, the energy output for canola production in Trakya region was higher than the energy output in the Mazandaran and Golestan provinces.

Table 3 presents the energy indices for canola production in these three regions. Average energy efficiency in Trakya province of Turkey, Golestan and Mazandaran provinces were 4.68, 3.02 and 1.44 respectively. The amount of energy efficiency of canola production was more than this index of energy for the production of canola in Mazandaran and Golestan provinces. The highest amount of canola yield belongs to canola production in Trakya province followed by canola production in Golestan province of Iran. The amounts of energy efficiency of canola production in Trakya province of Turkey, Golestan and Mazandaran provinces of Iran were 0.17, 0.12 and 0.066 kgMJ^{-1}, respectively. The share of direct and indirect energy of canola production is shown in Figure 2. The average direct and indirect forms of energy for canola production in Mazandaran province were 20% and 80%, respectively. The share of direct energy for canola production in Mazandaran province of Iran was less than that of canola production in Golestan province of Iran and Trakya province of Turkey.

GHG emissions of canola production

The GHG emissions of machinery input for canola production in Trakya province was more than that of producing canola in Mazandaran and Golestan provinces. In addition, GHG emissions from the diesel fuel for the same crop production in Trakya province was less than this volume in producing canola in the two northern provinces of Iran. The GHG emissions of chemical fertilizer from canola production in Golestan province was less than that of producing canola in Mazandaran and Trakya provinces.

Total GHG emissions of canola production were calculated as 562.85, 652.86 and 887.30 kgCO_2eq ha^{-1}, respectively (Table 4). The GHG emissions from canola production in Mazandaran province were high. The highest share of GHG emissions belongs to diesel fuel followed by chemical fertilizer in these three areas. Diesel fuel was accounted for 56% of total GHG emissions. The amounts of GHG emissions from canola production in Trakya province of Turkey, Golestan and Mazandaran provinces of Iran for 1000 MJ generation of energy were calculated as 6.58, 10.46 and 21.52 kgCO_2eq respectively. Moreover, the amounts of GHG emissions from canola production in these three regions were

Table 3: Energy indicators and forms for canola production in Iran and turkey

	Unit	Trakya province (Turkey)	Golestan province (Iran)	Mazandaran province (Iran)
Energy efficiency	-	4.68	3.02	1.44
Energy productivity	kg MJ^{-1}	0.17	0.12	0.066
Specific energy	MJ kg^{-1}	5.90	8.27	15.1
Net energy	MJ ha^{-1}	67259.36	36012.09	12524.69
Direct energy	MJ ha^{-1}	4516.82	7585.56	8674.74
Indirect energy	MJ ha^{-1}	13780.78	10200.80	20030.57
Renewable energy	MJ ha^{-1}	172.06	952.12	28478.54
Non-renewable energy	MJ ha^{-1}	18125.55	16861.24	226.77

[1] *DE: direct energy and ID: indirect energy*

Table 4: Greenhouse gas emissions of canola production in Iran (Golestan and Mazandaran provinces) and Turkey (Trakya province) ($kgCO_2eq\ ha^{-1}$)

	Trakya		Golestan		Mazandaran	
		Percentage		Percentage		Percentage
Agricultural machinery	76.04	13.51	69.95	10.71	66.07	7.45
Diesel fuel	219.28	38.96	280.08	42.90	496.81	56
Chemical Fertilizer						
N	246.48	43.79	144.83	22.18	307.87	34.70
P_2O_5	5.98	1.06	10.13	1.55	14.22	1.60
K_2O	-	-	2.68	0.41	-	
Biocide	15.06	2.67	13.69	2.10	2.33	0.26
Electricity	-	-	118.57	18.16	-	
Water for irrigation	-	-	12.92	1.98	-	
Total GHG emissions	562.85		652.86		887.30	

Table 5: GHG emission of canola production in Iran and Turkey ($kgCO_2eq$ per $unit^{-1}$)

	1000 kg	1000 MJ
Trakya province of Turkey	181.57	6.58
Golestan province of Iran	303.38	10.46
Mazandaran province of Iran	467	21.52

calculated as 181.75, 303.38 and 467 $kgCO_2eq$ $Tone^{-1}$, respectively (Table 5).

Analysis of energy and GHG emissions from potential feedstock production of biodiesel

Energy consumption for potential feedstock production for one kg production of biodiesel fuel in these provinces was calculated as 14.76, 20.66 and 37.77 MJ, respectively. The amounts of greenhouse gas emissions for potential feedstock production as one kg production of biodiesel were obtained 0.45, 0.76 and 1.17 $kgCO_2eq$ for Trakya, Golestan and Mazandaran provinces, respectively. According to the study was done by Dyer and Desjardins (2003), the GHG emissions for the combustion of each liter of diesel fuel are 2.76 $kgCO_2eq$. If we are going to use the canola to produce biodiesel, we should consider the inputs of canola cultivation systems in terms of environmental management.

CONCLUSIONS

Based on the present study the following conclusions are drawn:

- Diesel fuel and chemical fertilizers had the largest share in energy use and greenhouse gas emissions from canola production in Trakya province of Turkey, Golestan and Mazandaran provinces of Iran.
- Despite the higher energy use of machinery in Trakya province of Turkey, the energy use of diesel fuel in this province is less than the energy consumed in the two northern provinces of Iran.
- The energy consumption and GHG emissions from canola production in Trakya province were less than that of producing canola in Mazandaran and Golestan provinces.
- The amounts of greenhouse gas emissions for potential feedstock production for one kg production of biodiesel were obtained 0.45, 0.76 and 1.17 $kgCO_2eq$ for Trakya, Golestan and Mazandaran provinces, respectively.

REFERENCES

1- Acaroğlu, M., & Aydoğan, H. (2012). Biofuels energy sources and future of biofuels energy in Turkey. *Biomass and Bioenergy,* 36, 69-76.

2- Aghaalikhani, M., Kazemi-Poshtmasari, H., & Habibzadeh, F. (2013). Energy use pattern in rice production: A case study from Mazandaran province, Iran. *Energy Conversion and Management*, 69, 157-162.

3- Ahmad, AL., Yasin, N.H.M., Derek, C.J.C., & Lim, J.K. (2011). Microalgae as a sustainable energy source for biodiesel production: A review. *Renewable and Sustainable Energy Reviews*, 15,584-593.

4- Asgharipour, M.R., Mondani, F., & Riahinia, S. (2012). Energy use efficiency and economic analysis of sugar beet production system in Iran: A case study in Khorasan Razavi province. *Energy*, 44,

1078-1084.
5- Aytav, E., & Kocar, G. (2013). Biodiesel from the perspective of Turkey: Past, present and future. *Renewable and Sustainable Energy Reviews*, 25, 335-350.
6- Banaeian, N., Omid, M., & Ahmadi, H., (2011). Energy and economic analysis of greenhouse strawberry production in Tehran province of Iran. *Energy Conversion and Management*, 52, 1020-1025.
7- Beheshti-Tabar, I., Keyhani, A., & Rafiee, S. (2010). Energy balance in Iran's agronomy (1990–2006). *Renewable and Sustainable Energy Reviews*, 14, 849-855.
8- Çetin, B., & Vardar, A. (2008). An economic analysis of energy requirements and input costs for tomato production in Turkey. *Renewable Energy*, 33, 428-433.
9- Dyer, J.A., & Desjardins, R.L. (2003). Simulated farm fieldwork, energy consumption and related greenhouse gas emissions in Canada. *Biosystems Engineering,* 85, 503-513.
10- Dyer, J.A., & Desjardins, R.L. (2006). Carbon Dioxide Emissions associated with the manufacturing of tractors and farm machinery in Canada. *Biosystems Engineering,* 93, 107-118.
11- Erdal, G., Esengün, K., Erdal, H., & Gündüz, O. (2007). Energy use and economical analysis of sugar beet production in Tokat province of Turkey. *Energy*, 32, 35-41.
12- Food and Agriculture Organization (FAO). (2012). FAOSTAT, Available on: www.fao.org
13- Ghobadian, B. (2012). Liquid biofuels potential and outlook in Iran. *Renewable and Sustainable Energy Reviews*, 16, 4379-4384.
14- Heidari, M.D., & Omid, M. (2011). Energy use patterns and econometric models of major greenhouse vegetable productions in Iran. *Energy*, 36, 220-225.
15- Kant, S., Burch, D., Ehrenberger, W., Bitter, R., Rüger, S., Mason, J., Zimmermann, U., & Spangenberg, G. (2014). A novel crop water analysis system: identification of water stress tolerant genotypes of canola (*Brassica napus* L.) using non-invasive magnetic turgor pressure probes. *Plant Breeding*: doi:10.1111/pbr.12187.
16- Khoshnevisan, B., Rafiee, S., Omid, M., Yousefi, M., & Movahedi, M. (2013). Modeling of energy consumption and GHG (greenhouse gas) emissions in wheat production in Esfahan province of Iran using artificial neural networks. *Energy*, 52, 333-338.
17- Kuswardhani, N., Soni, P., & Shivakoti, G. P. (2013). Comparative energy input–output and financial analyses of greenhouse and open field vegetables production in West Java, Indonesia. *Energy*, 53, 83-92.
18- Lal, R. (2004). Carbon emission from farm operations. *Environment International,* 30, 981-990.
19- Liu, B., Wang, F., Zhang, B., & Bi, J. (2013). Energy balance and GHG emissions of cassava-based fuel ethanol using different planting modes in China. *Energy Policy,* 56, 210-220.
20- Ministry of Jihad-e-Agriculture of Iran (MAJ). (2012). Annual Agricultural Statistics. Retrieved from http://www.maj.ir (In Persian).
21- Mohammadi, A., & Omid, M. (2010). Economical analysis and relation between energy inputs and yield of greenhouse cucumber production in Iran. *Applied Energy*, 87, 191-196.
22- Mohammadi, A., Tabatabaeefar, A., Shahin, S., Rafiee, S., & Keyhani, A. (2008). Energy use and economical analysis of potato production in Iran a case study: Ardabil province. *Energy Conversion and Management,* 49, 3566-3570.
23- Mousavi-Avval, S.H., Rafiee, S., Jafari, A., & Mohammadi, A. (2011). Energy flow modeling and sensitivity analysis of inputs for canola production in Iran. *Journal of Cleaner Production*, 19, 1464-1470.
24- Nikkhah, A., Khojastehpour, M., Emadi, B., Taheri-Rad, A., & Khorramdel, S. (2015). Environmental impacts of peanut production system using life cycle assessment methodology. *Journal of Cleaner Production*, doi:10.1016/j.jclepro.2014.12.048
25- Ozkan, B., Akcaoz, H., & Fert, C. (2004). Energy input–output analysis in Turkish agriculture. *Renewable Energy*, 29, 39-51.
26- Pishgar-Komleh, S.H., Sefeedpari, P., & Ghahderijani, M. (2012). Exploring energy consumption and CO [sub 2] emission of cotton production in Iran. *Journal of Renewable and Sustainable Energy*, doi: 10.1063/1.4727906
27- Rajaeifar, M.A., Ghobadian, B., Safa, M., & Heidari, M.D. (2014). Energy life-cycle assessment and CO_2 emissions analysis of soybean-based biodiesel: a case study. *Journal of Cleaner Production*, 66, 233-241.
28- Ramedani, Z., Rafiee, S., & Heidari, M.D. (2011). An investigation on energy consumption and sensitivity analysis of soybean production farms. *Energy*, 36, 6340-6344.
29- Rempel, C.B., Hutton, S.N., & Jurke, C.J. (2014). Clubroot and the importance of canola in Canada. *Canadian Journal of Plant Pathology*, 36, 19-26.
30- Safieddin-Ardebili, M., Ghobadian, B., Najafi, G., & Chegeni, A. (2011). Biodiesel production potential from edible oil seeds in Iran. *Renewable and Sustainable Energy Reviews*, 15, 3041-3044.

31- Salehi, M., Ebrahimi, R., Maleki, A., & Ghasemi Mobtaker, H. (2014). An assessment of energy modeling and input costs for greenhouse button mushroom production in Iran. *Journal of Cleaner Production*, 64, 377-383.

32- Samavatean, N., Rafiee, S., Mobli, H., & Mohammadi, A. (2011). An analysis of energy use and relation between energy inputs and yield, costs and income of garlic production in Iran. *Renewable Energy,* 36, 1808-1813.

33- Sims, R.E.H., & Sayigh, A. (2003). Bioenergy options for a cleaner environment: in developed and developing countries. Elsevier Science.

34- Soltani, A., Rajabi, M.H., Zeinali, E., & Soltani, E. (2013). Energy inputs and greenhouse gases emissions in wheat production in Gorgan, Iran. *Energy*, 50, 54-61.

35- Taheri-Garavand, A., Asakereh, A., & Haghani, K. (2010). Energy elevation and economic analysis of canola production in Iran a case study: Mazandaran province. *International Journal of Environmental Sciences*, 1, 236-242.

36- Unakitan, G., Hurma, H., & Yilmaz, F. (2010). An analysis of energy use efficiency of canola production in Turkey. *Energy*, 35, 3623-3627.

37- Yilmaz, I., Akcaoz, H., & Ozkan, B. (2005). An analysis of energy use and input costs for cotton production in Turkey. *Renewable Energy*, 30,145-155.

17

Use of Computer and Internet in Agricultural Extension as Perceived by Extension Workers

***Gholamreza Pezeshki Rad* [1], *Zahra Haji Hashemi* [2*] *and Mohhamad Chizari* [3]**

Abstract

The purpose of this study was to determine computer and Internet use in agricultural extension by Extension Workers (EWs). This study used a descriptive-correlational design. Population for the study consisted of all extension workers (N = 320) in Isfahan Province, Iran. A stratified sampling technique and census was used to select EWs (n = 200). Overall, findings indicate that EWs have access to computers both at work and home. On average, they use computers approximately 6 – 8 hours per week and majority of them use the Internet on average of 1 – 4 hours per week. A majority of them reported "fair to high level" of skills in computer usage. For example, writing CDs, word processing, using computers for file management, and basic computer skills. However, extension agent's report "no to a low level" of skills relative to using excel, installing software, power point and micro soft access, Photoshop, statistical software, were the skills that EWs did not have. It was proved necessary to conduct a systematic assessment of training needs relative to computer and Internet use. Systematic training should be conducted in the areas that Extension Workers perceived to be less skill.

Keywords:
Extension, Agriculture, Information Technology, Applications, Internet

[1] Associate Professor of Agricultural Extension and Education Department of Tarbiat Modares University, Peykanshahr, Tehran, Iran.

[2] PhD. Student of Bu Ali Sina University, Hamedan.

[3] Professor of Agricultural Extension and Education Department of Tarbiat Modares University, Peykanshahr, Tehran, Iran.

** Corresponding author's email: Haji.hashemii@yahoo.com*

INTRODUCTION

The development of computers and advancement in telecommunications provide numerous opportunities to obtain a variety of information – economic, agriculture, news and to use it for decision making (Omidi Najaf Abadi *et al.*, 2009; Bheenick and Brizmohum, 1998). A host of researchers have examined the use of computers and Internet by extension workers in different parts of the world (Ahmadpour and Mirdamadi, 2010; Badragheh *et al.*, 2010; Bheenick and Brizmohum, 1998; Chieochan *et al.*, 2000; Gregg and Irani, 2004). Key findings from these studies suggest that computer and Internet use in agriculture, especially extension services has increased tremendously. This increase is mainly because of the advancement made in information technology.

Access to information by farmers and extension personnel and administers are critical to the agricultural development process and Iran is no exception. However, many challenges face developing nations relative to using information technology. The challenges include: 1) to equip extension centers with necessary facilities such as phone line and computer; 2) establishing AKIS Network – application of information technology; and 3) provide online networks, computer communications and digital interactive multimedia to facilitate dissemination of agricultural technologies (Fami, 2006).

Ommani and Chizari (2006) examined the use of information technology training needed by extension agents in Khuzestan province of Iran. They found that extension agents needed training in using tasks e-mail, the Internet, developing PowerPoint presentations, word processing and SPSS. In addition, they found significant relationship among age, income, position and attitudes toward using information technology. These variables explained almost 78% of the variance in the dependent variable.

Fallah Haghighi (2006) examined factors that influenced the use of information technology in agricultural extension. He identified four factors that influenced attitudes toward using information technologies in Iran. These included: effectiveness of extension, organizational environment, the qualities of the information material provided, and cost effectiveness of technology used in extension.

Ahmadpour and Mirdamadi (2010) identified major challenges in the application of E-learning in agricultural extension services in Iran. By using ordinal factor analysis; they classified factors into six major challenges in the following areas: financial, technical, supporting services, regulatory, cultural and human factors. Finally they discussed challenges and suggested solu-

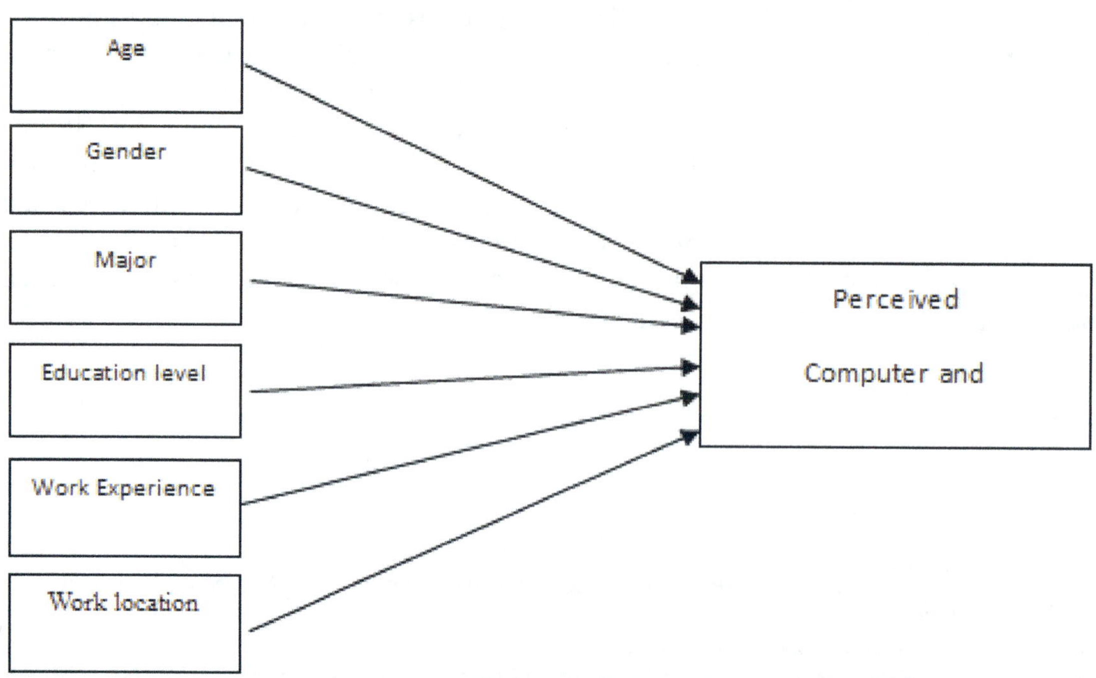

Figure 1: Conceptual Framework for the study

tions. In another study Omidi NajafAbadi *et al.*, (2009) Identified Requirements and Challenges of Information and Communication Technology System to Train Private Agricultural Insurance Brokers in Iran, too.

A number of studies have examined relationships among demographic and work characteristics and use of the computer and Internet (Ommani and Chizari, 2006; Fallah Haghighi, 2006; Akpabio, 2007). Based on this review of literature, a conceptual framework for the study was developed (see Figure 1). This framework identifies key demographic and work variables related to computer and Internet use. These variables include: age, gender, educational level, work experience, work location, and employment status.

Recognizing these factors is critical in expanding the use of information technology. A need exists to identify critical challenges. In Iran, increasing attention is being paid to information technology as an avenue to transfer technology in the agriculture sector. This increased attention is reflected in the recent Rural Information Communication Technology Strategic Plan.

Table1: Frequency and Percentage for Demographic and Job's Characteristics

Demographics Profile	f	%
Gender		
Male	154	87.0
Female	23	13.0
Total	177	100.0
Age*		
30 years and less	42	23.5
31 – 40 years	64	35.8
41 – 50 years	63	35.2
51 years and more	10	5.6
Total	179	100.0
Highest education completed		
High school diploma	18	10.2
Bachelors of art	33	18.6
Bachelors of science	104	58.8
Masters of science	22	12.4
Total	177	100.0
Major area of study		
Extension and education	17	9.5
Agronomy	50	27.9
Animal science	21	11.7
Other agricultural majors	64	35.8
Not related to agricultural major	27	15.1
Total	179	100.0
Work experience in agricultural extension**		
5 years and less	75	41.2
6 – 15 years	68	37.4
16 years and more	39	21.4
Total	182	100.0
Work location		
Provincial level	36	19.7
District (Shahrestan) level	51	27.9
County (Dehestan) level	96	52.5
Total	183	100.0
Employment		
Tenured	87	48.9
Tenured track	91	51.1
Total	178	100.0

*Mean of age = 37.83, SD = 8.16 and range between 23 to 55 years old.

** Mean of work experience = 9.42, SD = 7.57 and range between 1 to 29 years of experience.

Table 2: Frequency and percentage for computer and internet use

Computer and Internet Experience	f	%
Access to computer (including Internet and Web)		
Home	17	9.3
Work	38	20.8
Both	119	65
Laptop	5	2.7
No Access	4	2.2
Total	**183**	**100**
Total number of hours/week use computer (home + work)		
0.0 hour	4	2.2
1 to 4 hours	67	36.8
4.1 to 7 hours	41	22.5
7.1 hours & more	70	38.5
Total	**183**	**100**
Total number of hours/week use Internet (home + work)		
0.0 hour	28	16.6
1 to 4 hours	124	73.4
4 to 7 hours	12	7.1
7 hours & more	5	3
Total	**169**	**100**

Table 3: Means, Standard Deviations for Computer Skills

Competencies	n	Mean*	S.D	Rank
Writing CD	178	3.45	1.18	1
Using Computer and file management	179	3.42	o.97	2
Word processing	172	3.40	1.12	3
Basic Understanding of ICT	179	3.36	0.87	4
Excel	176	3.05	1.29	5
Install software	172	1.15	3.02	6
Presentation such as PowerPoint	171	1.20	2.74	7
Access	170	1.18	2.46	8
Basic problem solving with Software issues	169	1.25	2.23	9
Basic problem solving with Hardware issues	172	1.15	2.07	10
Photo Shop	173	1.05	2.03	11
Statistical software	172	0.99	1.89	12
Summated Scores	**179**	**10.63**	**32.18**	

* Mean computed on a scale 1 = none, 2 = little, 3 = fair, 4 = high

Purpose and objectives

The purpose of this study was to determine computer and Internet use in agricultural extension by Extension Workers (EW) in Isfahan Province of Iran. Specific objectives were:

1. To determine EWs job and demographic characteristics;
2. To determine EWs' computer and Internet use in extension;
3. To determine competencies (computer and Internet) among EW; and
4. To determine differences, if any, between perceived competencies and selected demographic characteristics.

MATERIALS AND METHODS

This study used a descriptive-core_ relational design. The population for the study consisted of all extension workers (N = 320) in Isfahan Province, Iran. Among whom 42 individuals were located in the agricultural organization of Isfahan province and the branch of the county management (N=70), were studied by census. For the selection of the required sample of the centers of

Table 4: Means, Standard Deviations for Internet Skills

Competencies	n	Mean*	S.D	Rank
Send E-mail	173	3.07	1.39	1
Understand E-mail operation	172	3.05	1.39	2
Search (Getting general information from sites)	171	3.01	1.42	3
Search (Getting information from special sites)	173	2.98	1.34	4
Attach File to E-mail	171	2.79	1.46	5
Internet Chat	172	2.34	1.41	6
Group News (Usenet)	172	1.88	1.03	7
Online Discussion	172	1.78	1.01	8
Membership in Internet Libraries	171	1.77	1.08	9
Develop Website	171	1.68	1.07	10
Summated Scores	**175**	**24.03**	**10.73**	

* Mean computed on a scale 1 = none, 2 = little, 3 = fair, 4 = high

providing agricultural services, the appropriate stratified sampling method was used that 200 answered questionnaire were returned, totally. A survey was developed to determine computer competencies and Internet use among EWs.

The survey was assessed for content and face validity by a panel of experts consisting of faculty members in the Departments of Agricultural Education and Extension, Penn State University, and extension officers in the Isfahan Province Agricultural Jihad Organization. The instrument was pilot tested using EWs (n = 30) in Isfahan Province who were not included in the study. Minor changes were made to improve the clarity and readability of the instrument. Cronbach's alpha, an internal consistency measure, was used to estimate the reliability. The reliability coefficient for part two of the instrument was found to be acceptable (alpha = .0.86)

Data were collected via normal mail. Of the 200 surveys mailed, 110 (55 %) were returned within four weeks. A follow up mailing was sent to non-respondents four weeks after the initial mail. As a result, an additional 75 responses were received. In all, 183 (95 %) surveys were returned. Data were analyzed using frequencies, percentages, means, and standard deviation. For the fourth research objective, analysis of variances (ANOVA) was used to compare demographic variables and computer skills.

RESULTS

Objective 1: Demographic Profile of Respondents.

The majority of the respondents were male (see Table 1) and the respondents' age was distributing evenly among three groups (23.5% under 30 years; 36.0% between 31-40 years and 35.0 % between 41-50 years). Three of five respondents (59%) had completed the Bachelor of Science degree. Nearly 28% reported agronomy as their major of study followed by a major other than agriculture (15.1%). 11.7 % had studied in animal sciences and only 9.5 % had agricultural extension and education degree. A little over one-third of the respondents (35.8%) indicated their major area of study in other agricultural subjects such as farm machinery, plant pathology, food science, irrigation engineering, and horticulture. On average, respondents had 9.4 years of work experience and a little over one-half of the respondents (52.5%) had worked in offices at county level.

Objective 2: Use of Computers, and the Internet in Extension

Use of computers and the Internet by respondents was examined using three factors – access to computers, total number of hours per week of computer use and total number of hours per week on the Internet (Table 2). Almost three out of every four respondents (73.4%) used the Internet one-to four hours per week.

Objective 3: Computer and Internet Competencies

Competencies of respondents in relation to computer skills was assessed using a five-point scale that ranged from 1 (none) to 5 (very high), see Table 3. Overall, respondents indicated a "fair to high" level of computer skills (Table 3). However, respondents indicated that their level

Table 5. ANOVA table for comparing demographic variables and computer skills**

Variable	n	M*	SD		
Age	41				
30 years old and <	64	36.58A	9.36		
31 to 40 years old	70	33.09A	11.36		
41 years old and >	175	28.44B	9.58		
Total		32.05	10.68		
Source	**SS**	**df**	**MS**	**F**	**Sig**
Between groups	1823.877	2	911.938	8.70	<0.001
Within groups	18028.660	172	104.818		
Total	19852.537	174			
Education					
High school diploma/B.A.	49	26.82	8.99		
Bachelors of science	104	32.72	10.52		
Masters of science	22	40.22	8.87		
Total	175	32.01	10.67		
Source	**SS**	**df**	**MS**	**F**	**Sig**
Between groups	2859.853	2	1429.927	14.52	<0.001
Within groups	16938.1244	172	98.477		
Total	19797.977	174			
Major of study					
Agronomy major	67	34.97	10.41		
Other agricultural major	83	32.11	10.15		
Not agricultural major	25	25.96	10.31		
Total	175	32.32	10.62		
Source	**SS**	**df**	**MS**	**F**	**Sig**
Between groups	1485.510	2	742.755	7.04	<0.001
Within groups	18156.924	172	105.564		
Total	19642.434	174			
Work experience					
5 years of experience and <	73	35.90	10.46		
6 to 15 years of experience	66	30.36	9.99		
16 years of experience and >	39	28.51	10.18		
Total	178	32.23	10.60		
Source	**SS**	**df**	**MS**	**F**	**Sig**
Between groups	1754.211	2	877.106	8.39	<0.001
Within groups	18295.345	175	104.545		
Total	20049.556	177			
Work location					
Provincial office	34	35.35	10.86		
Region office	51	31.63	9.95		
County office	94	31.34	10.80		
Total	179	32.18	10.63		
Source	**SS**	**df**	**MS**	**F**	**Sig**
Between groups	424.124	2	212.062	1.89	0.153
Within groups	19692.793	176	111.891		
Total	20116.916	178			

* Mean computed on a scale 1 = none, 2 = little, 3 = fair, 4 = high, and 5 = very high

** Mean could range from a low of 12 to a high of 60 with a theoretical midpoint of 38

Table 6. ANOVA table for comparing demographic variables and internet skills**

Variable	n	M*	SD		
Age					
30 years old and <	41	30.17A	11.37		
31 to 40 years old	63	24.40	11.08		
41 years old and >	67	19.46	8.45		
Total	171	23.85	10.73		
Source	**SS**	**df**	**MS**	**F**	**Sig**
Between groups	2946.506	2	1473.253	14.88	<0.001
Within groups	16631.541	168	98.997		
Total	19578.047	170			
Education					
High school diploma/B.A.	46	17.85	7.22		
Bachelors of science	103	24.23	10.45		
Masters of science	22	36.45	7.56		
Total	171	24.09	10.81		
Source	**SS**	**df**	**MS**	**F**	**Sig**
Between groups	5157.887	2	2578.944	29.47	<0.001
Within groups	14713.797	168	87.582		
Total	19871.684	170			
Major of study					
Agronomy major	67	25.81	11.25		
Other agricultural major	80	24.01	10.82		
Not agricultural major	24	19.87	8.30		
Total	171	24.13	10.79		
Source	**SS**	**df**	**MS**	**F**	**Sig**
Between groups	623.816	2	311.908	2.73	<0.001
Within groups	19186.090	168	114.203		
Total	1989.906	170			
Work experience					
5 years of experience and <	73	27.62	1.27		
6 to 15 years of experience	66	21.41	10.16		
16 years of experience and >	35	21.46	9.60		
Total	174	24.02	10.76		
Source	**SS**	**df**	**MS**	**F**	**Sig**
Between groups	1624.008	2	812.004	7.55	<0.001
Within groups	18401.901	171	107.613		
Total	20025.908	173			
Work location					
Provincial office	34	25	10.83		
Region office	50	22.58	10.27		
County office	91	24.46	10.97		
Total	175	24.03	10.73		
Source	**SS**	**df**	**MS**	**F**	**Sig**
Between groups	154.062	2	77.031	0.667	0.515
Within groups	19872.795	172	115.540		
Total	20026.857	174			

** Mean could range from a low of 12 to a high of 60 with a theoretical midpoint of 38

of computer skills was somewhat limited.

Respondents were asked to indicate their level of skills using various aspect of the Internet, see Table 4.

Objective 4: Differences in Perceived Competencies and Demographic Characteristics.

Significant differences were found between select demographic characteristics (age, education level, major area of study, work experience, and work location) and use of computer and Internet skills.

ANOVA results indicated significant differences [F (2, 172) = 8.70; p = .001] between the age and computer and Internet skills, see Tables 5 and 6. Younger extension workers (under 40 years old) tended to possess higher skills than those educators who were 42 and older.

In the ANOVA results between the demographic characteristic and Internet skills, gave the same results of Tables 5. In this case, indicated significant differences [F (2, 171) = 14.88; p = .001] between the age and Internet skills, too. Younger extension workers (under 40 years old) tended to possess higher skills than those educators who were 42 and older.

CONCLUSIONS AND RECOMMENDATIONS

Overall, findings indicate that extension workers have access to computers both at work and home. On average, they use computers approximately 6 – 8 hours per week and majority of them use the Internet on average of 1 – 4 hours per week.

Significant differences were found between key demographic variables (age, highest education level, major of study, work experience, and work location) and computer and Internet skills. Scheffé post hoc analysis revealed the following difference:

Younger extension workers (under 30 year) exhibited high level of both computer and Internet skills than older extension workers (41 year and older). This finding mirror with other findings reported in the literature (Table 5) and confirms findings Ommani and Chizari, 2006; Fallah Haghighi, 2006; Akpabio, 2007.

Extension workers with master's degrees reported significantly higher levels of computer and Internet skills than workers who had completed bachelor's degrees and agents with high school diplomas. Extension workers with bachelor's degrees reported higher skills when compared to workers with high school diploma.

More experienced extension workers reported lower levels of computer and Internet skills than less experienced (5 year or less) extension workers. This finding mirrors several studies reported in the literature regarding experience and computer skills such as: Ommani and Chizari, 2006; Fallah Haghighi, 2006; Akpabio, 2007.

Based on the findings and conclusion of this study the following recommendations are offered to make informed policy and programmatic decisions relative to computer and Internet use by extension in Isfahan Province of Iran.

First, need exists to conduct a systematic assessment of training needs relative to computer and Internet use extension workers. Such identification will help in characterizing a training program for extension workers to enhance their skills in computer and Internet use.

Second, systematic training should be conducted in the areas that extension workers perceived that they possessed less skill. For example, training should be offered in the areas of photo shop, power point, and statistical software.

Third, careful assessment of training needs relative to computer and Internet should be undertaken kept in mind the work-related characteristics such as work location, experience, etc.

Fourth, offering training/workshops, on computer and Internet emphasis should be given to age, education level and experience of extension workers. In order for training to be effective, staff development in Iran should consider separate tracks for training. For example, offer training (from basic to advance for older and experienced extension workers).

Finally, as computer and Internet dominate the information explosion, periodic upgrading of skills, software and hardware should be provided. Such effort will reduce the regional imbalances relative to equipment, both software and hardware.

This finding should be shared with all regional and provincial administration to make informed decisions relative to allocates resources for enhancing computer and Internet in Iran.

REFERENCES

1- Ahmadpour, A., & Mirdamadi, M. (2010). Determining challenges in the application of e-learning in Agricultural Extension Services in Iran. *American-Eurasian Journal of Agriculture and Environmental Science*, 9(3), 292-296.

2- Akpabio, I. A. (2007). Agricultural extension officials and utilization of information and communication technologies in the Niger Delta, *Nigeria [Abstract]*, Retrieved 2008, January 25 from http://jesonline.org/current.htm

3- Badragheh, A., Chizari, M., Hosseini, J.F., & Daneshfard, K .A. (2010). Challenges and necessary applying of Management Information System (MIS) in Agricultural Education and Extension System of Iran. *American- Eurasian Journal. Agriculture and Environmental Science*, 8(5), 561-569. Retrieved from: http://www.idosi.org/ aejaes/jaes8(5)/ 14.pdf

4- Bheenick, K. J., Brizmohum, R. (1998), November. The scope of information technology applications in agricultural extension in Mauritius. *Paper presented at the Third Annual Meeting of Agricultural Scientists*, Leduil, Mauritius. Retrieved 2008, February 12, from http://www.uom.ac.mu/faculties/fao/AIS/AMAs98/Index.htm

5- Chieochan, O., Lindley, D., & Dunn, T. (2000). Factors affecting the use of Information technology in Thai agricultural cooperatives: a work in progress. *The Electronic Journal of Information Systems in Developing Countries*, 2(1), 1-15

6- Fallah Haghighi, N. (2006). *An Analysis of Factors Influenced in the Use of IT in Agricultural Extension, Gilan Province, Iran.* Unpublished master's thesis, University of Tehran, Iran.

7- Fami, H. S. (2006). Enhancement of Extension Systems in Agriculture-Asian Productivity Organization, Tokyo, 5, 116-125, Retrieved 2008, January 10, from http://www.apo-tokyo.org/00e-books/AG-16_EnhanceExtSystem/AG-16_EnhanceExtSystem.pdf

8- Gregg, J. A., Irani, T. A. (2004). Use of information technology by county extension agents of the Florida cooperative extension service. Research in Brief, Retrieved 2008, January 20, from http://www.joe /2004june/rb.shtml

9- Omidi Najafabadi, M., Hosseini, J. F., & Mirdamadi, M. (2009). A Case Study of Requirements and Challenges of Information and Communication Technology System to Train Private Agricultural Insurance Brokers. *American- Eurasian Journal. Agriculture and Environmental Science*, 6 (2), 152-159. Retrieved from http://www.idosi.org/aejaes/ jaes6 (2)/5.pdf

10- Ommani, A. R., Chizari, M. (2006). Use of information technology by extension agents in Khuzestan province of Iran. *Paper presented at the Association for International Agricultural and Extension Education (AIAEE)*. 22th Annual Conference, Clearwater Beach, Retrieved 2008, February 12 from http://www.aiaee.org/2006/index.html

18

Teaching Methods and Tools Used In Food Safety Extension Education Programs in the North Central Region of the United States

Vikram Swaroop Chandra, Koundinya[1] and Robert A. Martin[2]

Abstract

One of the ways to ensure food safety is to educate the public. Of the organizations providing food safety education in the United States (U.S.), the Cooperative Extension System (CES) is one of the most reliable. The effectiveness CES programs depends not only on what is being taught but also on how it is taught. Both a needs-based curriculum and how that curriculum is delivered are equally important. This descriptive cross-sectional study using a disproportional stratified random sample identified the teaching methods and tools being used by food safety extension educators of the CES of North Central Region (NCR). A Likert-type scale administered to extension educators revealed that they were adopting a balanced use of teaching methods and tools, and using learner-centered teaching methods in their programs. However, distance education, case studies and podcasts, which are commonly used in education programs, were not being used extensively. We recommend that food safety extension educators of NCR should increase the use of these two teaching methods and tool while continuing to use the current ones. This study has implications for improving food safety education delivery to clients in the NCR and for designing inservice education for food safety extension educators.

Keywords:
Food safety education, extension educators, teaching methods, and teaching tools.

[1] Postdoctoral Research Associate, Global Extension Programs & Department of Agricultural Education and Studies, Iowa State University, Ames, IA – 50011.
[2] Professor, Department of Agricultural Education and Studies, Iowa State University, Ames, IA – 50011.
* *Corresponding author's email:* vikram@iastate.edu

INTRODUCTION

Food safety is an important national and global concern with serious health and trade implications. Ensuring the safety of food is a challenging task to governments despite rigorous regulatory practices. The incidence of food-related illnesses is increasing throughout the world (Motarjemi & Kaferstein, 1999), and foodborne illnesses represent a major health threat in all countries, from the most to the least developed (De Waal & Robert, 2005). In 2005 alone, 1.8 million people died of diarrheal diseases caused due to lack of food and water safety worldwide (World Health Organization (WHO), 2007). In the United States (U.S.), every year, one in six Americans suffers a foodborne illness with 48 million reported cases of foodborne diseases resulting in 128,000 hospitalizations and 3,000 deaths (Centers of for Disease Control and Prevention, 2011).

Food is needed for survival, but "Eating could very well be called America's national pastime" (Pennsylvania Impact, 1999, p.1). To ensure that public gets safe food, the American government invests vast sums of money and effort; yet, food-related illnesses remain prevalent (Ellis, 2006; Nyachuba, 2010). Aside from health, foodborne illnesses also have severe economic implications. A single outbreak of a foodborne illness in a metropolitan area can cause losses of up to $7 million for a chain of foodservice operations (Guion *et al.,* 2004). For a small business one such outbreak could challenge its very survival (Guion *et al.,*). Consequently, food safety is a legitimate problem in the U.S. (Barton & Barbeau, 1992). One of the ways to address this problem is to educate public on food safety.

It is found that the average American is not knowledgeable about food safety (Altekruse *et al.,* 1995). Consequently, effective food safety education programs are needed (Nyachuba, 2010). Of the organizations that offer food safety educational programs in the U.S., the Cooperative Extension System (CES) is considered the most reliable owing to its research support from land-grant universities (Feller *et al.,* as cited by McDowell, 2001). CES's reliability is further validated by research findings that indicate the impact of CES's food safety educational programs in bringing about changes in food safety behaviors (Gentry-Van Laanen & Nies, 1995; Wardlaw, 1999), improvement in knowledge (Laminack *et al.,* 2008), and adoption of recommended food safety practices (Dean *et al.,* 2008).

The effectiveness of such education programs conducted by the CES depends not only on what is being taught but also on how it is taught. Hence, a needs-based curriculum alone cannot ensure the success of an educational program but how that curriculum is delivered to clients is equally important. Cole (1981) stated that the primary role of extension educators is teaching, which means that they should learn more about the teaching methods and tools they use in their educational programs (Jayaratne & Martin, 2003). Further, Apantaku *et al.,* (2008) found that the teaching methods used by extension educators have an effect on the clients' outcomes. In addition, food safety has been identified as a scientifically complex subject matter whose underlying principles are not properly understood by the general public (Barton & Barbeau, 1992). Therefore, selection of appropriate teaching methods and tools is essential for conducting effective food safety education programs.

Verner (1959) stated that the methods used in an educational program indicate the ways in which people involved with the program are organized, and that it also establishes a relationship between learners and the agency offering the educational program (as cited by Conti & Kolody, 2004). He further stated that certain devices (identified as teaching tools in this study) facilitate the learning. Conti and Kolody stated that proper selection of methodology in the teaching-learning process results in professionalism, which contributes to the overall quality of the information being delivered. Therefore, food safety extension educators must give due consideration to the teaching methods and tools they use in their educational programs. This study identified and analyzed the teaching methods and tools being used by extension educators of the 12 member states of the North Central Region (NCR) of the U.S. in their food safety educational programs. There are no other known research studies that have analyzed the teaching methods and tools used in food safety extension programs in the CES of NCR.

PURPOSE AND OBJECTIVES

The purpose of this study was to identify and analyze the teaching methods and tools used by food safety extension educators of the NCR of the U.S. The study had the following objectives:

1. To identify the demographic characteristics of extension educators,
2. To identify the extent of use of the selected teaching methods by extension educators, and
3. To identify the extent of use of the selected teaching tools by extension educators.

MATERIALS AND METHODS

This study adopted a descriptive cross-sectional survey design. The population for this study was all the County Extension Directors (CEDs) and all extension educators in the program areas of Family and Consumer Sciences (FCS) and Agriculture and Natural Resources (ANR). CEDs were included because most had job responsibilities in more than one program area. While 4-H extension educators offer food safety educational programs to youth, they were not included as this study focused primarily on adult educational processes.

A disproportional stratified random sample was drawn from the total population. The formula for calculating sample size as suggested by Ary *et al.,* (2006, p.419) was used to arrive at a sample size of 384. Anticipating a 50% return rate (Ary *et al.,*), 64 extension educators were randomly selected from each of the 12 member states of the NCR yielding a sample size of 768. The 12 member states of the NCR include Illinois, Indiana, Iowa, Kansas, Michigan, Minnesota, Missouri, Nebraska, North Dakota, Ohio, South Dakota and Wisconsin. An electronic questionnaire developed using SurveyMonkey® and modeled on questionnaires used by Creswell (1990), Jayaratne (2001), Kwaw-Mensah (2008), and Walczyk and Ramsey (2003) was used for this study. This format for the survey questionnaire was used because the researchers had previous success conducting a similar study with agriculture teachers (Koundinya & Martin, 2010).

A list of 12 teaching methods and 14 teaching tools was developed based on a review of literature that focused on extension educational processes in livestock waste management, sustainable agriculture, water quality management and private pesticide training, and also based on input from Extension experts at the Iowa State University. The identified teaching methods and tools depicted those most commonly used in typical extension programs in the NCR. Additionally, a provision was given to extension educators to identify any additional teaching methods and tools they were using in their food safety educational programming. The extent of use of different teaching methods and teaching tools was identified on a Likert-type scale that ranged from 0 = Not Used (NU) to 4 = Always Used (AU).

The questionnaire was validated by an expert panel for face and content validity. The panel consisted of professors from the departments of agricultural education, food science and human nutrition; CES state program leaders in the program areas of FCS and ANR; and the Director of Extension at the Iowa State University. The questionnaire was pilot-tested with extension educators, and the resulting data was used to establish the reliability of the questionnaire. The extension educators that participated in the pilot test were excluded from the survey population to prevent a contaminated sample. For reliability of the questionnaire, Cronbach's α was computed from the pilot test data. Values of .768 and .893 were reported for teaching methods and tools sections, respectively, which are categorized as "acceptable" and "good" reliability, respectively, by George and Mallery (2003).

Selected extension educators were emailed a letter informing them the purpose of this research study. This letter sought their cooperation, and it was made clear that their participation in this study was voluntary and that they could withdraw at any time. They were also informed that any changes in the study's objectives would be communicated to them. Extension educators did not receive any monetary incentive for participating in this study. After this introductory email, the survey questionnaire was emailed to them. Extension educators' consent for the study was assumed if they filled out and returned the questionnaire. Four follow-ups were conducted at suitable time intervals. This study was a part of a bigger study that focused on the food safety educational processes in the CES of NCR of the U.S.

A potential limitation to this study was a comparatively lower response rate. According to Linder, Murphy, and Briers (2001), any response rate of less than 85% could result in significant differences between early and late respondents, thus affecting the external validity of the study. One method to account for nonresponse error is to compare early and late respondents (Dooley & Lindner, 2003; Miller & Smith, 1983). An independent samples t-test was conducted for comparing early and late respondents, and no statistically significant differences were recorded on the extent of use of all teaching methods and tools, except on use of the following tools: PowerPoint®, posters and videos. Late respondents had significantly higher mean scores than the early respondents on the extent of use of PowerPoint® and posters, whereas early respondents recorded significantly higher scores on the extent of use of videos in their food safety educational programs. Therefore, it was decided not to generalize the findings on these items to non respondents and the total population.

RESULTS AND DISCUSSION

Four hundred sixteen of the 768 extension educators responded to the survey for an initial response rate of 54.16%. However, only 325 questionnaires were usable, yielding a response rate of 42.31%. The findings for each objective are presented below.

Objective 1: To identify the demographic characteristics of extension educators

The respondents had a work experience of 14.86 years on an average, with a standard deviation (SD) of 10.04. Work experience ranged from 1 to 40 years. The respondents ranged from 24 to 73 years of age. The mean age of the respondents was 48.62 years with a SD of 10.85. Since outliers were detected in the age distribution, a median was calculated to account for the skewed data. The median age of the respondents was 51 years, indicating a negatively skewed age distribution. Of the respondents, 56% were female and 61.88% had earned a master's degree.

The demographic distribution was in line with the findings of Camara (2006); Creswell (1990); Jayaratne (2001); Kwaw-Mensah (2008); Radhakrishna and Thomson (1996) regarding demographic characteristics like age, educational level, and work experience; and with the findings of Webster *et al.,* (2001) regarding age, but differed in gender distribution. All these six studies found that a majority of the extension educators who participated in their studies were males. Braiser *et al.,* (2009), and Selby *et al.,* (2005), however, found that a majority of extension educators in their research samples were females. This difference could have been due to the differences in the states in which extension educators were working.

Objective 2: To identify the extent of use of the selected teaching methods by extension educators

The mean and standard deviation data on the extent of use of teaching methods by extension educators indicated that discussion (M = 2.88) was the most used among the 12 identified teaching methods followed by demonstration (M = 2.76) and lecture-discussion (M = 2.62), whereas learning contracts was the least used method (M = 0.57) followed by distance education (M = 1.35), quizzes (M = 1.46), and case studies (M = 1.79) (Table 1).

Kwaw-Mensah (2008) found discussions and lecture-discussions to be among the more commonly used teaching methods by extension educators of the NCR in livestock waste management education. On a similar vein, Shinn (1997) found that in the U.S., teachers of agriculture used demonstrations more often followed by discussion. From the learners' perspective, Reisenberg and Gor (1989) found that farmers preferred demonstrations to learn about innovations. Discussion has been identified as an inclusionary and participatory teaching method that facilitates critical thinking skills in learners (Brookfield, 2004), which is an essential component for teaching safe food practices and behaviors. Also, this teaching method is the most respectful of learners, and places the educator and learners on equal footage as it assumes that everyone has a useful contribution to make to the educational program (Brookfield). Subsequently, discussion is identified as a learner-centered teaching method (Jarvis, 2004).

Table 1: Mean and standard deviation scores of extension educators based on the extent of use of teaching methods in food safety education programs

S. No	Teaching method	M	SD	n
1	Discussion	2.88	0.90	321
2	Demonstration	2.76	0.90	320
3	Lecture-discussion	2.62	0.99	318
4	Lecture	2.51	1.07	320
5	Questioning	2.44	1.07	316
6	Small group work	2.16	1.06	316
7	One-on-one instruction	2.12	1.18	320
8	Problem Solving	2.10	0.98	317
9	Case Studies	1.79	1.04	320
10	Quizzes	1.46	1.16	315
11	Distance Education	1.35	1.09	315
12	Learning Contracts	0.57	0.92	305

Scale used: 0= Not Used, 1= Rarely Used, 2= Sometimes Used, 3= Frequently Used, 4= Always Used.

Demonstrations have their own significance for teaching skills by providing hands-on learning experiences. One of the basic principles in extension work is 'learning by doing' (Reddy, 1993) and the doing part is well provided by demonstrations. Essential specific skills related to safe food practices can be readily learned by demonstration. From the extension educators' perspective, demonstrations make economical use of time, equipment and materials (Gilley, 2004). Similarly, lecture-discussions have their own unique role to play in food safety education. A brief lecture for identifying and clarifying concepts is necessary in educational programs (Farrah, 2004). There are some complex scientific facts that learners need to know about food safety (Barton & Barbeau, 1992), and a brief lecture before opening the floor for discussion can accomplish that purpose.

Furthermore, it was found that none of the teaching methods were "Frequently Used" or "Always Used." This might suggest that food safety extension educators in the NCR were not relying excessively on any one particular method, and thereby adding variability to the adult education settings (Garton, 1999). Lynn (1996) suggested that a variety of strategies should be used in food safety educational programs (as cited by Costello *et al.,* 1997). The use of multiple teaching methods is also supported by Hall *et al.,* (2004); Ota *et al.,* (2006); and Rodewald (2001).

On the other hand, Shinn (1997) found that learning contracts, distance education and case studies were not being used much by agricultural educators in the U.S. which is in line with the findings from this study. Berger *et al.,* (2004) suggested that learning contracts can be used in a variety of settings with diverse adult learners. Also, they indicated that learning contracts are useful in the acquisition of skills as well as content, and they can be individualized to the learners. But, learning contracts demand more of the educator's time (Berger *et al.,*), which could be the reason for its lesser use by food safety extension educators of the NCR, and researchers have found that extension educators do have time constraints (Camara, 2006; Jayaratne, 2001). Koundinya and Duttala (2002) found that teaching through distance education contributed to a statistically significant gain in farmers' knowledge of peanut and mango cultivation practices in India. In the U.S., Dooley *et al.,* (1999) found that distance education methods like videoconferencing were effective in disseminating food safety updates. They reported an increase in the knowledge of Food Protection Management instructors who participated in a food safety instructor training via distance education methods. Shanley *et al.,* (2009) concurred that distance education was indeed an effective method for food safety education.

Another lesser used teaching method was 'case studies' which was being used only rarely to sometimes (M = 1.79) in food safety extension educational programs. This is in-line with the finding of Creswell (1990), who found that case studies were not heavily used by agricultural extension professionals in training private pesticide applicators in Iowa, Nebraska, North Dakota and Wisconsin. Case studies have been identified as one of the more effective teaching methods that promote active learning. A case study approach gives in-depth information (University of Idaho, 2006), can be used to answer how and why questions, and is effective in studying contemporary issues in real-life context (United States Department of Veteran Affairs, 2008; Yin, 2003). Case studies as a teaching method have been found to improve practical thinking, and help learners formulate and solve problems (Marsick, 2004) which are the skills that need to be developed in food safety educational programs.

Objective 3: To identify the extent of use of the selected teaching tools by extension educators

The mean and standard deviation scores computed on the extent of use of the identified teaching tools showed the same trend as that of the teaching methods with none of the tools falling under "Frequently Used" or "Always Used" categories. Brochures (M = 2.32), pamphlets (M = 2.31), newsletters (M = 2.28) and

Table 2: Mean and standard deviation scores of extension educators based on the extent of use of teaching tools in food safety education programs

S. No	Teaching tool	M	SD	*n*
1	Brochures	2.32	1.08	309
2	Pamphlets	2.31	1.10	306
3	Newsletters	2.28	1.19	304
4	PowerPoint®	2.22	1.54	314
5	Internet/websites	2.11	1.17	309
6	Research Publications	1.89	1.19	305
7	Charts	1.82	1.15	308
8	Compact Discs	1.62	1.16	303
9	Videotapes	1.53	1.16	308
10	Posters	1.28	1.26	312
11	Text Books	1.08	1.21	302
12	Podcasts	0.56	0.85	299
13	WebCt	0.54	0.91	306
14	Interactive Whiteboard	0.41	0.82	307

Scale used: 0= Not Used, 1= Rarely Used, 2= Sometimes Used, 3= Frequently Used, 4= Always Used.

PowerPoint® (M = 2.22) were the most commonly used teaching tools, while interactive whiteboard (M = 0.41), WebCt (M = 0.54) and podcasts (M = 0.56) were the least used tools in conducting food safety educational programs (Table 2).

These findings were comparable to related extension education studies. Roberts *et al.,* (2009) reported that adult learners preferred newsletters for receiving information on food safety. Jenkins *et al.,* (2000) concurred with this finding, stating that newsletters are effective in reaching a wide audience. Further, Marrotte (2000) found that learners preferred PowerPoint® presentations. In addition, podcasts have also been found to be useful in extension educational programs (Smith & Davis, 2008; Xie & Gu, 2007); however, the food safety extension educators from this study were not utilizing podcasts that much in their education programs. The same was true for interactive teaching media and other materials found to be useful in food safety educational programs (Bednar *et al.,* 2003; Trepka, 2008). In the same vein, WebCt is being used as a teaching tool in many distance education programs at U.S. land grant institutions, but the extension educators from this study were not using it to a great extent in their food safety educational programs.

In addition to the teaching methods and tools identified in this study, the respondents claimed they "Frequently Used" facilitated dialogue as a teaching method in their food safety programs. In addition to the ones identified for this study, webinars, real-life props and settings, and news columns were commonly used teaching tools. Webinars and news columns were "Frequently Used," whereas real-life props and settings were "Always Used". Thomson, Abel, and Maretzki (2001) indicated facilitated dialogue as a method of educating people on food, farm and community. Similarly, webinars were being used as an educational tool in a livestock insurance website (Small *et al.,* 2008).

CONCLUSIONS AND RECOMMENDATIONS

Four major conclusions were drawn based on the results of the study. First, the typical extension educator who participated in this study was a middle-aged woman with substantial years of work experience and a master's degree. Second, extension educators seemed to use a variety of teaching methods and tools in their food safety educational programs as evidenced by the reports of balanced use of the teaching methods and tools. The extension educators of the NCR were doing a good job of using a combination of different teaching methods and tools in their food safety educational programs and are encouraged to continue doing so.

The third conclusion was that extension educators were not making extensive use of 'distance education', 'case studies', and 'podcasts' in their food safety educational programs, which are commonly used and effective methods and tools in educational programs. The fourth and final conclusion was that extension educators were adopting learner-centered teaching approaches, as evidenced by the maximal use of discussions, demonstrations and lecture-discussion in their food safety educational programs. These three methods require and encourage learner involvement in the educational process. Hence, it is recommended that extension educators continue using these methods. However, given the effectiveness of distance education and case studies, and podcasts, it is also recommended that extension educators utilize these two teaching methods and tool more in their food safety educational programs in addition to the ones they are already using.

Implications

The findings have implications for improving the food safety education delivery to clients in the North Central Region. This study identified a lesser use of some of the commonly used teaching methods such as distance education and case studies and teaching tools such as podcasts in the food safety extension programs. The reasons for not using these teaching methods and tool to the desired extent need to be researched in future studies and duly addressed. The findings could also have implications for inservice education of extension educators as these educators may not be well trained in using the latest online resources for distance education. There is also a possible lack of resources to carry out online education, and these potential limitations can be found out in future research. Since food safety is a very important topic regardless of the country, the findings from this study have implications for international extension education settings, also. Hence, similar research needs to be conducted in other parts of the world. This study also has implications for designing teaching methods and tools for food safety education programs conducted in formal settings like colleges and universities.

REFERENCES

1- Altekruse, S. F., Steet, D. A., Fein, S. B., & Levy, A. S. (1995). Consumer knowledge of foodborne microbial hazards and food-handling practices. Journal of Food Protection, 59(3), 287-294.

2- Apantaku, S. O., Olufade, O., Adebayo, K., & Oyekanmi, A. (2008). Analysis of effects of extension teaching methods on farmers' level of cassava and maize production in Ogun State, Nigeria [Abstract]. Ghana Journal of Agricultural Science, 41(1), ISSN: 0855-0042.

3- Ary, D., Jacobs, L. C., Razavieh, A., & Sorenson, C. (2006). Introduction to research in education (7th ed). Belmont, CA: Thomson Wadsworth.

4- Barton, J. A., & Barbeau, W. E. (1992). Is Extension ready for food safety education in the '90s?. Journal of Extension, 30(1), Article number 1FRM2.

5- Bednar, C., Kwon, J., Baker, M., & Kennon, L. (2003). Survey of food safety education needs for low-income consumers. Journal of the American Dietetic Association, 103(9), pp 111-112.

6- Berger, N. O., Caffarella, R. S., & O'Donnell, J. M. (2004). "Learning contracts". In Galbraith, M. W. (eds.), Adult learning methods. A guide for effective instruction (3rd Ed) (pp. 289-319). Malabar, FL: Krieger Publishing Company.

7- Braiser, K., Barbercheck, M., Kiernan, N. E., Sachs, C., Schwartzberg, A., & Trauger, A. (2009). Extension educators' perceptions of the educational needs of women farmers in Pennsylvania. Journal of Extension, 47(3). Article number 3FEA9.

8- Brookfield, S. D. (2004). "Discussion". In Galbraith, M. W. (eds.), Adult learning methods. A guide for effective instruction (3rd Ed) (pp. 209-226). Malabar, FL: Krieger Publishing Company.

9- Camara, M. (2006). Agricultural extension educators' perceptions regarding the teaching and learning process related to water quality issues in the North Central Region of the United States (Doctoral Dissertation). Iowa State University, Ames.

10- Centers for Disease control and Prevention (2011). CDC Estimates of Foodborne Illness in the United States. Retrieved March 23, 2011 from, http://www.cdc.gov/foodborneburden/2011-foodborne-estimates.html

11- Cole, J. M. (1981). Selecting extension teaching methods. Journal of Extension, 27-32.

12- Conti, G. J., & Kolody, R. C. (2004). "Guidelines for selecting methods and techniques". In Galbraith, M. W. (eds.), Adult learning methods. A guide for effective instruction (3rd Ed) (pp. 181-192). Malabar, FL: Krieger Publishing Company.

13- Costello, C., Gaddis, T., Tamplin, M., & Morris, W. (1997). Evaluating the effectiveness of two instructional techniques for teaching food safety principles to quick service employees. Journal of Foodservice Systems, 10, 41-50.

14- Creswell, J. L. (1990). An identification and assessment of Extension educational delivery systems for training of private pesticide applicators (Doctoral Dissertation). Iowa State University, Ames.

15- Dean, K. W., Reames, E. S., Tuuri, G., Keenan, M. J., Bankston Jr, J. D., Friendship, D. Y., … Tucker, E. H. (2008). Improved knowledge and adoption of recommended food safety practices by food recovery agency personnel and volunteers participating in the serving food safely program. Journal of Extension, 46(4), Article number 4RIB4.

16- De Waal, C. S., & Robert, N. (2005). Global and local: Food safety around the world. Center for Science in the Public Interest, 1875 Connecticut Avenue, N.W., Suite 300, Washington, D. C. 20009-5728.

17- Dooley, L. M., & Lindner, J. R. (2003). The handling of nonresponse error. Human Resource Development Quarterly, 14(1), 99-110. doi: 10.1002/hrdq.1052

18- Dooley, K. E., Van Laanen, P. G., & Fletcher, R. D. (1999). Food safety instructor training using distance education. Journal of Extension, 37(3), Article number 3FEA5.

19- Ellis, J. D. (2006). Food safety training as adult education: Determining prior knowledge in the service of scientific conceptual change (Doctoral Dissertation). Iowa State University, Ames.

20- Farrah, S. J. (2004). "Lecture". In Galbraith, M. W (eds.), Adult learning methods. A guide for effective instruction (3rd Ed) (pp. 227-252). Malabar, FL: Krieger Publishing Company.

21- Garton, B. L. (1999). "Adult teaching methods". In Birkenholz, R. J (eds.). Effective adult learning (pp. 43-56). Danville, IL: Interstate Publishers, Inc.

22- Gentry-Van Laanen, P., & Nies, J. I. (1995). Evaluating Extension program effectiveness: Food safety education in Texas. Journal of Extension, 33(5). Article number 5FEA4.

23- George, D., & Mallery, P. (2003). SPSS for windows step by step. A simple guide and reference. 11.0 update (4th ed). New York, NY: Pearson Education, Inc.

24- Gilley, J. W. (2004). "Demonstration and simulation". In Galbraith, M. W. (eds.), Adult learning methods. A guide for effective instruction (3rd Ed) (pp. 361-381). Malabar, FL: Krieger Publishing Company.

25- Guion, L. A., Simonne, A., & Easton, J. (2004). Youth perspectives on food safety. Journal of Extension. 42(1), Article number 1RIB5.

26- Hall, J. B., McKinnon, B. R., Greiner, S. P., & Whittier, W. D. (2004). Teaching complex, indepth programs. Journal of Extension, 42(3), Article number 3FEA2.

27-Jarvis, P. (2004). Adult education and lifelong learning. Theory and practice (3rd ed.). New York, NY: RoutledgeFalmer.

28- Jayaratne, K. S. U. (2001). Agricultural extension educators' perceptions regarding the teaching and learning processes as related to sustainable agriculture: Implications for agricultural extension education (Doctoral Dissertation). Iowa State University, Ames.

29- Jayaratne, K.S.U & Martin, R. A. (2003). Agricultural extension educators' perceptions regarding teaching methods and tools for educating farmers about sustainable agricultural practices. Proceedings of AEEE central region agricultural education research conference, St Louis, Missouri.

30- Jenkins, J. H., Newman, M. E., Castellaw, J. C., & Lane, Jr, C. D. (2000). Partnering with the local livestock market in educational programs. Journal of Extension, 38(3), Article number 3RIB2.

31- Koundinya, V., & Duttala, R. (2002). Impact of Annadata-velugubata, a distance education program on the knowledge gain in groundnut and mango cultural practices of the televiewers in Chittoor district of Andhra Pradesh. Journal of Research ANGRAU, 30(2), 127-129.

32- Koundinya, V. & Martin, R. A. (2010). Food safety inservice educational needs of agricultural teachers. Journal of Agricultural Education, 51(4), 82-91. doi: 10.5032/jae.2010.04082

33- Kwaw-Mensah, D. (2008). Perceptions of agricultural extension educators regarding livestock waste management education in the North Central Region (Doctoral Dissertation). Iowa State University, Ames.

34- Laminack, J., Dainello, F., Degenhart, S. H.,Vestal, T. A., & Wingenbach, G. (2008). Increasing positive perceptions of food irradiation: Appealing to one's affective domain. Journal of Extension, 46(4), Article number 4RIB3.

35- Lindner, J. R., Murphy, T. H., & Briers, G. H. (2001). Handling nonresponse in social science research. Journal of Agricultural Education, 42(4), 43-53.

36- Marrotte, E. L. (2000). The evolution in using high tech teaching tools in outreach programs: From stone age to high tech. Journal of Extension, 38(5). Article number 5TOT2.

37- Marsick, V. J. (2004). "Case study". Galbraith, M. W. (eds.), In Adult learning methods. A guide for effective instruction (3rd Ed) (pp. 383-404). Malabar, FL: Krieger Publishing Company.

38- McDowell, G. R. (2001). Land grant universities and extension into the 21st century: Renegotiating or abandoning a social contract. Ames, IA: Iowa State University Press.

39- Miller, L. E., & Smith, K. L. (1983). Handling nonresponse issues. Journal of Extension, September/October, 45-50.

40- Motarjemi, Y., & Kaferstein, F. (1999). Food safety, hazard analysis and critical control point and the increase in foodborne diseases: a paradox? Food Control, 10, 325-333.

41- Nyachuba, D. G. (2010). Foodborne illness: Is it on the rise? Nutrition Reviews, 68(5), 257-269.

42- Ota, C., DiCarlo, C. F., Burts, D. C., Laird, R., & Gioe, C. (2006). Training and the needs of adult learners. Journal of Extension, 44(6), Article number 6TOT5.

43- Pennsylvania Impact (1999). Building food safety into our food system. College of Agricultural Sciences, Teaching, Research and Cooperative Extension Programs Work for Pennsylvania. U.Ed. AGR99-78 5M299PS23980. Retrieved August 3, 2010, from http://paimpact.cas.psu.edu/pdf/agr9978.pdf

44- Radhakrishna, R, B., & Thomson, S. (1996). Extension agents' use of information sources. Journal of Extension, 34(1), Article number 1RIB2.

45- Reddy, A. A. (1993). Extension Education (5th ed). Bapatla, A.P: Sree Lakshmi Press.

46- Reisenberg, L. E., & Gor, C. O. (1989). Farmers' preferences for methods of receiving information on new or innovative farming practices. Journal of Agricultural Education. Fall, 7-13.

47- Roberts, K. R., Remig, V. M., Bryant, T. J., & Snyder, G. (2009). Older adults' technology preferences related to food safety education. Journal of the American Dietetic Association, 109(9), pg A64.

48- Rodewald, A. D. (2001). Delivery systems--Is the "latest" technology the greatest? Journal of Extension, 39(4), Article number 4TOT2.

49- Selby, K. A., Peters, J. L., Sammons, D. J., Branson, F. F., & Balschweid, M. A. (2005). Preparing extension educators for a global community. Journal of Extension, 43(4). Article number. 4RIB1.

50- Shanley, E. L., Thompson, C. A., Dziura-Duke, C., & Rodriguez, N. (2009). Innovative food safety education for low literacy food service employees using traditional versus distance learning methods. Journal of the American Dietetic Association, 109(9), pg A63.

51-Shinn, Y. (1997). Teaching strategies, their use and effectiveness as perceived by teachers of agriculture: A national study (Doctoral Dissertation). Iowa State University, Ames.

52- Small, R. M., Waterbury, J. A., & Mark, D. R. (2008). Livestock insurance web site: A learning tool for producers, insurance agents, and educators. Journal of Extension, 46(3), Article number 3TOT4.

53- Smith, S., & Davis, D. (2008). The use of radio broadcast, internet, and podcasting in a family life education program, 46(2), Article number 2IAW1.

54- Thomson, J. S., Abel, J. L., & Maretzki, A. N. (2001). Edible Connections: A model to facilitate citizen dialogue and build community collaboration. Journal of Extension, 39(2), Article number 2FEA5.

55- Trepka, M. J. (2008). Interactive multimedia can be used effectively in food safety education. Journal of the American Dietetic Association, 108(6), pg 985.
56- United States Department of Veteran Affairs (2008). Methodological challenge. Using qualitative research methods. Retrieved from http://www.colmr.research.va.gov/mgmt_research_in_va/methodology/qualitative_research.cfm#2
57- University of Idaho (2006). Research methods. How psychologists conduct research. Retrieved June 23, 2010 from, http://www.class.uidaho.edu/psyc101/ResearchMethods6.pdf
58- Walczyk, J. J. & Ramsey, L. L. (2003). Use of learner-centered instruction in college of science and mathematics classroom. Journal of Research in Science Teaching. 40(6), 566-584.
59- Wardlaw, M. K. (1999). Empowering consumers: Hamburger safety. Journal of Extension, 37(3), Article number 3IAW5.
60- Webster, J., Rogers, D. L., & Mariger, S. C. (2001). Utah extension educators' perceived satisfaction with and needs for agricultural health and safety information. Journal of Extension, 39(2), Article number 2RIB3.
61-World Health Organization (2007). Food safety and foodborne illnesses. WHO Media centre. Retrieved from http://www.who.int/mediacentre/factsheets/fs237/en/
62- Xie, K., & Gu, M. (2007). Advancing Cooperative Extension with podcast technology, Journal of Extension, 45(5), Article number 5TOT2.
63- Yin, R. K. (2003). Case study research. Design and methods (3rd ed). Applied Social Research Methods. Vol 5.

19

Perceptions of Constraints Affecting Adoption of Women-in-Agriculture Programme Technologies in the Niger Delta, Nigeria

Iniobong A. Akpabio, Nsikak-Abasi A. Etim[1] and Sunday Okon*

Abstract

The study focused on constraints affecting the adoption of innovative agricultural technologies disseminated by the Women-in-Agriculture (WIA) unit of the Akwa Ibom Agricultural Development Programme (AKADEP) to its women clientele. The study also ascertained the awareness and adoption levels of such introduced technologies. Findings revealed that respondents were aware of 61.9% of introduced technologies, while only 33.3% were fully adopted. The study also identified seven factors responsible for the non-adoption of women farmers' related technologies. The three highest ranking constraining factors were revealed as; high cost of inputs, low income level of women farmers and lack of regular contact with WIA extension agents. Reasons have been proffered for the relatively low technologies' adoption levels. Recommendations have also been made to enhance the technology adoption level. These include the necessity to introduce only socio- economically and culturally compatible technologies to WIA clientele, a wholesale focus on follow-up activities after initial group based technology introduction activities, and the attachment of a credit scheme to the WIA program.

Keywords:
Constraints to Adoption, Awareness, Agricultural Technologies, Women farmers.

[1] *Department of Agricultural Economics and Extension, University of Uyo, PMB 1017, Uyo, Akwa-Ibom State, Nigeria.*
* *Corresponding author's email: etimbobo@yahoo.com*

INTRODUCTION

Agriculture constitutes a large share of national output and employs a majority of the labour force in most developing countries; hence the sector has been integrated into any thinking about development (World Bank, 2003). However, whereas agriculture-led growth played an important role in slashing poverty and transforming the economies of many Asian and Latin American countries, the same has not occurred in Africa, including Nigeria (Diao *et al.*, 2007). According to Baker (2005), technical change is the engine of long-term growth and it becomes technically important through diffusion. This is more so for agricultural production, where the prospect of enhanced production offered by improved agricultural technologies is recognized, according to the World Food Program, as essential to improving the household food security of small scale farmers, raising rural incomes and creating national surplus that can improve the basis for economic growth (WFP, 1998).

Baker (2005) took a retrospective view at Africa's lack of robust economic growth and dearth of modern technology and concluded that technology (especially agricultural technology) diffusion appears to have failed therein. Eicher (1992) revealed that nearly 100% of the increase in food production in the West African sub-region, since 1960, has come from expanded harvest area, rather than improvements in technology, a trend which Sanders (1996) has deemed, inefficient and with negative long term prospects.

Jafry (2000), Brown *et al.,* (2001) and the Directorate for International Development (DFID, 2004), among many other authors and research scientists, revealed that women are the key farmers, food producers and natural resource managers, in most countries of sub-Sahara Africa. This is because they provide 65 – 89% of food, provide nearly half of farm labour, shoulder over 90% of domestic responsibilities and work twice as many hours as men. Akpabio (2005) also reported an African study which revealed that women carry over 80 tonnes of fuel, water and farm produce for a distance of more than one kilometer over the course of a year. Despite all these contribution, the Technical Centre for Agricultural and Rural Cooperation, asserted that women are still restricted in their roles as farmers by unequal rights and unequal access to and control over resources, especially land (CTA, 2000). Women also carry out their work without much help from agricultural support mechanisms such as extension agencies, input suppliers and credit institutions (FAO, 2000).

The Women-in-Agriculture (WIA) sub-component of the Agricultural Development Programme (ADPs) was instituted in 1988 to address gender specific agricultural problems. The focus is on food nutrition, processing, storage and utilization of crop and livestock produce, in order to raise women's income and living standards through business oriented farming and processing strategies. Ever since the introduction of the WIA programme in Nigeria, and with the current emphasis on participatory extension, various efforts have been made to elicit various types and levels of information on the activities and effectiveness of the programme in specific limited areas (states) of Nigeria and the Niger Delta. Akpabio (2005b) reported that the WIA programme in Akwa Ibom State remains less than effective, in terms of its contribution to the upliftment of the economic and socio-psychological status of rural women while Adetoun (2000) in South Western Nigeria, and Eshiett (2007) with reference to Akwa Ibom State, revealed that only a few of the technologies disseminated to WIA clientele have been fully adopted.

The importance of agricultural technologies in the development process cannot be overemphasized. It is against this background that this study sought to ascertain clientele perceptions on reasons for the reported low trend of adoption of agricultural technologies. This study however covers the larger South – South (Niger Delta) region of Nigeria, hence it was decided first of all to ascertain on a wider scale of the Niger Delta, the validity of earlier reports of Akpabio (2005b) and Eshiett (2007). In essence, the study sought to answer pertinent questions relating to: (i) the level of women farmers' awareness of specified innovations introduced through the WIA programme and (ii) respondents' perceptions of constraints affecting adoption of technologies disseminated through the WIA programme in the Niger Delta.

MATERIALS AND METHODS

Study Area

The Niger Delta is located in the Southern part of Nigeria. It spreads over a total land mass of about 75,000 square kilometers. It is inhabited by an estimated 30 million population. The people are distributed into forty ethnic groups in about 13,329 communities/settlements in nine states. It is characterized by wetlands and water bodies, with creeks and rivers criss-crossing the entire Southern parts and is often regarded as the largest wetland in Africa and the third largest in the world. The region is however endowed with natural resources. It has the third largest mangrove forest, with the most extensive fresh water swamp forest and tropical rainforest characterized by great biological diversity. Alongside its immense potentials for agricultural revolution, the study area also hosts vast reserves of non-renewable natural resources, particularly hydro-carbon deposits in oil and gas.

The population for the study comprised all the leaders/representatives of different WIA groups who attended the various one-day interactive fora organized by the ADPs in all states in the region. Relevant data could be collated for five states. These were Akwa Ibom, Cross River, Delta, Edo, and Rivers. All the 267 participants were purposefully utilized for the study, although responses from 250 respondents were eventually utilized for data analysis (viz, table 1). A pre-tested and validated structured Interview Schedule and Focus Group Discussions were utilized to elicit relevant information from the selected sample. These activities were performed with the aid of trained enumerators.

Table 1: Selected Sample

S/N	State	Population	Sample
1	Akwa Ibom	51	51
2	Cross River	60	53
3	Delta	53	48
4	Edo	56	54
5	Rivers	47	44
	TOTAL	**267**	**250**

To ascertain the level of women farmers' awareness and adoption of specified innovations introduced through the WIA program, a list of technologies disseminated through the WIA programme was obtained, after which awareness and adoption scores were computed for each technology. Scores of 0 and 1 were recorded for awareness and non-awareness of disseminated technologies while scores of 2, 1 and 0 were recorded for adopted, discontinued and non-adopted technologies, respectively. A mean cut-off score of 0.5 was adopted to demarcate between technologies for which respondents were either 'aware' or 'not aware', while a cut-off mean score of 1.0 was utilized to differentiate between technologies which have either been 'adopted' or 'not adopted'. In essence, respondents were deemed to be aware of a technology with a mean score of 0.5 and above, while they were not aware of technologies with mean scores of less than 0.5. Similarly, technologies which recorded mean scores of 1.0 and above were perceived as adopted by respondents, unlike technologies with mean scores of less than 1.0, which were regarded as not adopted.

To determine respondents' perceptions of constraints affecting adoption of WIA programme technologies, a list of possible constraints that may hinder the adoption of disseminated technologies was drawn up with the aid of interviews and literature search. A 3-point Likert continuum of agreed (3) undecided (2) and disagreed (1) was employed to compute responses on reasons for non-adoption of WIA technologies. A cut-off mean score of 2.5(3+2+1/3 +0.5) was utilized to differentiate between 'major' and 'minor' factors for non –adoption, where a score of 2.5 and above, was depicted as a 'major' factor for non-adoption, while items with scores below 2.5 were adjudged minor factors.

RESULTS AND DISCUSSION

Awareness and adoption levels of WIA technologies

Tables 2 and 3 show that women farmers were aware of 61.9% (13 of 21) introduced

Table 2: Distribution of Respondents based on the extent of Awareness of WIA technologies

	AKADEP Technologies Food Crops	Aware (1)	Not aware (0)	Means	Remarks
1	Cassava/maize/melon planting	230 (92) *	20 (8)	0.92	Aware
2	Yam mini set	124 (49.6)	126 (50.4)	0.49	Not Aware
3	Dry season vegetable	170 (68)	80 (32)	0.68	Aware
4	Wet season vegetable	210 (84)	40 (16)	0.84	Aware
5	Rice cultivation	86 (34.4)	164 (65.6)	0.34	Not Aware
	Processing & Utilization				
6	Soya bean milk/ flour	140 (56)	110 (44)	0.56	Aware
7	Odorless fufu/garri	250 (100)	0 (0)	1.00	Aware
8	Fruit drinks	116 (46.4)	134 (53.6)	0.46	Not Aware
9	Plantain chips processing	210 (84)	40 (16)	0.84	Aware
10	Pineapple chips processing				
	Input use	200 (80)	50 (20)	0.80	Aware
11	Fertilizer use	180(72)	70(28)	0.72	Aware
12	Improved crop varieties e.g. maize	160 (64)	90 (36)	0.64	Aware
13	Agro chemicals e.g. Pesticides	84 (33.6)	166 (66.4)	0.33	Not Aware
14	Improved animal breeds	130 (52)	120 (48)	0.52	
	Agroforestry technology				
15	Snail rearing	124 (49.6)	126 (50.4)	0.49	Aware
16	Plantain /Cocoyam intercropping	196 (78.4)	54 (21.6)	0.78	Aware
17	Afang cultivation	210 (84)	40 (16)	0.84	Aware
18	Bee raising	78 (31.2)	172 (68.8)	0.31	Aware
	Tree crops planting				
19	Improved oil palm seedlings	160 (64)	90 (36)	0.64	Aware
20	Rubber seedlings	90(36)	160 (64)	0.36	Not Aware
21	Improved cocoa seedlings	70 (28)	180(72)	0.28	Not Aware

*-Percentages in parentheses

technologies, while only 33.3% (7 of 21) of the technologies were eventually adopted. It was also observed on table 3, that respondents adopted 53.9% (7 of 13) of the technologies for which they were aware. A related finding in the course of study revealed that only 59.2% respondents received information on improved agricultural technologies from extension officials of the WIA program, while 20.8% and 20% respondents received information from relatives/friends and husbands, respectively. There is a cause for concern here. This is because an extension program deliberately targeted at women farmers reaches only 59.2% of intended clientele. Many reasons have been proffered for this undesirable situation. These include; long distance from meeting venues and concomitant non-attendance at group meetings (Adetoun, 2000) lack of interpersonal contact, arising from lack of follow-up after group meetings (Udoh, 2001) and lack of relevance of disseminated messages to the amelioration of female farmers livelihood constraints (Reij and Waters-Bayer, 2001) among many others, might have led to clientele' loss of interest in extension offerings.

Table 3 also shows that none of the technologies disseminated to respondents under "input use" and "tree crops planting" classifications was adopted. Odourless fufu/garri (fresh/dried cassava paste) x = 1.96; cassava/maize/melon crops combination (x = 1.76) and intercropping (x = 1.45) were the most adopted technologies. This result corroborates Baker's (2005) and Swinkels and Franzel's (1997) assertion that compatible technologies and technologies that differed very little from the old technologies would diffuse faster since there would be less of an information problem associated with them. Pannell (1999) described four conditions necessary for farmers to adopt innovative technologies, two of which are "awareness of the technology" and "perception that technology promotes farmers objectives". It may be inferred that farmers will adopt more of the technologies for which they are aware. In essence, awareness of technology is a motivating factor for the adoption of technological packages. Hardarker, Huirne and Anderson (1997) however

Table 3: Distribution of respondents based on the extent of adoption of WIA technologies.

	AKADEP Technologies (Food Crops)	Not Adopted (0)	Discontinued (1)	Adopted (2)	Means**	Remarks
1	Cassava/maize/melon planting	0 (0) *	20 (8)	210 (84)	1.76	Adopted
2	Yam mini set	0 (0)	88 (35.2)	36 (14.4)	0.64	Not Adopted
3	Dry season vegetable	60 (24)	60 (24)	48 (19.2)	0.62	Not Adopted
4	Wet season vegetable	0 (0)	40 (16)	170(68)	1.52	Adopted
5	Rice cultivation	172 (34.4)	0 (0)	0 (0)	0.00	Not Adopted
	Processing & Utilization					
6	Soya bean milk/ flour	10 (4)	70 (28)	60 (24)	0.76	Not Adopted
7	Odourless fufu/garri	0 (0)	10 (4)	240 (96)	1.96	Adopted
8	Fruit drinks	110 (44)	6 (2.4)	0 (0)	0.02	Not Adopted
9	Plantain chips processing	4 (1.6)	16 (6.4)	190 (76)	1.58	Adopted
10	Pineapple chips processing	14 (5.6)	10 (4)	176 (70.4)	1.45	Adopted
	Input Use					
11	Fertilizer use	40 (16)	52 (20.8)	88 (35.2)	0.91	Not Adopted
12	Improved crop varieties e.g maize	24 (9.6)	60 (24)	76 (30.4)	0.84	Not Adopted
13	Agro chemicals eg. pesticides	34 (13.6)	46 (18.4)	4 (1.6)	0.21	Not Adopted
14	Improved animal breeds	40 (16)	20 (8)	60 (24)	0.56	Not Adopted
	Agroforestry technology					
15	Snail rearing	16.(6.4)	24 (9.6)	84 (33.6)	0.76	Not Adopted
16	Plantain /Cocoyam intercropping	12 (4.8)	4 (1.6)	180 (72)	1.45	Adopted
17	Afang cultivation	30 (12)	16 (6.4)	164 (65.6)	1.36	Adopted
18	Bee raising	78 (31.2)	0 (0)	0 (0)	0.00	Not Adopted
	Tree crops planting					
19	Improved oil palm seedlings	140 (56)	20 (8)	0 (0)	0.08	Not Adopted
20	rubber seedlings	82 (32.8)	8 (3.2)	0 (0)	0.03	Not Adopted
21	Improved cocoa seedlings	66 (26.4)	4 (1.6)	0 (0)	0.01	Not Adopted

*-Percentages in parentheses
** Mean Scores calculated, based on total no. of respondents- regardless of the no. of actual recorded responses per innovation.

cautioned that high level awareness of technologies does not necessarily translate into higher adoption levels. This is because farmers will only adopt those innovations which are adjudged useful and beneficial to their particular situation. A disaggregated analysis of table 2 reveals relatively high frequencies of "non-awareness" scores that were recorded for some technological offerings which were generally perceived (mean scores) as being "aware" of by respondents (viz; items 3, 6, 12, 14 and 19). Bunch (1982) and Baker (2005) harped on the importance of critical mass in the adoption of innovations. The researchers contended that that there is a higher level of adoption and less discontinuance for a new technology in which the whole community or a critical mass (proportionately larger than average) of farmers are aware of and in which they are interested. They explain reasons for this in terms of traditional communities being accustomed to living in an environment of consensus, and that schemes which entail community risk sharing are more easily imbibed than otherwise.

Constraints affecting adoption of WIA programme technologies

Results as shown on table 3 reveals that respondents perceived 7 of the 21 identified items as possible reasons for the relatively low adoption levels of agricultural technologies introduced through the WIA programme. These are: high cost of inputs ($x = 3.0$), low income level of women farmers' ($x = 2.97$) lack of regular contact with extension agents ($x = 2.82$) old age of women farmers in the study area ($x = 2.73$) poor attitude towards risk and change ($x = 2.55$) and complexity of introduced technologies ($x = 2.55$). The above revealed findings find relevance in related literature. Rogers (1995) identified five key characteristics of innovations that determine their adoption potential, including: relative advantage, trialability, compatibility, observability and complexity. Reed (2001) iden-

Table4: Distribution of respondents based on perception of factors affecting non-adoption of WIA technologies.

	Reasons for Non Adoption/ Discontinuace	Disagreed (1)	Undecided (2)	Agreed (3)	Mean (X)	Remarks
1	High cost of inputs	0(0) *	0(0)	250 (100)	3.00	Major Factor
2	Lack of supporting inputs	88(35.2)	62(24.8)	100(40)	2.05	Minor Factor
3	Problem of diseases / pests	66 (26.4)	34(13.60)	150 (60)	2.33	Minor Factor
4	Non-appropriateness of the technological package to the Local environment	60 (24.1)	68 (27.2)	122 (48.8)	2.25	Minor Factor
5	Non-availability of the improved package	84 (32.80)	16 (6.4)	152 (60.8)	2.28	Minor Factor
6	Non-Profitability of the new technology	188 (75.2)	40 (16)	22 (8.8)	1.16	Minor Factor
7	Superiority of the old technology to the newly introduced one.	174 (69.6)	48 (19.2)	28 (11.2)	1.42	Minor Factor
8	Incompatibility of the new technology with the norms and customs of the local environment	144 (57.6)	60(24)	46 (18.4)	1.60	Minor Factor
9	Lack of clear understanding of the newly introduced package	12(4.8)	104(14.6)	134 (53.6)	2.48	Minor Factor
10	Low level of educational attainment by women farmers in the study area.	48 (19.2)	106 (42.4)	96(38.4)	2.19	Minor Factor
11	Low level of income of women farmers in the area	2 (0.8)	4 (1.6)	244 (97.6)	2.97	Major Factor
12	Insufficient Programs designed to convince and encourage farmers to change	12 (4.8)	144 (59.6)	94 (37.6)	2.32	Minor Factor
13	Women farmers perception of the old technology as better than the new one.	138 (55.2)	68 (29.2)	44 (17.6)	1.62	Minor Factor
14	Inconsistence of the innovation with the existing farming system, values and needs of women farmers in the area	156 (62.4)	36(14.4)	58 (23.2)	1.61	Minor Factor
15	Inadequate information about the newly introduced technological package.	50(20)	58 (23.2)	142(59.2)	2.44	Minor Factor
16	Complexity of the introduced innovation.	44(17.6)	24 (9.6)	182 (92.8)	2.55	Major Factor
17	Failure of some demonstration plots set –up by the extension agents.	138 (55.2)	58(23.2)	54 (21.6)	1.66	MinorFactor
18	Lack of regular contact with extension agents	20(8)	4(1.6)	226(90.4)	2.82	Major Factor
19	Poor attitude of women farmers towards change and risk	8 (3.2)	60 (24)	182 (72.8)	2.70	Major Factor
20	Age of women farmers in the study area	20 (8)	28 (11.2)	202 (80.8)	2.73	Major Factor
21	Lack of access and control over production resources such as land and credit facilities.	8(3.2)	96(38.4)	146(58.4)	2.55	Major Factor

*-Percentages in parentheses

tified the most significant of these characteristics, as: high relative advantage, high compatibility and low complexity. Swinkels and Franzel (1997) agreed with the submission above, but also opined that for the female gender, additional incentives for adoption may include factors like, suitability to accepted gender roles, cultural acceptance and compatibility with other enterprises.

High cost of inputs for introduced technologies and low income of respondents' were revealed as the greatest constraints to adoption of introduced technologies. Obinne (1994) and Arokoyo (1996) mentioned low income level of farmers and high cost of inputs as constraints to technology adoption, especially among low income farmers. In that wise, Baker (2005) and Hebinck, Franzel and Richards (2007) asserted that the most successful programmes of agricultural change are those that tie adoption to credit programmes. Udoh (2001) and Eshiett (2007) maintained that contact with extension agents, especially with respect to interpersonal contacts, relate favorably to the adoption of new farm practices and concomitant improved agricultural production. Obinne (1994) and Baker (2005) opined that poor attitude to risk, in terms of excessive risk aversion may severely limit adoption of technological innovations especially among female rural farmers, while Baker (2005) opined that technologies that differed very little from the old technologies would diffuse faster than unrelated technologies, while the older generation may credibly block adoption even if the younger generation co-ordinates. Dove (1991) contended that individuals with insecure tenure will generally be less likely to invest in new technologies that require complementary immobile inputs, while Due, Mudenda and Miller (1993) asserted that although women want to increase the productivity of the resources they control, they face greater obstacles to change. One of such obstacles, according to Reij and Waters-Bayer (2001) is lack of relevance of disseminated messages to the amelioration of female farmers' livelihood constraints. (PLEASE, THIS IS NOT LITT REVIEW. THIS IS SIMPLY A COMPARISON OF FINDINGS WITH RESULTS FROM PREVIOUS STUDIES)

It is obvious that although poor female farmers in the study area are conscious of innovating in order to overcome their present precarious socio-economic situation, they are however precluded from benefiting from opportunities open to them due to various constraining factors, as have been identified above.

CONCLUSION AND RECOMMENDATIONS

It has been revealed that the WIA program, as being implemented in Akwa Ibom State does not reach out to a large number of its intended clientele base. This has resulted in an average level of awareness and concomitant relatively lower level of adoption of innovative technologies disseminated by WIA extension officials. The study also identified seven factors which combine to hinder the adoption of disseminated WIA technologies. The major constraints were: high cost of inputs, low income level of women farmers and lack of regular contact with WIA extension agents. Many reasons, backed by literature, have been proffered for this trend, including the fact that only 59.2 percent respondents regarded WIA extension officials as their source of information on innovative agricultural technologies. In order to enhance the success of the WIA programme in Akwa Ibom State, attempts should be made to ameliorate constraints which hinder extension officials' access to their potential clientele. To aid in this direction, adequate logistic support should be

provided to WIA extension agents so as to help enhance the process of contacting their expectant clientele Technologies slated for dissemination should be compatible to clientele socio-economic and cultural base and emphasis should be focused on follow-up activities, after initial group meetings. This would help to practicalize disseminated technologies on the farms and in the homes of potential adopters of technological innovations. It may also be necessary to attach credit schemes to the WIA program, in terms of linking the various women groups to various credit agencies.

REFERENCES

1- Adetoun, B. A. (2000) Organization and Management of Agricultural Extension Services for Women Farmers in South-Western Nigeria. Centre for Sustainable Development and Gender Issues. Abner Publishing, Ibadan, (Abstract).

2- Akpabio, I.A (1997) Extension Communication Methods as a Strategy for Agricultural Development in Akadep In: Issues in Sustainable Agricultural Development. Fabiyi, Y.L. and M.G. Nyienakuna (eds). Vol. 1. August. Dorand Publishers. Uyo. P. 122.

3- Akpabio, I.A. (2005). Human Agriculture: Social Themes in Agricultural Development. Abaam Pub. Co. Uyo. 219 pp.

4- Akpabio, I.A. (2005b). 'Beneficiary Perceptions of the Gender Specific Extension Delivery Service in Akwa-Ibom, Nigeria.' Gender and Behaviour 396-405.

5- Anderson, J. R and P. B. Hazell (1989). Variability in Grain Yields: Implications for Agricultural Research and Policy in Developing Countries. Baltimore: John Hopkins University Press, p. 23

6- Arokoyo, T. (1996). Agricultural Technology Development and Dissemination. A Case Study of Ghana and Nigeria. CTA. The Netherlands, p. 5.

7- Baker, E. (2005) Institutional Barriers to Technology Diffusion in Rural Africa. University of Massachusetts, Amherst. 30pp.

8- Bunch, R. (1982). Two Ears of a Corn. USA: World Neighbours, p. 11.

9- Brown, L., Feldstein, H., Haddad, L.C. & Pena, A. Quisumbing (2001). Generating Food Security in the year 2020: Women as Producers, Gatekeepers and Shock Absorbers" (In) The Unfinished Agenda. Pinstrup-Andersen and Pandya-Lorch (eds) IFPRI. Washington D.C. pp. 205-209.

10- Diao, X., P. Hazell, D. Resnick and J. Thurlow (2007). The Role of Agriculture in Development: Implications for SSA. Research Paper 153. IFPRI. Washington DC, p.5
11- Department for International Development (DFID 2004) "Gender" Fact Sheet. Sept. London.
12- Dove, M. R. (1991) Forester's Beliefs about Farmers: An Agenda for Social Science Research. In: Social Forestry- EAPI Working Paper (Abstract).
13- Due, J., Mudenda, T., & Miller, C. (1983). How Do Rural Women Perceive Development? A Case Study in Zambia. Report no 83-E-265. University of Illinois. Dept of Agricultural Economics, Urbana, IL.
14- Eicher, C. (1992) "African Agricultural Development Strategies. In: Alternative Development Strategies in sub-Saharan Africa. Stewart, F., L. Sanjaya and S. Wengwen (eds) Macmillan. London.
15- Eshiet, E. (2007) "Level of Participation of Women Farmers in the Women in Agriculture Programme in Akwa Ibom State". Unpublished M. Sc thesis. Dept of Agricultural Economics and Extension, Universilty of Uyo, Nigeria. 88 pp.
16- FAO, (2000). Rural Women and Development Agencies. Rome. P. 23
17- Hardarker J. B., Huirne R.B.M & Anderson, J.R. (1997). Coping with Risks in Agriculture. CAB. Int. Oxon. UK.
18- Hebinck, P., Franzel, S., & Richards, P. (2007). Adopters, Testers or Pseudo Adopters? Dynamics of the Use of Improved Tree Fallow by Farmers in Western Kenya. AgroSystems 94(2): 509-515.
19- Jafry, T. (2000). Women, Human Capital and Livelihood – an Ergonomics Perspective. 0D1 Natural Resources Perspectives (Abstract).
20- Pannell, D.J. (1999). Social and Economic Challenges in the Development of Complex Farming Systems. Agroforestry Systems 45: 393-409.
21- Obinne, C. 1994). Fundamentals of Agricultural Extension. ABIC Publishers, Enugu, Nigeria.
22- Reid, M.S. (2001). Paticipatory Technology Development for Agro-forestry: An Innovation – Decision Approach. Centre for Enviornmental Management Working Paper R01/121. University of Leeds.
23- Reij, C., & Waters-Bayer, A. (2001). An Initial Analysis of Farmers. Innovations in Africa. in: A Source of Inspiration for Agricultural Development. Reij and Water Bayer eds. Earthscan Publications pp. 3-22.
24- Rogers, E. M. (1995). Diffusion of Innovations, 4th Edition. The Free Press, NewYork.
25- Swinkels, R., & Franzel, S. (1997). Adoption Potential of Hedgerow Intercropping in Maize Based Cropping Systems in the Highlands of Western Kenya. Economic and farmers' evaluation. Experimental Agriculture 33:211-223.
26- Technical Centre for Agricultural and Rural Cooperation (CTA-2000) The Economic Role of Women in Agricultural and Rural Development: Promoting Income-Generating Activities. C.T.A. Wageningen. The Netherlands, 56pp.
27- Udoh, A.J. (2001). Agricultural Extension Development Administration. Etofia Media Services Ltd. Uyo. p. 39.
28- Udokah, E. (2004). Information Source Preference of Small Scale Farmers in Akwa Ibom State. Un-Published M. Sc thesis. Dept of Agricultural Economics and Extension, University of Uyo, Nigeria, 71pp.
29- World Bank (2003). Sudan Country Economic Memorandum: Stabilisation and Reconstruction. Washington DC. World Bank, p. 3.
30- World Food Programme (1998). Food Aid to Support Technology Adoption Among Small Scale Agriculturists. WFP, Rome, Italy.

20

A Logistic Regression Analysis: Agro-Technical Factors Impressible from Fish Farming in Rice Fields, North of Iran

***Seyyed Ali Noorhosseini-Niyaki* [1*], *Mohammad Sadegh Allahyari* [2]**

Abstract

This study was carried out to identify Technical-Agronomic Factors Impressible from Fish Farming in Rice Fields. This investigation carried out by descriptive survey during July-August 2009. Studied cities including Talesh, Rezvanshahr and Masal set in Tavalesh region near to Caspian Sea, North of Iran. The questionnaire validity and reliability were determined to enhance the dependability of the results. Data were collected from 184 respondents (61 adopters and 123 non-adopters) randomly sampled from selected villages and analyzed using logistic regression analysis. Results showed that there was a significant positive relationship ($p<0.05$) between biological control of pests in rice fields and the fish farming in rice fields. Also, there was a significant negative relationship ($p<0.10$) between the fish farming in rice fields and variables of quantity using pesticide of Diazinon in rice fields and number of plows in rice fields.

Keywords:
Rice-Fish Farming, Technical-Agronomic Factors, Pest, Weed, Plow, Fertilizer

[1]*Department of Agronomy, Lahidjan Branch, Islamic Azad University, Lahidjan, Iran*
[2]*Department of Agricultural Management, Rasht Branch, Islamic Azad University, Rasht, Iran*
* *Corresponding author's email: Noorhosseini.SA@gmail.com*

INTRODUCTION

The earliest records of fish culture in rice-fields originate from China, circa 2000 years ago, followed by India, 1500 years ago. Other countries with a recorded history of rice-fish culture are Indonesia, Malaysia, Thailand, Japan, Madagascar, Italy, Russia, Vietnam (Rothuis, 1998), Egypt, Philippines, Bangladesh, Cambodia, Korea and other countries (Saikia and Das, 2008; Frei and Becker, 2005; Halwart, 1998). Also in northern Iran rice-fish farming is a new farming system (Karami *et al.,* 2006; Noorhosseini and Allahyari, 2010). Integrated rice-fish farming offers a solution to economic problem of farmers by contributing to food, income and nutrition. Not only the adequate supply of carbohydrate, but also the supply of animal protein is significant through rice-fish farming. Fish, particularly small fish, are rich in micronutrients and vitamins, and thus human nutrition can be greatly improved through fish consumption (Larsen *et al.,* 2000; Ahmed and Garnett, 2011; Noorhosseini and Mohammadi, 2010). Many reports suggest that integrated rice-fish farming is ecologically sound because fish improve soil fertility by increasing the availability of nitrogen and phosphorus (Giap *et al.,* 2005; Dugan *et al.,* 2006; Noorhosseini, 2012).The feeding behavior of fish in rice fields causes aeration of the water. Integrated rice-fish farming is also being regarded as an important element of integrated pest management (IPM) in rice crops (Berg, 2001; Hilbrands and Yzerman, 2004). At the farm level rice-fish integration reduces use of fertilizer, pesticides and herbicides in the field. Such reduction of costs lowers farmer's economic load and increases their additional income from fish sale (Noorhosseini, 2010; Noorhosseini and Radjabi, 2010). Also, integrated rice-fish farming gave higher rice yields and fetched higher gross margin than sole rice cropping system (Das *et al.,* 2002; Hossain *et al.,* 2005). Ahmed and Garnett (2011) Reported that higher yields can be achieved by increasing inputs in the integrated farming system. Integrated rice-fish farming also provides various socioeconomic and environmental benefits. Nevertheless, only a small number of farmers are involved in integrated rice-fish farming due to a lack of technical knowledge, and an aversion to the risks associated with flood and drought. In addition, Ahmed *et al.,* (2011) Reported that rice-fish farming is as production efficient as rice monoculture and that integrated performs better in terms of cost and technical efficiency compared with alternate rice-fish farming. However, a lack of technical knowledge of farmers, high production costs and risks associated with flood and drought are inhibiting more widespread adoption of the practice. Our objective was to identify technical-agronomic factors impressible from fish farming in rice fields in north of Iran.

MATERIALS AND METHODS

Studied Location and Survey: This study was carried out by survey during July and August 2009. Studied area including Talesh, Rezvanshahr and Masal set in Tavalesh region of Guilan province near to Caspian Sea, north of Iran (Figure 1). Respondents were selected from rural area and categorized into adopters and non-adopters of integrated rice-fish farming. Totally 184 farmers were selected by stratified random sampling technique using the table for determining the sample from given population developed by Bartlett *et al.*, (2001) that including 61 (33.15%) adopters and 123 (66.85%) non-adopters (Table 1). This survey was conducted in using a questionnaire with open-ended questions. The questionnaire was pre-tested by in-

Table 1: Total sample size used in the study area

	Talesh	Masal	Rezvanshahr	Total
IRFF Adopters Population	31	31	17	17
IRFF Adopters Sample Size	19	28	14	14
IRFF Non-adopters Sample Size	38	56	29	29

Source: Survey Results, 2009

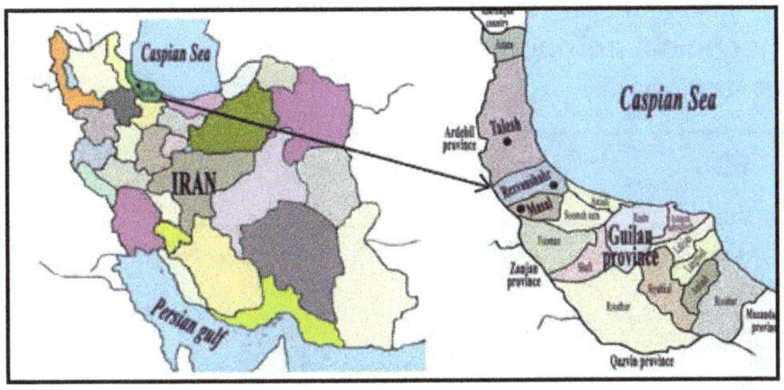

Figure 1: Site of study

terviewing three farmers (not included in the study). After some modifications, it was tested again with 10 other respondents.

Statistical Analysis: The technical-agronomic variables for the two groups were examined using logistic regression model. The dependent variable was the adoption of rice-fish farming. The dependent variable was dichotomized with a value 1 if a farmer was an adopter of integrated rice-fish farming and 0 if non-adopter. The definitions and measurement of variables are present in Table 2. AF, AH, MW, AP, BP and WI were entered in the model as dummy variables. The other variables namely QF, QD and NP were entered as continuous variables. Data analysis was conducted with Statistical Package for Social Sciences (SPSS 18).

The model was specified as follows;

$$Y=f(AF, QF, AH, MW, AP, BP, QD, NP, WI)$$

strength of the joint effect of the covariates on probability of adoption among farmers in the zone. The results also showed that the decision on adoption of rice-fish farming is determined by biological control of pests in rice fields (BP), quantity using Diazinon in rice fields (QD) and number of plows in rice fields (NP) which have significant influence. Also, the Wald indicating the relative contribution of individual variable to probability of adoption of rice-fish farming showed that BP (4.538) was the most important factor determining choice of adoption of rice-fish farming among the rice farmers. Generally, the results of logistic regression show that there was a significant positive relationship ($p<0.05$) between biological control of pests in rice fields and the fish farming in rice fields (Table 3). These results were similar to Frei and Becker (2005) and Kathiresan (2007). In other words, fish farming in rice fields reduced the use of chemical control methods that is light reason of rice-fish farming system sustainability. Also, there was a significant negative relationship ($p<0.10$) between the fish farming in rice fields and quantity using Diazinon in rice fields (Table 3). These results are consistent with Saikia and Das (2008), Salehi and Momen Nia (2006). The use of chemical control methods reduced with adoption of rice-fish farming which is also

Table 2: Definition of variables included in the regression model

Dependent variable **Y = Adoption**	**Adopters = 1, Non adopters = 0**
Independent variable	Yes = 1, No = 0
AF = Application of Chemical Fertilizers	Kg/ha
QF = Quantity Using Chemical Fertilizers	Yes = 1, No = 0
AH = Application of Herbicides	Yes = 1, No = 0
MW = Mechanical Control of Weed	Yes = 1, No = 0
AP = Application of Pesticides	Yes = 1, No = 0
BP = Biological Control of Pests	Kg/ha
QD = Quantity Using Diazinon	Number in year
NP = Number of Plows	Very mach = 5, Much = 4, Intermediate = 3, Little = 2, Very
WI = Accessibility to Water Supply for Irrigation	little = 1

RESULTS AND DISCUSSION

The results of the Logit likelihood regression model indicated that the overall predictive power of the model (70.1%) is quite high, while the significant Chi square ($p<0.01$) is indicative of

compatible with sustainable agriculture. Since many fish species feed partly on the aquatic fauna, it has been assumed that they can act as biological control agents in rice fields. Concurrent rice and fish culture decrease pesticides appli-

Table 3: Logistic regression coefficients of the technical-agronomic factors affecting adoption of rice-fish farming

	B	S.E.	Wald	Sig.
AF	-19.131	25320.477	0.000	0.999
QF	0.000	0.001	0.005	0.943
AH	-21.476	18858.287	0.000	0.999
MW	20.493	40192.879	0.000	1.000
AP	0.611	0.558	1.200	0.273
BP	0.996	0.468	4.538	0.033**
QD	-0.029	0.016	3.311	0.069*
NP	-0.717	0.396	3.273	0.070*
WI	0.112	0.167	0.450	0.502
Constant	20.822	51110.003	0.000	1.000

*** $p<0.01$, ** $p<0.05$ and * $p<0.10$
-2 log likelihood = 204.830
Chi square statistic = 28.943***
Overall Correct predictions = 70.1%

cation compare monoculture (Berg, 2002; Noorhosseini, 2011). In general, common carp, being an omnivorous feeder, seems to be the most promising species in controlling insects and snails [15]. In addition, there was a significant negative relationship ($p<0.10$) between fish farming in rice fields and number of plows in rice fields (Table 3). According to results, by adoption of rice-fish farming, number of plows was reduced. It seems that the decrease of tillage frequency among adopters, caused by farms occupied by fishes is the time dimension.

CONCLUSION

In general, our results suggest that biological control of pests in rice fields, quantity using Diazinon in rice fields, and numbers of plows in rice fields were the most important technical-agronomic factors impressible from fish farming in rice fields. In other words, adopters of rice-fish farming used less chemical materials in order to control pests and reduce the number of plows. Furthermore, society health and environment sustainability will be saved and they reach more profit that is economical. Also, since aquaculture requires resources such as pond, land, water and other inputs, poor farmers cannot afford the requirements. As a target to understand and meet their needs and to access the common water resources available in their rice-fields, rice-fish farming is the most appropriate technology in recent times.

REFERENCES

1- Ahmed, N. & Garnett, T.S. (2011). Integrated Rice-fish Farming in Bangladesh: Meeting the challenges of food security. Food Security. 3(1): 81-92.

2- Ahmed, N., Zander, K.K., & Garnett, S.T. (2011). Socioeconomic Aspects of Rice-fish Farming in Bangladesh: Opportunities, Challenges and Production Efficiency. The Australian Journal of Agricultural and Resource Economics, 55: 199–219.

3- Bartlett, J.E., J.W. Kotrlik and C.C. Higgins, (2001) Organizational Research: Determining Appropriate Sample Size in Survey Research. Information Technology, Learning and Performance Journal, 19(1): 43-50.

4- Berg, H. (2001) Pesticide Use in Rice and Rice-Fish Farms in the Mekong Delta, Vietnam. Crop Protection, 20: 897–905.

5- Berg, H. (2002) Rice Monoculture and Integrated Rice–Fish Farming in the Mekong Delta, Vietnam-economic and ecological considerations. Ecological Economics, 41: 95– 107.

6- Das, D.R., Quddus, M.A., Khan, A.H. & Nur-e-Elahi, M. (2002). Farmers Participatory Productivity Evaluation of Integrated Rice and Fish Systems in Transplanted Aman rice. Pakistan Journal Agronomi. 1(2-3): 105-106.

7- Dugan, P., Dey, M.M., & Sugunan, V.V. (2006). Fisheries and Water Productivity in Tropical River Basins: Enhancing Food Security and Livelihoods by Managing Water for Fish. Agricultural Water Management, 80: 262–275.

8- Frei. M., & K., Becker. (2005). Integrated Rice-Fish Culture: Coupled Production Saves Resources. Natural Resources Forum. 29: 135–143.

9- Giap, D.H., Yi, Y., & Lin, C.K. (2005). Effects of

Different Fertilization and Feeding Regimes on the Production of Integrated Farming of Rice and Prawn Macrobrachium Rosenbergii (De Man). Aquaculture Research, 36: 292–299.

10- Halwart, M. (1998). Trends in Rice-Fish Farming. FAO Aquaculture Newsletter, 18: 3-11.

11- Hilbrands, A., & Yzerman, C. (2004). On-Farm Fish Culture. Printed by: Digigrafi, Wageningen, the Netherlands. Second edition. 67p.

12- Hossain, S.T., Sugimoto, H., Ahmed, G.J.U., & Islam, M.R. (2005). Effect of Integrated Rice-Duck Farming on Rice Yield, Farm Productivity, and Rice-Provisioning Ability of Farmers. Asian Journal of Agriculture and Development. 2 (1): 79-86.

13- Karami E. A., Rezaei Moghadam K., Ahmadvand M., & Lari, M.B. (2006). Adoption of Rice- Fish Farming (RFF) in Fars province. Iranian Journal Agriculture. Extension & Education. 2(2): 31-43.

14- Larsen, T., Thilsted, S.H., Kongsbak, K., & Hansen, M. (2000). Whole Small Fish as a Rich Calcium Source. British Journal of Nutrition. 83: 191–196.

15- Noorhosseini-Niyaki, S.A. (2010). Fish Farming in Rice Fields Toward Sustainable Agriculture: The Case Study in Guilan Province. The First National Conference on Sustainable and Cleaner Product, Esfahan of Iran. Nov 10-11, 2010.

16- Noorhosseini-Niyaki, S.A. (2011). Ecological and Biological Effects of Fish Farming In Rice Fields. Regional Congress of Sustainable Management Science-based in Agriculture and Natural Resources. Gorgan University of Agricultural Sciences and Natural Resources, Iran. 21-22 May 2011. pp: 244-250.

17- Noorhosseini-Niyaki, S.A. (2012) Production of Fish Varieties in Rice Fields Simultaneously. National Conference on Food Industries, Young Researchers Club, Islamic Azad University, Quchan Branch, Iran.

18- Noorhosseini-Niyaki, S.A., & Allahyari, M.S. (2010). Problems of Fish Farming in Rice Fields: The Case Study in Guilan Province. The First National Conference of Aquatic Sciences, Islamic Azad University, Booshehr Branch, Iran. 23-24 Feb, 2010. 15p.

19- Noorhosseini-Niyaki, S.A., & Mohammadi, N. (2010). Environmental and Social Benefits of Fish Farming in Rice Fields. The 4th Conference and Exhibition on Environmental Engineering, Tehran University, Iran.

20- Noorhosseini-Niyaki, S.A., & Radjabi, R. (2010). Decline Application of Insecticide and Herbicides in Integrated Rice-Fish Farming: the Case Study in North of Iran. American-Eurasian Journal Agriculture & Environ Science. 8(3): 334-338.

21- Rothuis, A. (1998). Rice-Fish Culture in the Mekong Delta, Vietnam: Constraint Analysis and Adaptive Research. Thesis Submitted for the Award of the Degree of Doctor of Science. Department of Biology, Laboratory of Ecology and Aquaculture, Katholieke Universiteit Leuven, Vietnam. 113p.

22- Saikia. S.K., & Das. D.N. (2008). Rice-Fish Culture and its Potential in Rural Development: A Lesson From Apatani Farmers, Arunachal Pradesh, India Journal Agriculture Rural Development. 6(1&2): 125-131.

23- Salehi, H, Momen-Nia, M. (2006) The Benefits of Fish and Rice Integrated Culture in Iran. Iranian Journal Fish Science. 15(3):97-108.

24- Kathiresan, R.M. (2007). Integration of Elements of a Farming System for Sustainable Weed and Pest Management in the Tropics. Crop Prot 26: 424-429.

, 2012.

21

Relationship between Meta-Technology Ratios and Varietal Differences in Date Production (Case Study of Sistan and Baluchestan, Iran)

Shahram Saeedian[1*], Mahmoud Sabouhi Sabouni[2], Javad Shahraki[3], Ebrahim Moradi[4], Mohammad Sepahi[5]

Abstract

This paper uses the concept of Meta-frontier to investigate the relationship between Meta Technology Ratios (MTR) and different varieties in datepalms including; Mazafati, Rabi and Zardan that are produced widely in Sistan and Baluchestan province of Iran. We used a stratified sampling of 300 date farmers of this province in 2010, 2011 and 2012 and estimated a stochastic production frontier using pooled data. Then, stochastic production frontiers were estimated for each variety, separately. At the end, the meta-frontier parameters were obtained to estimate the Meta technology ratio. Results showed that the estimated mean values of technical efficiency for the pooled frontier, variety group frontiers and Meta frontier across all data were 0.558, 0.543 and 0.0014, respectively. The value of Meta technology ratio was 0.407 for Mazafati variety, 0.432 and 0.507 for Rabi and Zardan varieties respectively. Mazafati had the lowest MTR while, Zardan and Rabi varieties are more closed together. Surprisingly, Mazafati had the lowest technical efficiency with respect to group frontier and Meta-frontier. The estimated average technical efficiencies with respect to group frontiers for Mazafati, Zardan and Rabi are: 0.518, 0.520 and 0.592, respectively; while Mazafati is kind of date which is the most exported variety.

Keywords:
Meta -frontier, Meta technology ratio, Datepalms, Sistan & Baluchestan

[1] *Graduate Student of Agricultural Economics, Economics Department, Sistan and Baluchestan University.*
[2] *Associate Professor, Agricultural Economics Department, Zabol University.*
[3] *Assistante Professor, Economics Department, Sistan and Baluchestan University.*
[4] *Ph.D. Student of Economics, Economics Department, Sistan and Baluchestan University.*
[5] *Graduate Student of Agricultural Economics, Economics Department, Sistan and Baluchestan University.*
** Corresponding author's email: saeedian1989@gmail.com*

INTRODUCTION

Good performance is an important factor for improving productivity in agricultural economics of developing countries. These countries can use the results of studies involved with this context broadly for their benefits because of resource and opportunity shortages they are faced for accepting and developing new technologies. Results of these studies show that it is still possible to improve the productivity with efficiency increment without using more basic resources or expanding new technologies. In addition; these research outcomes will be helpful in level of policy making, for choosing between two options; accepting new technologies or increasing the efficiency of existent technology (Sabouni, 1997).

A few farmers produce solely one variety of date, but many of them turn out two or more kinds of dates. Animals are sometimes fed by Zard variety. Mazafati and Rabi varieties are just for selling in the country or exporting abroad. One of important issues related to date production that should be noticed by farmers is technical efficiency. Although much has been written to measure technical efficiency of crops in Iran (see Najafi and Zibaee, 1994; Najafi and Abdollahi, 1997; Bakhsude *et al.,* 2001; Karbasi *et al.*, 2004, and Karim koshte *et al.*, 2004 for surveys), relatively little attention has been paid to the varietal or regional differences in agricultural field. O'Donnell *et al.,* (2008), put it well, "Firms in different industries, regions and/or countries face different production opportunities. Technically, they make choices from different sets of feasible input-output combinations and these so-called technology sets differ because of differences in available stocks of physical, human and financial capital (e.g., type of machinery, size and quality of the labor force, access to foreign exchange), economic infrastructure (e.g., number of ports, access to markets), resource endowments (e.g., quality of soils, climate, energy resources) and any other characteristics of the physical, social and economic environment in which production takes place. Such differences have led efficiency researchers to estimate separate production frontiers for different groups of firms", (O'Donnell, *et al.*, 2008).

Battese and Rao (2002), Battese *et al.* (2004), and then O'Donnell *et al.* (2008) introduced Meta frontier (MF) approach to address this issue. Battese and Rao (2002) and also Rao *et al.* (2003) showed that, we should put aside the similar production technology assumption for all groups of an industry. In agricultural sector Mehrabi *et al.* (2006) estimated the technical efficiency of wheat farmers of Kerman province with translog production function. They obtained the technology gap, between five main regions which were producing wheat. Results showed a higher technology gap ratio respected to meta-frontier in drought regions than others. Mehrabi *et al.* (2007) and Villano *et al.* (2010) appear to be the only authors that have published using the Battese *et al.* (2004) procedure for varietal differences in agricultural crops. Mehrabi *et al.* (2007) reported a significant technology gap between three varieties of pistachio with using a selected sample random of 475 pistachio farmes of Kerman in 2003 and 2004 which was included three main varieties of pistachio. Results emphasized the importance of taking into account the differences in frontiers imposed by different tree varieties. Villano *et al.* (2010) discussed about factors which contribute to not using the meta-frontier production technology by most farmers except in the long run. Physical conditions, environmental constraints, capital scarce, production cycle span, specialy for perennial crops and unprofitability of changing the current varieties are the most important reasons expressed by them that should be taken into account, if farmers would like to reach the meta-frontier. So far, to the best of our knowledge, no study had tried to measure productive efficiency of the datepalm farms while taking into consideration the varietal differences and their relationship with technology gap. In this paper we are going to examine the unsimilarity of technologies in three main datepalm varieties by stochastic production analysis. The rest of this paper organizes as follows: section 2 describes the econometric modeling from basic stochastic frontier analysis to meta-frontier parameters linear programming. Section 3, elaborates the data and empirical models. Section 4 presents the empirical results and discussion and section 5, includes summary and conclusion.

Analytical frame work

Stochastic frontier analysis:

Stochastic frontier analysis was developed by Hayami (1969), Hayami and Ruttan (1970) to introduce the meta-production function idea as a function which envelops the neoclassical production functions. Thus, it is a common underlying production function that is used to represent the input-output relationship of a given industry (Lau and Yotopoulos, (1989)). The basic Meta production concept discusses the hypothesis that all the groups in an industry have potential access to the same technology. However, each producer may choose to operate on a different part of it depending on circumstances such as the natural endowments, relative prices of inputs, and the economic environment (Lau and Yotopoulos, 1989). Battese and Rao (2002) extended the Meta production approach which is reviewed here. Suppose a number of groups like different varieties of datepalms are defined. If for the j-th group there is data for Nj firms then the stochastic frontier model is written as (Battese *et al.* (2004); O'Donell *et al.* (2008):

$$y_{it}^{j} = f(x_{1it}, x_{2it}, \ldots, x_{lit}; \beta^{j}) e^{v_{it}^{j} - u_{it}^{j}} \tag{1}$$

Where y^{i}_{jt} implies the output of the i-th firm in the j-th group for the t-th period; x^{i}_{Nit} denotes the N-th input of the i-th firm in the j-th group for time period t; β^{j} is the vector of unknown parameters to be estimated, v^{i}_{jt} represents statistical noise (random variables) and is assumed to be distributed as N $(0, \sigma^{2}_{vj})$ that is independent on u^{i}_{jt} which is distributed as N $(z^{j}_{it}\theta + \omega^{j}_{it}, \sigma^{2}_{uj})$ and indicates inefficiency where the effects of different factors on technical efficiency (U_{iyj}) can be showed like this:

$$u_{it}^{j} = \delta_{it}^{j}\theta + \omega_{it}^{j} \qquad \text{s.t.} \qquad \omega_{it}^{j} \geq -\delta_{it}^{j}\theta \tag{2}$$

Where ω^{j}_{it} is a stochastic variable with normal distribution of zero mean and σ^{2} variance. Having Assumed that the exponent of the frontier production function is linear in the parameter vector, β^{j}, then the Eq. 3 is equivalent to:

$$e^{x'_{it}\beta^{j} + v_{it}^{j} - u_{it}^{j}} \tag{3}$$

Where x_{it} is the transformed vector of inputs for the *i*-th firm in the *j-th* group for the time period t. Given data on inputs and output for group j, Maximum likelihood (ML) estimates of β^{j} vector for this frontier are attained. Using the above notations, we can calculate TE for the *i-th* firm in the *t-th* period with respect to group *j* frontier as:

$$TE_{it}^{j} = \frac{y_{it}}{e^{x'_{it}\beta^{j} + v_{it}^{j}}} = \exp(-u_{it}^{j}) = \exp(-\delta_{it}^{j}\theta - \omega_{it}^{j}) \tag{4}$$

Frontier software is programmed to estimate the predictor of the TE proposed by Battese and Coelli (1988). After estimating the group frontier parameters separately, it is time to examine whether the different groups have an identical frontier or not. Likelihood ratio test can verify this issue, where L (H_0) is the value of log likelihood function associated with pooled data of all groups and L (H_1) will be the sum of the values of log likelihood functions exploited from the individual group frontiers. The degrees of freedom (df) for the test of significance of LR would be the difference between the number of parameters being estimated under the alternative hypothesis (eg., three times the total number of parameters being estimated for each model, if there are three groups, and number of parameters being estimated under the null hypothesis (the number of parameters in any model)). For example the df will be thus twice the number of the parameters in each model (for the pooled data, say) if we have three groups[1]. If the null hypothesis was rejected, it means that the groups have different frontiers and the Meta-frontier should be estimated; otherwise, the pooled frontier is enough and it makes sense not to estimate Meta-frontier parameters. (Battese *et al.*,

[1] *The LR statistic is calculated by:* $\lambda = -2\,[L(H_0) - L(H_1)]$. $L(H_0)$, *is the log likelihood function of null hypothesis and* $L(H_1)$ *is the log likelihood function under alternative hypothesis.* λ, *has a chi-square distribution.*

2004). Therefore, if the null hypothesis was rejected, then the Meta-frontier which is a deterministic parametric frontier and will have a larger predicted value from all group frontiers and also envelops all those grouped frontiers smoothly (Battese *et al.,* 2004). Meta frontier is defined as below:

$$y^*_{it} = f(x_{1it}, x_{2it}, \dots, x_{lit}; \beta) \equiv e^{x'_{it}\beta} \tag{5}$$

Where y^*_{it} is the Meta frontier output and β is an unknown column vector of Meta frontier parameters satisfying the inequality. 5: (O'Donnell *et al.* (2008)):

$$x'_{it}\beta^f \geq x'_{it}\beta^j \quad \text{For all} \quad j=1, 2, \dots, J \tag{6}$$

A Meta frontier has to envelop all the group frontiers, but the above constraints may do not do this. For obtaining the Meta frontier parameters which envelops all the group frontiers, this optimization is to be solved:

$$\min_{\beta^f} \sum_{i=1}^{m}\sum_{t=1}^{T}[\ln f(x_{1it}, x_{2it}, \dots, x_{Lit}; \beta^f) - \ln f(x_{1it}, x_{2it}, \dots, x_{Lit}; \hat{\beta}^j)]$$

$$\text{s.t. } \ln f(x_{1it}, x_{2it}, \dots, x_{Lit}; \beta^f) \geq \ln f(x_{1it}, x_{2it}, \dots, x_{Lit}; \hat{\beta}^j) \tag{7}$$

Where $\hat{\beta}^j$ is the column vector of coefficients associated with the group – *j* stochastic frontier model. Moreover, if the Ln $f(x_{1it}, x_{2it}, \dots, x_{Lit}; \beta^f)$ is log linear in parameters then, an equivalent form of the linear problem can be shown as:

$$\min_{\beta^f} \bar{x}'_{it}\beta^f$$

$$\text{s.t.} \quad x'_{it}\beta^f \geq x'_{it}\beta^j \quad \text{For all } i \text{ and } t\,. \tag{8}$$

Where $\bar{x}$ is the arithmetic average of the x_{it} -vectors for all firms in all periods (O'-Donnell *et al.* (2008)). After solving the LP problem, in order to estimate the Meta technology ratios, we use equation (9) that includes three parts. The first term is the technical efficiency which estimated by eq. (3). The second term is the Meta technology ratio (technology gap) for the *i-th* firm in the *t-th* period (in the *j-th* group):

$$y_{it} = e^{-u^j_{it}} \times \frac{e^{x'_{it}\beta^j}}{e^{x'_{it}\beta^f}} \times e^{x'_{it}\beta^f + v^j_{it}} \tag{9}$$

Relationship between Meta technology ratio and Technical efficiency of different groups is verified as:

$$\hat{TE}^f_{it} = \hat{TE}^j_{it} \times \hat{MTR}^j_{it} \tag{10}$$

Where, $\hat{TE}^j_{it}$ is the technical efficiency of *i-th* firm relative to a specific group in the *t-th* period, and $\hat{TE}^f_{it}$ is the technical efficiency of the firm in period with respect to Meta frontier. $\hat{MTR}^j_{it}$ is the Meta technology ratio which according to O'Donnell *et al.* (2008) indicates that a higher value in this measure, showes a lower technology gap between a specific group frontier and Meta frontier. Meta technology ratio for the *i-th* firm in *t-th* period of the group *j-th* is measured simply by substituting the β^j and β^f in Eq. 11. (O'Donnell *et al.* (2008)).

$$\hat{MTR}^j_{it} = \frac{e^{x'_{it}\beta^j}}{e^{x'_{it}\beta^f}} \tag{11}$$

Data and Empirical models

The five largest producers of date in the world, are Egypt, Iran, Saudi Arabia, Iraq and United Arab Emirates, but Egypt, China, Bahrain, Gaza strip and Qatar have the largest yield by hectare respectively, (FAO, 1992-2010). Date is kind of fruit and food that is eaten around Sistan and Baluchestan by people and also animals. Datepalms have been cultivated because they have provided almost all rural people's necessities. Now this province is the second largest producer of date in Iran because of it's weather condition; 76 kinds of dates are produced and sold from 30 regions of the province, 46000 hectares of Sistan and Baluchestan agricultural lands are under datepalms with 200000 tones dates produced in 2012 (Ministry of Jihad-e- Agriculture, 2012). As shown in the figure 1, Production of date reduced in 2006 and 2008 0.7 and 4 percent relative to the past year, respectively, in Iran, this reduction took place because of

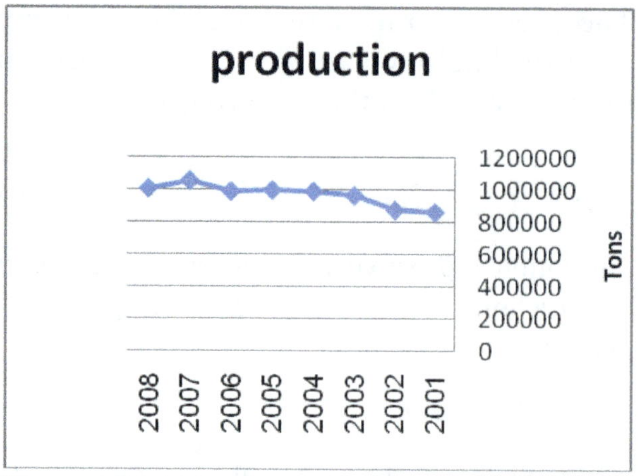

Figure 1: Historical trends in Iran dates production
Sources: Ministry of Jihad-e-Agriculture, 2012

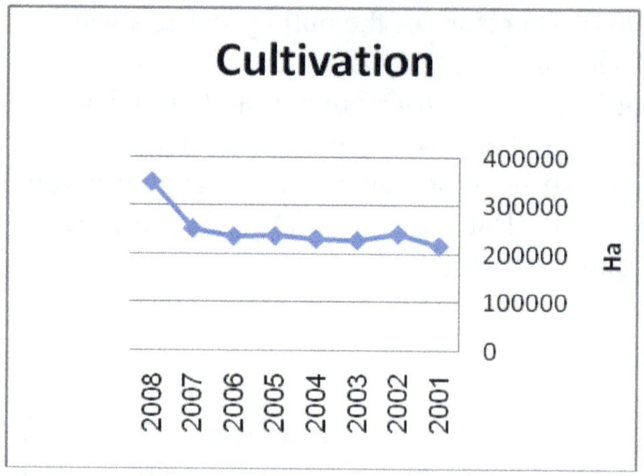

Figure 2: Historical trends in Iran palms cultivation

sequent droughts, salinity of irrigation water, storms in eastern and southern regions and palming phenomenon.

In this research, t he sample included 341 palm farmers which involved 126 Mazafati farmers, 113 Zard farmers and 102 Zard farmers. With using stratified random sampling approach, finally 300 respondents (100 Mazafati farmers, 100 Zard farmers and 100 Rabi farmers) were specified. Data was collected by questionnaires in the region of Saravan, Sistan and Baluchestan, over 3 agricultural years: 2009/2010, 2010/2011 and 2011/2012.The sample statistics for the input and output variables are reported in Table 1 briefly. Seven inputs were considered in this research, i.e., labor use (man-days), sown area (hectare), fertilizer (kg), tree age (years), density (trees per hectare), water (m^3), toxin (liter), and one output that is the physical product of dates in each period of harvesting. Hayami and Ruttan. (1970) measured fertilizer by thousands of tonnes of nitrogen, phosphate and potassium. In our sample, farmers use the animall dungs as fertilizer not chemical fertilizer, so we only measured the fertilizer by Kilogramms of animal dungs used in farms. Also there are two dummy variables; D_1 and D_2 for fertilizer and toxin, because fertilizer and toxin variables had zero values for some of farmers. For considering the zero observations of toxin and fertilizer, following Battese (1997), for each dummy variable, we used this modification:

$D_i = 1$, if $x_i = 1$ and $D_i = 0$, if $x_i > 0$; and $x^*_i =$ max (x_i, D_i) and we use x^*_i instead of x_i in our production function. This implies that when the x_i variable has a positive value, then $x^*_i = x_i$; but if x_i has value zero then $x^*_i = 1$ (Battese,1997). Our production function involving seven inputs and one output is defined by:

$$y_{it} = \beta_0^j + \sum_{l=1}^{3} \beta_l^j x_{it} + 0.5 \sum_{l=1}^{3} \sum_{k=1}^{3} \beta_{lk}^j x_{lit} x_{kit} + \sum_{d=1}^{2} \lambda_d^j D_d + \alpha_t^j T_{it} + v_{it}^j - u_{it}^j \quad (12)$$

Where subscript j represents the *j-th* group and T_{it} denotes the time trend which captures the technological change effects during time for each *i-th* firm in period *t-th*. The Eq. 12 expresses the translog functional form of stochastic production function, but it should be noted that the approperiate functional form has to be specified by LR test based on the values of log likelihood function. After opting out the best functional form (Cob-Douglas or Translog) for SFA model, three hypothesis tests were examined regardin U_{it} and V_{it} distributions (Coelli, 1995; Battese, 1993). At first, the existence of inefficiency effects was examined by this null hypothesis: H_0: $\gamma = \delta_i = 0$. Rejecting this hypothesis, confirms the existence of inefficiency effects. An additional test is done regarding the H_0: $\gamma = 0$, if this Hypothesis is accepted, this would indicate that the U_{it} term should be removed from the model, leaving a specification with parameters that can be consistently estimated using ordinary least squares (Coelli, 1996) otherwise the maximum likelihood estimates are prefered. Moreover, rejecting the hypothsis: H_0: $\delta_i = 0$ insinuates that the social variables like education or having second job, are influencing technical efficiencies. By omitting the second-order interaction terms from translog model, the Cob-Douglas model can be written as:

$$y_{it} = \beta_0^j + \beta_{x_1}^j x_1 + \beta_{x_2}^j x_2 + \beta_{x_3}^j x_3 + \beta_{x_4}^j x_4 +$$

$$\beta_{x_5}^j x_5 + \beta_{x_6}^j x_6 + \beta_{x_7}^j x_7 + \lambda_1^j D_1 + \lambda_2^j D_2 + \alpha_t^j T$$

$$+v_{it}^j - u_{it}^j \qquad (13)$$

Dependent and explanatory variables' quantities were stated in natural logarithms.

RESULTS

Table1 represents that the mean production of Mazafati and Rabi is 900 kg higher than Zard, but Mazafati has the highest standard deviation. Results also denote that the value of labor used in Mazafati is the highest, because Mazafati variety needs more tendency either in harvesting period or before. Furthermore, the average water used in Mazafati is 73.96 M3 higher than Zard variety and 74.32 M3 compared to Rabi variety.In addition, sown area for Mazafati variety is comparatively higher than sown area usage for Zard and Rabi variety as, 0.11 and 0.13 hectare. There is no significant difference between varieties in density. Fertilizer is more utilized in Zard variety than Mazafati and Rabi, as 13.79 and 21.84 kg, respectively. Farmers applied more toxins for Rabi variety than other varieties. Trees in Rabi variety are notably older than other two varieties. As we see in table 1, most of inputs are used in Mazafati are higher than other varieties, and this vindicates the higher production of this variety. Mazafati variety has a higher price in markets, so this induces farmers to use more inputs for getting more output from Mazafati variety. It can be seen in table 1 that, on average, farmers have 14 years experience in agriculture as a profession and almost 43 percent of them have a second job. This is because of inadequate income they have. Unfortunately, the average year of education is almost 8 years. (Table 1). This might be an important reason for inefficiency of most farmers. More than half of farmers in all three groups are member of agricultural cooperative according to table1 and 35 percent of them have participated in extension classes. This section proceeds by at first, fitting stochastic production frontier for pooled data including all dates' varieties and individual dates, separately. Then, the specification tests are discussed and finally we analyze technical efficiencies with respect to group frontiers and with respect to Meta-frontier. It is necessary

Table 1: Input and output variables summary

Variable	Symbol	Unit	All dates	Mazafati	Zard	Rabi
Production	y	10^3 kg	1.96(1.8)	1.99(1.9)	1.9(1.7)	1.99(1.8)
Sown area	x_1	Hectare	1.73(1.39)	1.81(1.43)	1.70(1.39)	1.68(1.39)
Water	x_2	M^3	123.74(77.4)	173.17(91.55)	99.21(84.55)	98.85(84.18)
Labor	x_3	Man-days	1.52(1.11)	1.57(1.05)	1.53(1.2)	1.46(1.07)
Toxin	x_4	Liter	15.43(24.52)	15.64(32.10)	14.40(14.17)	16.25(24)
Fertilizer	x_5	Kg	138.74(299)	136.81(302.5)	150.66(392)	128.76(155.77)
Tree age	x_6	Year	24.34(13.3)	23.4(13.23)	25.04(13.5)	60.09(24.06)
Density	x_7	Tree per ha	60.83(24.5)	62.24(25.45)	60.17(24.2)	60.09(24.06)
Experience	δ_1	Year	14.69	14.62	14.73	14.73
Education	δ_2	Year	8.41	8.14	8.35	8.74
Having 2nd job	δ_3	Percentage	43	40	38	50
Co-member	δ_4	Percentage	65	61	67	68
PIEXC	δ_5	Percentage	35	37	38	31

Notes: Number of observations, is 100 for Mazafati variety, 100 for Zard variety and 100 for Rabi variety; Values in brackets are standard errors of variables. Co-member is the brevity of being the the agricultural cooperative member and PIEXC implies participating in Extension classes. Agricultural Cooperatives and Extension programms are planned by Ministry of Jihad-e-agriculture to improve farmers knowledge about results of new agricultural researches in country. Density of datepalms was measured by tree per hectare which is shown by "Tree per ha".

Table 2: LR test results for choosing better functional form

Groups	LLF of CD	LLF of TR	LR	χ2 statistic(df)	Decision
Pooled	-873.828	-812.031	123.594	49.578**(29)	Translog
Mazafati	-304.862	-282.11	45.504	49.578**(29)	Cobb-Douglas
Zard	-278.059	-235.025	86.068	49.578**(29)	Translog
Rabi	-275.357	-224.772	101.17	49.578**(29)	Translog

**P < 0.01

to specify the appropriate functional forms for SPF model, before measuring technical efficiencies. The translog (TR) and Cobb-Douglas (CD) forms were estimated for pooled data and each group, separately. By LR test results which are reported in table 2, just for Mazafati variety, Cobb-Douglas functional form is opted out, but it is obvious that translog form is prefered. So we applied translog form for all varities.

We also have to test the hypothesis whether or not the pooled frontier model is prefered to separate group frontiers (all varieties adapt the same stochastic production frontier or no) with the separate estimation of the three varieties as the alternative model. The likelihood ratio statistic is distributed as chi-square with a degree of freedom as 94[2]. The alternative hypothesis is accepted by 1% of significance. So the data for different varieties cannot be pooled and we have to estimate meta-frontier to measure the technical efficiencies of three groups with respect to it. The estimates of stochastic frontier coefficients which can be interpreted as production elasticities, because of using corrected to geometric mean data are reported in table 3. As shown in table 3, most of first-order coefficients for Mazafati variety are highly significant except Toxin, fertilizer and density. But, fertilizer has a more considerable effect on production of Mazafati variety than other varieties separately. Interestingly, Tree age had a positive value for Mazafati, demonstrating that older Mazafati datepalms will increase the production. Mazafati production is remarkably affected by toxin as 0.4081. This is justified by high sensitivity of Mazafati variety to pests. The coefficient of t (time) for Mazafati, shows that this variety was faced a higher negative technical change during time than other varieties. Tree age has the largest positive coefficient among other inputs, which portends that a percent change in tree age will mutate the production of Mazafati variety more strongly than other inputs. Intercept which is an index for the level of technology, is highly significant as 1.02** for Mazafati variety. And Mazafati variety has a higher technology level compared to other datepalm's varieties has a higher technology level compared to other datepalms. The results show that the coefficients for water, fertilizer, toxin, density and the interactions of these inputs are negative in the instance of Mazafati. This is because, Mazafati farmers use overplus inputs.

Zardan variety is somehow different from Mazafati. It is kind of dry date whose coefficients of fertilizer and density are negative and almost all its first-order coefficients are highly significant with except to toxin. The function coefficient of Zardan variety is less than one. In fact, it implies that if farmers increase the size of farm, the production of Zardan variety will decrease (decreasing returns to scale). As we aforementioned for Mazafati, t (time) coefficient for Zardan variety is negative too and negative technical change is incurred with Zardan variety. The level of technology for Zardan variety is dramatically lower than Mazafati and Rabi because the price of Zardan date is not as high as Mazafati and Rabi, so, farmers do not use high technologies for Zardan variety. Rabi variety is sort of moist date which has a less sweet taste in comparison to Mazafati. According to table 2, sown area and density coefficients for Rabi are negative. The intercept (technology level) of Rabi is higher than Zardan variety's intercept but lower than Mazafati's intercept. Density has

[2] *The degree of freedom is calculated by twice the number of the parameters in each group model. In this case it is as 2 47=94.*

Table 3: Estimates of stochastic frontier parameters

Variable	Pooled	Mazafati	Zard	Rabi	MF
Intercept	0.5570	1.02**	0.405**	0.7302***	1.0054
X_1	0.3024***	0.301***	0.3368***	-0.3702*	0.3387
X_2	-0.0083**	-0.2339**	0.2198*	0.2471*	0
X_3	- 0.0841**	-0.0798**	-0.2316**	0.1811	0
X_4	-0.0622***	-0.2592*	-0.0579**	0.0805**	0
X_5	0.04671	0.4081	-0.0181	-0.0013	0.3544
X_6	-0.4833	0.5408**	-0.4334***	- 0.6882***	0.1469
X_7	0.9306***	-0.9860	0.9076**	1.1408***	0
X_1^2	0.2578	0.3501	0.1871*	0.4778***	0.1141
X_2^2	0.01116***	0.00617*	-0.0516***	-0.0894	0
X_3^2	0.6458**	0.6692*	0.6185	0.4533*	0
X_4^2	0.06995**	0.1043	0.08794	0.1012**	0.0036
X_5^2	-0.00427	-0.0138**	0.0118	-0.0315*	0
X_6^2	1.1162	1.5630	1.5908*	1.4343***	1.2378
X_7^2	0.3102***	-0.1608	0.2702	1.3095***	1.0647
X_1X_2	-0.0199**	-0.1827	-0.4853	0.0685	0
X_1X_3	-0.9163*	-0.4514*	0.0175**	-0.6015*	0
X_1X_4	0.1003	0.1581	0.3725	-0.0172	0
X_1X_5	-0.0499	-0.0022	0.0088	0.2346***	0
X_1X_6	-0.2687	-0.4826	-1.1363*	0.4163	0
X_1X_7	-0.9487	-1.6052**	-0.7389*	-0.2421	0
X_2X_3	-0.1072**	-0.0721	0.1352	0.5788**	0
X_2X_4	-0.9590	-0.1062	-0.1303	-0.3641***	0
X_2X_5	0.0066	0.1017	-0.07964*	0.0511	0
X_2X_6	0.3847	0.1801	0.5013	0.6993**	0
X_2X_7	0.4313	1.1037*	0.8028	-1.5122*	0
X_3X_4	0.0158	0.1954	-0.1772	-0.2553*	0.2486
X_3X_5	-0.0413	-0.0027	0.1191*	0.1694**	0
X_3X_6	-0.2908	-0.3886*	-0.3206	-0.4388	0.1017
X_3X_7	-0.2381	-0.0249	-0.2076	-0.6153	0
X_4X_5	-0.02892*	-0.0257	-0.3873	-0.0522**	0
X_4X_6	0.1484*	0.3271	0.0200	0.0707	0
X_4X_7	-0.4221	-0.3797*	-0.5298	-0.8793***	0
X_5X_6	-0.009186	-0.1047	0.0660	0.0085	0.0279
X_5X_7	0.1440**	0.2322	0.1363	0.5374***	0
X_6X_7	-0.2124*	-1.2611	-1.652	-1.3711***	0
D_1	-2.1248	-0.1376	0.0635**	0.0201	0
D_2	0.0241	-0.1110	0.0063	-0.05103	0
T	-0.0116***	-0.2626*	-0.117*	-0.1724***	0
ω	0.0420**	0.0113***	-0.1271***	0.2651**	0.06921**
δ_1	-0.0260**	-0.0026**	0.0045***	-0.0557**	0.00530**
δ_2	-0.1711**	-0.121*	-0.0280*	-0.0123	0.01160*
δ_3	-0.0302*	0.1330**	-0.1205	0.0589**	-0.00580
δ_4	0.1666	0.0499	0.2340	-0.0158*	-0.01453*
δ_5	0.0028***	0.0015**	0.0101**	-0.0082	-0.00093*
σ^2	0.2883***	0.3114***	0.2422**	0.1788***	-a
γ	0.9406***	0.9587***	0.9022***	0.9999***	-a
FC	0.0642	-0.309	0.7232	0.5898	0.84
LLf	-812.031	-282.110	-235.025	-224.772	-a

Note: LLF represents log likelihood function, ***P<0.001, **P<0.05, *P<0.1. FC denotes on function coefficient. –a, implies that those parameters are not measured, because the meta frontier parameters were estimated by mathematical programming, not by stochastic frontier function.

the largest coefficient between inputs in the case of Rabi. It shows that the Rabi farms are so diffused and it is still possible to plant more datepalms per hectare. Lots of interaction coefficients are significant in Rabi farms compared to other varieties. Rabi variety has dropped off, technically with a lower t coefficient than other varieties. Function coefficient of Mazafati is -

Table 4: Hypothesis test

Groups	Null hypothesis	LR	$\chi 2$ (df)0.01	Decision
Pooled	H_0: $\gamma = \delta_i = 0$	461.726	17.755(7)*	reject
	H_0: $\gamma = 0$	346.439	18.475(7)	reject
	H_0: $\delta_i = 0$	461.720	16.811(6)	reject
Mazafati	H_0: $\gamma = \delta_i = 0$	180.188	17.755(7)*	reject
	H_0: $\gamma = 0$	146.665	18.475(7)	reject
	H_0: $\delta_i = 0$	180.197	16.811(6)	reject
Zard	H_0: $\gamma = \delta_i = 0$	155.152	17.755(7)*	reject
	H_0: $\gamma = 0$	122.142	18.475(7)	reject
	H_0: $\delta_i = 0$	155.153	16.811(6)	reject
Rabi	H_0: $\gamma = \delta_i = 0$	194.857	17.755(7)*	reject
	H_0: $\gamma = 0$	139.518	18.475(7)	reject
	H_0: $\delta_i = 0$	194.858	16.811(6)	reject

Note: Asterisk *, indicates that the values have a mixed chi-square distribution extracted from Table 1 of Kodde, D.A., and F.C. Palm (1986).

0.309, the lowest among three species of dates, again indicating decreasing returns to scale. Table 2 also portrays the results of regression model for pooled data and meta-frontier parameter estimates attained by linear programming of Eq. 8. First-order coefficients are significant with except to toxin and tree age and production is affected by density more than other inputs for pooled data. The level of technology ascents over Zard but it is lower than Mazafati and Rabi. Beside, the technical change is falling over three agricultural years by -0.0116 as coefficient of t, for pooled data. Finally, Table 2 includes the Meta-frontier parameters which all of them are positive and a function coefficient of 0.84. Toxin has the largest coefficient and the level of technology is 0.4484 higher than the level of technology of pooled data.

Moreover, the technical change coefficient is zero denoting that the technology is constant in meta-frontier. After choosing the best functional form, for estimating the production function, some hypothesis tests have to be done. The results of likelihood ratio tests are portrayed in Table 3. Rejecting the hypothesis H_0: $\gamma = \delta_i = 0$ confirms the existence of inefficiency effects. This hypothesis is examined by comparing the LR test of the one-sided error reported in Frontier Software output. Amazingly, LR one-sided error values for all three varieties are highly significant and reject the lack of inefficiency effects. The highest value of this index belongs to Rabi as 194.857 and the lowest is for Mazafati as 146.665. Accepting the null hypothesis H_0: $\gamma = 0$ vouches that the conventional production function is enough, so Ordinary least square is prefered to

Table 5: Technical efficiencies and Meta technology ratios

	Groups	Average	SD	Minimum	Maximum
Meta technology ratio	Mazafati	0.407	0.1725	0.02796	1
	Zard	0.507	0.1703	0.00404	0.8592
	Rabi	0.432	0.1927	0.00339	1
Technical efficiency with respect to group frontiers	Mazafati	0.518	0.270	0.460	0.999
	Zard	0.520	0.254	0.010	0.941
	Rabi	0.592	0.229	0.046	0.932
	All dates	0.558	0.239	0.012	0.939
Technical efficiency with respect to meta frontier	Mazafati	0.210	0.09	0.125	0.360
	Zard	0.263	0.011	0.011	0.480
	Rabi	0.255	0.011	0.031	0.395
	All dates	0.200	0.011	0.010	0.250

Source: Research findings

Table 6: Spearman correlation coefficients for TE measures of translog stochastic frontier and meta-frontiers

Variety / Model	TL-frontier	MF- frontier
Mazafati		
TL-frontier	1	
MF-frontier	0.28	1
	Translog	**Meta-frontier**
Zard		
TL-frontier	1	
MF-frontier	0.11	1
	Translog	**Meta-frontier**
Rabi		
TL-frontier	1	
MF-frontier	0.05	1

TL-frontier: TE measured with respect to translog production frontier, MF-frontier: TE measured with respect to meta-frontier.

Maximum likelihood estimation and closer γ to 1 means that there are more significant inefficiency effects (Battese and Coelli, 1992). According to Table 3 the parameter γ revolves from 0.009 for Rabi variety to 0.006 for Mazafati variety. Statistical tests exhibit that all varieties have stochastic frontiers at the 1% level of significance (Table 3). Rejecting the hypothesis H_0: $\delta_i = 0$ explains that the socio-economic characteristics like education, having second job or participating in extension classes which are considered in empirical model are influencing the efficiency scores of farmers. This null hypothesis was strongly rejected too for all varieties.

Meta technology ratios and technical efficiencies

Meta technology ratios (MTR) and technical efficiencies (TE) with respect to group frontiers and Meta-frontier are summarized in table 5. MTR is the ratio of a specific farmer's output from a particular group to the meta-frontier output with the same inputs. Therefore, the higher MTR, the less technology gap between individual frontier and Meta frontier and if the MTR is unity, then the individual frontier is palced on Meta frontier, so the MTR cannot be greater than one. The mean estimated MTR for the Mazafati instance is 43.2% varying from a minimum of 2.7 % to a maximum of 100.0%. The average estimated MTR for Zard variety is 40.7% and goes from a low of 0.4% to 85.9%; and the average MTR for Rabi variety is 50.7%, ranges from 0.3% to 100.0%. The highest average MTR belongs to Rabi variety which means that Rabi farmers are closer to the expected output of Meta frontier. The lowest MTR is related to Zard farmers which have the lowest gross incomes compared with other farmers because of low prices of Zardan date. This analysis emphasizes that Rabi farmers use higher technologies than Zardan and Mazafati farmers. The average technical efficiencies with respect to Meta-frontier and group frontiers are 7.5% and 51.8% for Mazafati, 7.1% and 59.2% for Zard, and 5.9% and 52% for Rabi (Table 5). There is not a significant difference between TE measures of Zardan (26.3%) and Rabi (25.5%) with respect to Meta-frontier, although they are remarkably higher than for Rabi (5.9%). A comparison of the average TE among the three varieties, estimated from the pooled stochastic frontier, portends that there is no main difference (Mazafati = 56.1%, Zard = 54.8% and Rabi = 56.5%). This issue again corroborates usage of Meta-frontier for these three varieties, i.e., a reasonable frontier which envelops all the three different frontiers to compare their performance. Victor, *et al.* (2010) found similar TE measures in their research on three South American regions.

It is useful to compare the TE measures obtained from translog model with those of meta-frontier model for each group. So the Spearman correlation coefficients of TE measures of translog stochastic production and meta-frontiers are reported in table 6. As shown in table6, all pair wise comparisons show positive correlations, ranging from 0.05 between the TL-frontier and the MF-frontier for Rabi variety and 0.28 for Mazafati. These results are also similar to those of Moreira *et al.* (2010), and imply that the TEs measured by the two models are fairly consistent.

TL-frontier: TE measured with respect to translog production frontier, MF-frontier: TE measured with respect to meta-frontier.

CONCLUSION

Agriculture is the mainstay economic sector of rural areas in Iran that it has great potential for development. Datepalms are cultivated in dozens southern provinces of Iran, such as: Khuzestan, Kerman, Hormozgan, Fars and Sistan and Baluchestan. This study provides some interesting results on the datefarmers technical efficiencies in Sistan and Baluchestan province, one of the leading producers of different varieties of dates in Iran. The objective of this paper was to compare technical efficiency (TE) for different varieties of dates in Sistan and Baluchestan province using the Meta-frontier method extended by Battese and Rao (2002) and advanced by Battese *et al.* (2004) and O'Donnell *et al.* (2008). The data are completely balanced panels including 300 farms and 3 periods for the various varieties of date including 100 observations for Mazafati, 100 for Zard and 100 for Rabi variety. Stochastic Production Frontier (SPF) models were estimated separately for each variety and pooled for all three varieties. Then, a meta-frontier model was estimated using linear programming with the pooled data. Translog functional form was prefered to Cobb-Douglas by specification test and the maximum likelihood estimation was used instead of ordinary least square. Besides, the socio-economic characteristics were strongly influencing the efficiency scores of date farmers and the inefficiency effects were revealed to exist significantly. The null hypothesis that the date farms from the three varieties operate on the same pooled production frontier was strongly rejected. So we estimated a meta-frontier which envelops all the three individual variety frontiers. The difference or gap between individual variety frontiers and meta-frontier is called Meta Technology Ratio (MTR). The average MTRs for Mazafati, Zard and Rabi are 0.407, 0.507 and 0.432, respectively. Thus, the Mazafati frontier is the most distant to the meta-frontier while the Zard frontier is the closest, and the Rabi is in an intermediate position. The average TEs with respect to group frontiers are 0.518, 0.520 and 0.592, respectively, for Mazafati, Zardan and Rabi. But these TEs are not comparable because they are not from the same frontier. So the TEs which obtained with respect to the Meta-frontier can be compared. The TEs with respect to Meta-frontier are 0.00134, 0.00144 and 0.00142, respectively for Mazafati, Zardan and Rabi. The average TEs for Zardan and Rabi variety are not significantly different from each other and both are higher than the value for Mazafati. All frontier models including Mazafati, Zard, Rabi and Meta-frontier had decreasing returns to scale (RTS), with average Function Coefficient (FC) values as, 1.309, 1.723, 0.589 and 0.840 respectively. These RTS indices imply that on average date farms in Sistan and Baluchestan are not operating at a sub-optimal scale. A noteworthy result of the analysis reported in this paper is that, Mazafati farmers had the lowest efficiency score among others while Mazafati variety is the most popular and expensive variety in the region. It is, therefore, felt, that if the Mazafati farmers apply more efficient technologies, they will definitely gain more profit than now.

REFERENCES

1- Bakhshoodeh, M. & Thomson, K.J. (2001). Input and Output Technical Efficiencies of Wheat Production in Kerman, Iran. Agricultural Economics, 24: 307-313.

2- Battese, G.E. (1997). A Note on the Estimation of Cobb-Douglas Production Functions When Some Explanatory Variables Have Zero Values, Journal of Agricultural Economics, 2: 250-252.

3- Battese, G.E. (1993). Frontier Production Function and Technical Efficiency: A Survey of Empirical Applications in Agricultural Economics. Journal of Agricultural Economics, 7: 185-208.

4- Battese, G.E. & Coelli, T.J. (1988). Prediction of Firm-Level Technical Inefficiencies with a Generalized Frontier Production Function and Panel Data. Journal of Econometrics, 38: 387-399

5- Battese, G., Malik, S.J., Broca, S. (1993). Production

Functions for Wheat Farmers in Selected Districts of Pakistan: An Application of a Stochastic Frontier Production Function with Time-varying Inefficiency Effects, The Pakistan Development Review, 32: 233-268.
6- Battese, G. & Rao, D.S.P. (2002).Technology Gap, Efficiency, and a Stochastic Meta-frontier Function. International Journal of Business and Economics, 1(2): 87-93.
7- Battese, G.E., Rao, D.S.P. & O'Donnell, C.j. (2004). A Meta-frontier Production Function for Estimation of Technical Efficiencies and Technology Gaps for Firms Operating under Different Technologies. Journal of Productivity Analysis, 21:91-103.
8- Coelli, T.J. (1996). A Guide to Frontier Version 4.1: A Computer Program for Stochastic Frontier Production And Cost Function Estimation. CEPA working Papers, No.7/96, Department of Econometrics, University of New England, Armidale, NSW 2351, Australia, Retrieved from http: // www. une.edu.au/econometrics/cepawp.htm.
9- Coelli, T.J. (1995). Recent Developments in Frontier Modeling and Efficiency Measurement. Australian Journal of Agricultural Economics, 3:219-24.
10- Food and Agriculture Organization of the United Nations (FAO), Retrieved from http:// faostat. fao. org/site/339/default.aspx/ 2012.
11- Hayami, Y. (1969). Sources of Agricultural Productivity Gap Among Selected Countries. American Journal of Agricultural Economics, 51:564-575.
12- Hayami, Y., Rutan, V.W. (1970) Agricultural Productivity Differences Among Countries. American. Economics Review, 60:895-911.
13- Karbasi, A.R., Karim Koshteh, M.H., Ashrafi, M. (2004). Technical Efficiency Analysis of Pistachio-Production in Iran (Khorasan Province Case Study). Paper Presented at the 4th Asia-Pacific Productivity Conference, Brisbane, Australia, July, 14-16. 2004.
14- Karim koshteh, M.H., Akbari, A. & Mehri, M.A. (2004). A Survey on Efficiency of Wheat Farms in Sistan Area, Paper Presented At the 4th Asia-Pacific productivity Conference, University of Queensland, Brisbane.
15- Kodde, D.A. & Palm, F.C. (1986). Wald Criteria for Jointly Testing Equality and Inequality Restrictions, Econometrica, 54: 1243-1248.
16- Lau, L.J. & Yotopoulos, P.A. (1989). The Meta-Production Function Approach to Technological Change in World Agriculture. Journal of Development Economics, 31: 241-269.
17- Mehrabi Boshrabadi, H., Villano, R. & Fleming, E. (2006). Technical Efficiency and Environmental-Technological Gaps in Wheat Production in Kerman Province of Iran: A Meta-Frontier Analysis. Retrieved from http:// www. une.edu.au/economics/publications/gsare/arewp06_6.pdf
18- Mehrabi Boshrabadi, H., Villano, R. & Fleming, E. (2007). Production Relations and Technical Inefficiency in Pistachio Farming Systems in Kerman Province of Iran, Forests, Trees and Livelihood, 17:2, 141-156. Retrieved From http://www.tandfonline.com/doi/abs/10.1080/14728028.2007.9752590#preview.
19- Ministry of Jihad-e-Agriculture, (2004). Cultivation and Production Database, Agricultural Crops Information, retrieved from http:// www. dbagri. maj. ir/ Zrt/year/2012.
20- Najafi, B. & Abdollahi, M. (1997). Survey on Technical Efficiency of Pistachio Farmers in Iran. Iranian Journal of Agricultural Economics and Development, 17: 25-42.
21- Najafi, B. & Zibaee, M. (1994). Survey On Technical Efficiency of Wheat Farmers In Fars Province of Iran. Journal of Agricultural Economics and Development, 7: 71-85.
22- O'Donnell, C.J., Rao, D.S.P. & Battese, G. (2008). Meta-Frontier Frameworks for the Study of Firm-level Efficiencies and Technology Ratios. Journal of Empirical Economics, 34: 231-255.
23- Rao, D.S.P., O'Donnel, C.j. & Battese, G. (2003). Meta-frontier Functions for the Study of Inter-regional Productivity Differences, Center for Efficiency and Productivity Analysis. School of Economics, University of Queensland, Australia, Working Paper Series, No, 01/2003.
24- Villano, R., Mehrabi Boshrabadi, H. & Fleming, E. (2010). When is Meta-frontier Analysis Appropriate, An Example of Varietal Differences in Pistachio Production in Iran. Journal of Agricultural Science and Technology, 12: 379-389.
25- Sabouni, S. M. (1997). Efficiency Measurement of Dairy Farms in Fars Province, (Master's thesis, University of Shiraz). Retrieved from http://www.ensani.ir/fa/content/20616/default,aspx.
26- Moreira, V., Bravo-Ureta, B. (2010). Technical Efficiency and Meta-Technology Ratios for Dairy Farms in Three Southern Cone Countries: A Stochastic Meta-Frontier Model. Journal of Productivity Analysis, 33(1): 33-45.

Comparative Analysis of Stochastic Frontier Partially non-parametric and Stochastic Frontier Parametric Methods Case Study: Measuring Cost Efficiency in Wheat Production in Iran

Ebrahim Moradi [1], Mosayeb Pahlavani [2], Ahmad Akbari [3] and Hossain Mehrabi Bashrabadi [4]*

Abstract

Most research on performance evaluation of agricultural crops in Iran analyzes technical efficiency with estimation of stochastic production function. In this paper in spite of past research we survey cost efficiency with panel data. Information about input prices, yield and production cost per hectare collected for 28 provinces in 10 years and stochastic frontier cost function estimated with panel data in two methods, parametric stochastic frontier and partially non-parametric stochastic frontier. The results show that according to parameter significance and discretion of production structure parametric methodology is more suitable than non- parametric methodology. Land rent (price) had maximum influence and chemical fertilizer price had minimum influence on frontier production cost per hectare. Cost efficiency of wheat production in Iran is suitable and over 90 percent. Surveyed period Khuzestan province had maximum cost efficiency and Yazd province had minimum cost efficiency in wheat production. In whatever province where efficiency is low, agricultural education and knowledge needed to improve further.

Keywords:
Cost Efficiency, Partially Non- Parametric Method, Stochastic Frontier Method, Iran

[1] *Ph.D. Student, University of Sistan and Baluchestan, Iran.*
[2] *Associate Professor, University of Sistan and Baluchestan, Iran.*
[3] *Professor, University of Sistan and Baluchestan, Iran.*
[4] *Professor, University of Bahonar Kerman, Iran.*
* *Corresponding author's email: ebmoradi31@gmail.com*

INTRODUCTION

Today, grain has unique and important role in consuming pattern at any country. Wheat, barley, rice and maize have cultivation at highest level in Iran as Wheat 73.17%, barley 18.45%, rice 5.90%, and maize 2.48% of the cultivation has devoted to themselves (Agricultural Statistics, Iran, 2010). Wheat planted area was estimated 6.65 million hectares in 2010, that 36.75 percent is produced by dry farming and 63.25 percent is irrigated farming. Khorasan razavi with 9.21 percent of total wheat of Iran lands allocated highest level planted wheat area to himself. After this province, Kordestan, Fars, Hamadan, eastern Azarbaygan, Zangan and Kermanshah with 8.27%,6.91%,6.75%,6.6% and 6.40 %, allocated second rank to seventh rank to themselves. Therefore, more than half (50.76%) of wheat lands picked up in these seven province. Despite that Fars province has Third place in planted area, it has been in first place with 36.1% of total wheat production and Khuzstan, Khorasan Razavi, Golestan, Kermanshah, Hamadan and Eastern Azarbaygan with 8.75%, 8.36%, 8.17%, 6.11%, 5.69% and 5.43 %, allocated second rank to seventh rank to themselves. (Costs of agricultural production, 2011).

Most of study in Iran in concerned with measuring the performance of agricultural crops production analyzed estimation of stochastic frontier production function and technical efficiency. Since the accuracy of input prices data is higher than amount of inputs in agricultural sector macro data, cost function is suitable in efficiency study. Comparison of the results of the parametric and non parametric methods can show that whether application of nonparametric method Compared to the parametric method is an advantage or not.

Mosavi and Khalilian (2005), with purpose of estimating technical efficiency of some wheat farm in Shahr Kord city, used Translog stochastic frontier production function and estimated model with panel data. whose results show that average technical efficiency is 78%. Gahani and Asghari (2006), surveyed mathematical structure of production wheat cost function in framework of panel data and non frontier model where they used non frontier Translog cost function with cost share equations. After model estimation, they analyzed substitute and complement relation among inputs. Results from this study shows that fertilizer and seed are complement, and machinery and labor are complement.

Zeram Nejad and Yousefi Hagiabad (2008), for survey technical efficiency of wheat production at different province of Iran have used DEA and stochastic frontier production function methods. They have estimated technical efficiency by using panel data and obtain that the mean of wheat production in Iran is 57% by parametric method as well as the mean of wheat production is 84% by non-parametric method. Moradi Shahrbabak (2008), for estimating the wheat production efficiency at Baft county of Kerman province preferred using cob-Douglass stochastic frontier production function than by duality principle was derivated cost function and got a result that avrage efficiency of technical, allocative and economic are 88%,84% and 74% ,Respectively. The method used for decomposition of efficiency in his paper is not suitable and compatible. Haghiri (2003) has used stochastic non-parametric frontier production function and parametric production function for measuring technical efficiency of north American dairy industry. His motivation for using stochastic non-parametric frontier estimates comes from the fact that there are problems inherent in the structure of stochastic parametric frontier models. The results show that the overall mean of technical efficiency obtained from translog function for all regions is higher than that of the corresponding values obtained from the non-parametric approaches. Both parametric and nonparametric methodologies indicated evidence of differences between the mean technical efficiency of dairy farms in all regions. Cesar Revoredo, *et al.* (2009) estimate indicators of farm efficiency for the period 1989 to 2008 by farm type and to analyses they used stochastic frontier cost function. The results indicate while mixed farms and lowland farms have maintain their levels of efficiency. LFAfarms have seen their efficiency reduced since approximately 2004 or 2005 (especially LFAsheep farm specialists). Also, the analysis shows that there seems to be an increase in the dispersion of farmers in terms of efficiency for some farm types in periods of change in agricultural policy.

Considering uncontrollable factors in agriculture, the DEA approach, because of the disregard random effects, has a higher error than stochastic frontier approach. Panel data estimation has higher accuracy than cross-sectional data. According to data quality in agricultural sector of Iran, using of cost function can provide better results.

Despite the fact that most past study in Iran surveyed technical efficiency by estimating production function, this study pay attention to frontier cost function and so stochastic frontier partially non- parametric is introduced. Next, the theoretical basis of the parametric and non-parametric methods of calculating cost efficiency was surveyed and Information about prices, inputs, yield and cost of production per hectare in a 10 year period has been collected for 28 provinces. Stochastic frontier cost function with panel data was estimated by both parametric and nonparametric methods and Analysis and comparison of the results of the estimates are analyzed and compared.

MATERIALS AND METHODES

We assume that we have observation a panel of producers through time periods. Also we assume that the deterministic kernel of the stochastic cost frontier takes cob-douglas form and initially cost efficiency is time invariant. Thus cost frontier model as

$$\ln E_{it} = \beta_0 + \beta_y \ln y_{it} \sum_n \beta_n \ln w_{nit} + v_{it} + u_i \quad (1)$$

Where v_{it} represents error term, represents time invariant cost efficiency, and $\sum_n \beta_n = 1$ ensures homogeneity of degree1 of cost frontier in input prices. We make the following assumption on the error components in the stochastic cost frontier model.

i. v_i □ *iid* $N(0, \sigma^2_v)$

ii. u_i □ *iid* $N^+(0, \sigma^2_v)$

iii. u_i and v_{it} are distributional independently of each other, and of regressors.

The log likelihood function for I a sample of producers, each observed for periods of time, becomes

$$\ln L = \text{constant-}\frac{I(T-1)}{2}\ln\sigma_v^2 - \frac{I}{2}\ln(\sigma_v^2 + T\sigma_u^2)$$
$$+\sum_i \ln\left[1-\Phi\left(-\frac{\mu_{*_i}}{\sigma_*}\right)\right]-\left(\frac{\varepsilon'\varepsilon}{2\sigma_v^2}\right)+\frac{1}{2}\sum_i\left(\frac{\mu_{*_i}}{\sigma_*}\right)^2 \quad (2)$$

Where and $\mu_{*_i} = T\sigma_u^2\bar{\varepsilon}/(\sigma_v^2+T\sigma_u^2)$ and $\sigma_2^* = \sigma_u^2\sigma_v^2/(\sigma_v^2+T\sigma_u^2)$. This log likelihood function can be maximized with respect to the parameters. To obtain maximum likelihood estimates of β, σ^2_v and σ^2_u. The conditional distribution of $(u|\varepsilon)$ is

$$f(u|\varepsilon) = \frac{1}{(2\pi)^{1/2}\sigma_*[1-\Phi(-\mu_*/\sigma_*)]}.\exp\left\{-\frac{(u-\mu_*)^2}{2\sigma_*^2}\right\} \quad (3)$$

Which is the density function of a variable distributed as $N^+(\mu*, \sigma^2*)$.either the mean or the mode of this distributions cab be used as a point estimator of cost efficiency, and we have

$$E(u_i|\varepsilon_i) = \mu_{*_i} + \sigma_*\left[\frac{\phi(-\mu_{*_i}/\sigma_*)}{1-\Phi(-\mu_{*_i}/\sigma_*)}\right] \quad (4)$$

And

$$M(u_i|\varepsilon_i) = \begin{cases}\mu_{*_i} & if\,\varepsilon_i \ge 0\\ 0 & otherwise\end{cases} \quad (5)$$

Either can be substituted into to $CEi = \exp\{-u_i\}$ obtain producer-specific estimate of time-invariant cost efficiency. An alternative estimator is provided by the minimum squared error predictor

$$E(\exp\{-u_i\}|\varepsilon_i) = \frac{1-\Phi(\sigma_* - (\mu_{*_i}/\sigma_*))}{1-\Phi(-\mu_{*_i}/\sigma_*)}.\exp\left\{-\mu_{*_i}+\frac{1}{2}\sigma_*^2\right\} (6)$$

Confidence intervals for any of these estimators can be calculated (Kumbhakar and Lovell, 2000).

Most methods of measuring efficiency based on the MLE estimation such as in “Pooled data model”, “Battese and Coelli model”, “pitt and lee model” and “True random effects Green model” In addition, ”, “Cornwell, Schimdit, and Sickles model” based on OLS estimation and non frontier function (Green,2011). In this study we used “pitt and lee model” this model is similar to “Battese and Coelli model” With the difference that one efficiency score in time period for every firm is estimated, therefore this model is faced with less error.

Another method that can be used to measurement cost efficiency is non-parametric method. The locally weighted scatter plot smoothing, or the locally regression model, is introduced by Cleveland (1979) and developed by Fan (1992; 1993) and Hastie and Loader (1993). The basic idea of using the LOWESS [1] approach is to find a point in the space of the predictors and then search the neighborhood points that are smoothed using surface smoothers to estimate the mean response function. For instance, we may consider any point x, a so-called local observation, in the space of the predictors. Estimating a local regression model can be specified through different approaches. In a local regression we attempt to find a neighborhood containing the initial point x in which the regression surface is well approximated by a function from a specific parametric point of view. Therefore, our specification from the local regression model leads to methods of fitting the response function. The method consists of smoothing the response as a function of the predictors. Our goal is to smooth $s(x_0)$ in which $s(.)$ depicts the scatter plot smoother functions, using nearest neighborhoods. First, we identify the nearest neighbors of x_0 which are denoted by $\Omega\ (x_0)$. In the second step, we determine the furthest near- neighbor observation from x_0 and compute the distance between these two points. That is, we calculate $\Delta(x_0)= \max_{\Omega(x0)} |x_0-x_i|$ next, using the *tri-cube* weight function (Cleveland *et al.,* 1993), we assign weights to each point in ξ_i as

$$\xi\left(\frac{|x_0 - x_i|}{\Delta(x_0)}\right) \tag{7}$$

Where

$$\xi\ (u) = (1-u^3),\ for\ 0 \leq u \leq 1 \tag{8}$$

In the last step, and by using the weights computed in the third step, we obtain the Scatter plot smoother $s(x_0)$ value at the initial point by x_0 applying the *weighted least-squares* (WLS) of response variable y to predictors x in the domain of $\Omega\ (x_0)$. Locally weighted scatter plot smoothers are popular among statisticians and econometricians for at least two reasons. First, these methods are able to produce robust results in respect to the outliers. Second, with the scatter plot smoothers, we are able to easily find the neighborhoods for the target point . Econometricians prefer using the former method since it is less biased, as compared to the latter method (Haghiri, 2003).

RESULTS AND DISCUSSION

Data for this study is collected from agricultural crops cost production bank and statistical yearbooks of agricultural ministry of Iran. Data related to every province from 2000 to 2010 was collected for 28 provinces at 10 years time period. The cost efficiency of wheat production was modeled in a six input and single output framework. Quantity of wheat produced per hectare at any province was used to measure output. Price of Six inputs were used for cost efficiency analyses which are fertilizer, seed,

Table 1: Statistical summary of the variables

Variable	Describe	Number of observation	Unit	Mean	S.D	min	Max
E_{it}	Production cost per hectare	280	Ten rails	425170.80	209862.4	80560	1111588
y_{it}	Yield per hectare	280	kg	3380.06	803.95	1172.91	5358.92
w_{it}	Weighted mean of price of one kg fertilizer	280	Ten rails	56.38	14.54	26	101
w_{2it}	Price of one kg wheat seed	280	Ten rails	201.73	82.15	79	384
w_{3it}	Price of one kg agricultural pesticides	280	Ten rails	3343.87	2531.18	153.36	13010.48
w_{4it}	Rent of one hectare land	280	Ten rails	129110.80	87808.27	11626	401061
w_{5it}	Water price	280	Ten rails	64314.67	51826.77	959	2388891
w_{6it}	Labor wage	280	Ten rails	6469.36	3925.02	1366.127	21166.62

Source: Agricultural Ministry of Iran and research calculations

[1] *Locally Weighted Scatter Plot Smoothing.*

agricultural pesticides, labor and water prices. In order to aggregate fertilizer and agricultural pesticides input prices we used the weighted average. Descriptive statistics of input price and output is given in table 1.

We were encountered with limitations in machinery and water price. The water price wasn't available; therefore we used water cost per hectare instead of water price. Unfortunately, suitable variable wasn't available for machinery price.

Cost efficiency was calculated by two methods namely parametric and partially non-parametric. And we had compared results from two methods. At first model was specified in translog form, unfortunately, we encountered with singular error, after elimination of this error and estimation model and were not significant. Therefore we used cob-Douglas specification form.

$$\ln E_{it} = \beta_0 + \beta_y \ln y_{it} + \sum_n \beta_n \ln w_{nit} + v_{it} + u_{it} \quad (9)$$

The cost function is homogenous degree one at input price. Therefore $c\ (y_{it}, \lambda w_{nit}; \beta) = \lambda c\ (y_{it}, \lambda w_{nit}; \beta)$ and $\lambda > 0$. We should impose $\beta_k = 1 - \sum_{n \neq k} \beta_n$ or we can change equation (9)

$$\ln(\frac{E_{it}}{w_{kit}}) = \beta_0 + \beta_y \ln y_{it} + \sum_n \beta_n \ln(\frac{w_{nit}}{w_{kit}}) + v_{it} + u_{it} \quad (10)$$

We used seed price for normalization variables and after model estimation with pitt and lee (1981) and LOWESS methods by Nlogit 5.0 software package and calculated cost efficiency of the type

$$CE_{it} = \exp E\ \{-u_{it}\} \quad (11)$$

A likelihood ratio test against the hypothesis of no inefficiency follows the variance estimates. Appropriate tables for the mixed chi squared test used here are given in Kodde and Palm (1986). LR test for inefficiency vs. OLS only is

$$Chi - sq = 2 \times [\log l\ (sf) - \log l(ls)] \quad (12)$$

Table 2: LR test for inefficiency vs. OLS only

statistics	Stochastic frontier based on pooled data	Stochastic frontier based on panel data
Chi-sq	4.634	119.964

Kodde-Palm C*: 95%: 2.706, 99%: 5.412

Table 2 shows that there is not inefficiency in wheat production hypothesis that cannot be accepted and we should estimate model in frontier manner.

The stochastic frontier cost function estimation results was showed in table 3. In cost function estimation based on pooled data and panel data only normalized price of agricultural pesticides hasn't significant effect on Production cost per hectare wheat and other coefficient are significant at 99%. Normalized land price (land rent) has maximum effect on Production cost per hectare wheat, as one percent increase at land price 0.24 percent increase cost production and normalized fertilizer price has minimum effect on production cost as one percent increase at land price lead to only 0.14 percent increase cost production.

The results of Cost function estimation based on panel data show that log likelihood is increased

Table 3: Models estimation results

variables	Cost function estimation based on pooled data	Cost function estimation based on panel data	Cost function estimation based on panel data
constant	1.33494***	1.39714***	1.63018***
$\ln y_{it}$	0.18564***	0.14955***	0.01533
$\ln (w_{1it} / w_{2it})$	0.13590***	0.15295***	0.09694**
$\ln (w_{3it} / w_{2it})$	-0.01873	0.01457	-0.03328
$\ln (w_{4it} / w_{2it})$	0.23606***	0.24174***	0.29466**
$\ln (w_{5it} / w_{2it})$	0.18504***	0.15232***	0.24241**
$\ln (w_{6it} / w_{2it})$	0.18965***	0.22276***	0.19089**
$\lambda = \sigma_u / \sigma_v$	1.27005***	2.57029**	0.78903
$\lambda = (\sigma_u^2 + \sigma_v^2)^{1/2}$	0.07739***	0.10813***	0.05401
σ_u	0.06081	0.10813	0.03345
σ_v	0.04788	0.04207	0.04240
Log likelihood	390.65911	448.32391	-------------

Note: ***, **, * ==> Significance at 1%, 5%, 10% level.

Table 4: Cost efficiency and province rank

Province	Cost efficiency Parametric method	Rank	Cost efficiency LOWESS method	Rank	Province	Cost efficiency Paramet-ric method	Rank	Cost efficiency LOWESS method	Rank
Khuzestan	0.995	1	0.986	1	Mazandran	0.907	15	0.975	20
Bushier	0.989	2	0.985	2	Golestan	0.900	16	0.976	16
Kurdistan	0.955	3	0.979	4	Azerbaijan sharghi	0.898	17	0.974	23
Elam	0.950	4	0.978	10	Azerbaijan gharbi	0.896	18	0.973	24
Kerman	0.940	5	0.979	8	Kermanshah	0.893	19	0.977	13
Lorstan	0.929	6	0.978	9	Khorasan	0.886	20	0.979	7
Tehran	0.928	7	0.979	5	Qom	0.882	21	0.975	19
Hormozgan	0.927	8	0.979	6	Ardabil	0.881	22	0.969	27
Fars	0.927	9	0.976	18	South kerman	0.877	23	0.976	14
Markazi	0.924	10	0.977	12	Semnan	0.876	24	0.975	21
Kohkilo	0.923	11	0.976	17	Sistan va Blochstan	0.875	25	0.976	15
Zanjan	0.922	12	0.978	11	Esfahan	0.867	26	0.971	25
Hamadan	0.918	13	0.980	3	Chaharmahal bakhtiyari	0.854	27	0.966	28
Ghazvin	0.918	14	0.974	22	Yazd	0.789	28	0.970	27

Source: Agricultural Ministry of Iran and research calculations

significantly, only normalized price of agricultural pesticides hasn't significant effect on Production cost per hectare wheat and other coefficients are significant at 99%. Normalized land price (land rent) has maximum effect on Production cost per hectare wheat, as one percent increase at land price lead to 0.24 percent increasing cost production similar to pooled data model.

The results of Cost function estimation based on LOWESS method is to some extent different from panel data method. Yield per hectare and agricultural pesticides price haven't significant effect on production cost while other coefficient are significant at 95%. Land price have maximum effect on production cost and fertilizer price have minimum effect and this is similar to panel data method.

After cost function estimation and calculation cost efficiency, provinces ranked based on cost efficiency calculated by panel data and LOWESS methods that present in table 4.

Khuzestan and Bousher had maximum cost efficiency in wheat production, and Yazd province had minimum cost efficiency in wheat production at panel data method (Table 5).

Mean of efficiency in parametric method is 90% and in non-parametric method (LOWESS) is 97%. Standard deviation in nonparametric method is low and efficiency cost deference among provinces is inconsiderable as in this method minimum cost efficiency is 97% and maximum cost efficiency is 99%.

The kernel density estimator is a device used to describe the distribution of a variable non-parametrically, that is, without any assumption of the underlying distribution (silverman, 1986). The kernel density function for a single variable is computed using

$$\hat{f}(x) = \frac{1}{nh}\sum_{i=1}^{n} k(\frac{x - X_i}{h})$$

$$\int_{-\infty}^{+\infty} k(x)dx = 1 \qquad (13)$$

We plotted kernel density for EUP and EUNP that is displayed in figure 1.

Figure 1 shows Difference between two methods. EUP distribution is more suitable than EUNP. Although in cross section data results from parametric and non-parametric method is more similar, in this study that we used panel data, the results of parametric method is more suitable than non-parametric method.

Table 5: Statistical characteristics of the estimation cost efficiency.

Variable	Describe	Number of observation	Mean	S.D	min	Max
EUP	Cost efficiency (Parametric method)	28	0.908	0.0412	0.789	0.995
EUNP	Cost efficiency (LOWESS method)	28	0.976	0.004	0.966	0.986

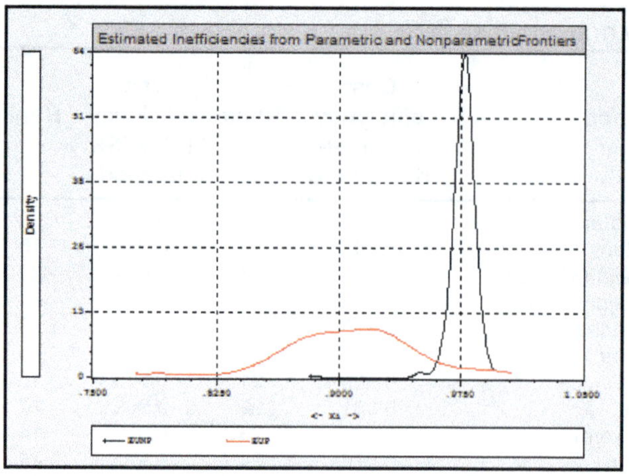

Figure1: cost efficiency kernel density of EUP and EUNP

CONCLUSION

Our purpose in this study was survey and comparative parametric and partially non-parametric methods in wheat production cost efficiency evaluation in Iran. We used specification and estimation stochastic frontier cost function with panel data. With Attention to quality of data in agricultural sector of Iran, the results of efficiency measurement by specification and estimation cost function is more suitable than by specification and estimation production function. Because input prices instead of inputs was used and quality of price data is higher than quality of input data at province levels in cost function estimation. The results show that agricultural pesticides price hasn't significant effect on frontier cost production. Share of agricultural pesticides is nugatory. Agricultural pesticides used when the farm is infected by disease therefore pesticides consume and yield per hectare are relevant and these reasons can be accounting for insignificant agricultural poison price.

In these two methods, wheat farmers Khuzestan province have the highest cost efficiency. Khuzestan province has vast fields as well as plenty of water for agriculture. The farm size in this province is high. Economics of scale seems to be one of the reasons for highest cost efficiency in Khuzestan province. Detailed studies in this field are offered and the main reasons for the high cost efficiency of irrigated wheat production in the province are found.

Parametric and nonparametric methods for estimating the model results indicate that from the Point of view of significant coefficients and differentiated production structure, parametric approach in each province is more appropriate. Therefore, in similar studies, especially when panel data is used one should be more discreet to use partially non-parametric models.

Parametric and nonparametric methods show that cost efficiency of wheat production in Iran is Suitable and over 90%. And this result emphasized on the fact that if farmers were given the facilities they would effectively use it.

It is necessary that department of agricultural promotion and education organize their forces and facilities based on the ranking in terms of cost efficiency of every province. So that in whatever province where the efficiency is low, agricultural education and knowledge is needed to improve further.

REFERENCES

1- Aigner, D.j., Lovell, C.A.K. & Schmidt, P. (1977). Formulation and Estimation of Stochastic Frontier Production Function Models, Journal of Econometrics, 6(1): 21-37.

2- Battese, G.E., & Corra, G.S. (1977). Estimation of Production Frontier Model: With Application to the Pastrol Zone off Eastern Australia, Australian Journal of Agricultural Economics, 21(3): 169-79.

3- Battese, G.E., & Coelli, T.J. (1992). Frontier Production Functions, Technical Efficiency and Panel Data: With Application to Paddy Farmers in India, Journal of Productivity Analysis 3(1/2): 153-69.

4- Cesar Revoredo, G., Catherine, E.M., Philip, M.L. & Woong, J.C. (2009). Efficiency of Scottish Farms: A Stochastic Cost Frontier Analysis. Agricultural Economics Review, 10(2): 17-35.

5- Cornwell, C., Schimdit, P. & Sickles, R.C. (1990). Production Frontier With Cross-sectional and Time-series Variation in Efficiency Levels, Journal of Economics 46(1/2): 185-200.

6- Cleveland, W.S., Grosse, E. & Shyu, W.M. (1993). Local Regression Models, in J. M. Chambers and T. J. Hastie (eds.) Statistical Model in S, London: Chapman and Hall, Pp: 309- 376.

7- Fan, J. (1992). Design Adaptive Nonparametric Regression, Journal of the American Statistical Association, 87: 998-1004.

8- Fare, R. & Lovell, C.A.K. (1978). Measuring the Technical Efficiency of Production, Journal of Economic Theory, 19: 150- 162.

9- Forsund, F.R., Lovell, C.A.K. & Schimidt, P. (1980). A Survey of Frontier Production Functions and of Their relationship to Efficiency Measurement. Journal of Econometrics, 13(1): 5-25.
10- Gahani, M. & Asghari, A. (2005). Wheat Cost Production Analysis by Used of One Output Translog Cost Function, Case Study: Arasbaran Region. Agricultural Economic Research Journal, Tehran University, 1(70): 233-262.
11- Greene, W. (2011). LIMDEP version 10, Econometric Modeling Guide, Volume 2, Econometric Software, Inc. Plainview, NY, USA.
12- Hastie, T.J. & Loader, C. (1993). Local Regression: Automatic Kernel Carpentry, Statistical Science, 8: 120- 143.
13- Jondrow, J., Lovell, C.A.K. Materov, I.S. & Schimdit, P. (1982). On the Estimation of Technical Inefficiency in the Stochastic Frontier Production Function Model. Journal of Econometrics, 19(2): 223-38.
14- Kumbhakar, S.C. & Knox Lovell, C.A. (2003). Stochastic Production Frontier. New York: Cambridge University Press. Second edition.
15- Kumbhakar, S.C. & Knox Lovell, C.A. (2000). Stochastic Frontier Analysis. New York: Cambridge University Press. First Published.
16- Kodde, D., & Palm, F. (1986). Wald Criteria for Jointly Testing Equality and Inequality Restrictions. Econometrica, 54: 1243–1248.
17- Lee, L.F. (1983). A Test for Distributional Assumptions for the Stochastic Frontier Functions. Journal of Econometrics, 22(3): 245-67.
18- Mundlak,Y. (1961). Empirical Production Function Free of Management Bias. Journal of Farm Economics 43(1): 44-56.
19- Haghiri, M (2003). Stochastic Non-Parametric Frontier Analysis in Measuring Technical Efficiency: A Case Study of The North American Dairy Industry. A Thesis Submitted to the College of Graduate Studies and Research In Partial Fulfillment of the Requirements For the Degree of Ph.D. in the Department of Agricultural Economics University of Saskatchewan Saskatoon.
20- Meeusen,W. & van den Broeck, J. (1977). Efficiency Estimation from cob-douglas Production Functions With composed Error. International Economic Review, 18(2): 435-44.
21- Musavi, H. & khaliliyan, S. (2005). Factors Affecting on Technical Efficiency of Wheat Production. Agricultural Economic and Development Journal, 1(52): 45-60.
22- Moradi shahrbabak, H.(2008). Economic Efficiency of Wheat Production in Kerman. Agricultural Journal, 1(2): 173-180.
23- Pitt. M. & Lee, L.F. (1981). The Measurement and Source of Technical Inefficiency in the Indonesian Weaving Industry. Journal of Development Economics, 1(9): 43-64.
24- Schimdit, p. & Sickles, R.C. (1984). Production Frontier and Panel Data. Journal of Business and Economic Statistics, 2(4): 367-74.
25- Stevenson, R.E. (1980). Likelihood Function for Generalized Stochastic Frontier Estimation. Journal of Econometrics, 13(1): 57-66.
26- Silverman, B.W. (1986). Density Estimation for Statistics and Data Analysis, Published in Monographs on Statistics and Applied Probability, London: Chapman and Hall.
27- Zeramnegad, M. & yousafi Hagiabad, R. (2008). Technical Efficiency Wheat Production Evaluation in Iran. Journal of Economics Research, 2(9): 145-172.

Permissions

All chapters in this book were first published in IJAMAD, by Islamic Azad University; hereby published with permission under the Creative Commons Attribution License or equivalent. Every chapter published in this book has been scrutinized by our experts. Their significance has been extensively debated. The topics covered herein carry significant findings which will fuel the growth of the discipline. They may even be implemented as practical applications or may be referred to as a beginning point for another development.

The contributors of this book come from diverse backgrounds, making this book a truly international effort. This book will bring forth new frontiers with its revolutionizing research information and detailed analysis of the nascent developments around the world.

We would like to thank all the contributing authors for lending their expertise to make the book truly unique. They have played a crucial role in the development of this book. Without their invaluable contributions this book wouldn't have been possible. They have made vital efforts to compile up to date information on the varied aspects of this subject to make this book a valuable addition to the collection of many professionals and students.

This book was conceptualized with the vision of imparting up-to-date information and advanced data in this field. To ensure the same, a matchless editorial board was set up. Every individual on the board went through rigorous rounds of assessment to prove their worth. After which they invested a large part of their time researching and compiling the most relevant data for our readers.

The editorial board has been involved in producing this book since its inception. They have spent rigorous hours researching and exploring the diverse topics which have resulted in the successful publishing of this book. They have passed on their knowledge of decades through this book. To expedite this challenging task, the publisher supported the team at every step. A small team of assistant editors was also appointed to further simplify the editing procedure and attain best results for the readers.

Apart from the editorial board, the designing team has also invested a significant amount of their time in understanding the subject and creating the most relevant covers. They scrutinized every image to scout for the most suitable representation of the subject and create an appropriate cover for the book.

The publishing team has been an ardent support to the editorial, designing and production team. Their endless efforts to recruit the best for this project, has resulted in the accomplishment of this book. They are a veteran in the field of academics and their pool of knowledge is as vast as their experience in printing. Their expertise and guidance has proved useful at every step. Their uncompromising quality standards have made this book an exceptional effort. Their encouragement from time to time has been an inspiration for everyone.

The publisher and the editorial board hope that this book will prove to be a valuable piece of knowledge for researchers, students, practitioners and scholars across the globe.

List of Contributors

Adetumbi, Saheed.Ige, Olaniyi, Olumuyiwa., Akin and Adewale, Jacob., Gbemiga
Department of Agricultural Extension and Rural Development, Ladoke Akintola University of Technology, P.M.B. 4000, Ogbomoso, Nigeria

Gholamreza Pezeshki Rad
Associate Professor of Agricultural Extension and Education Department of Tarbiat Modares University, Peykanshahr, Tehran, Iran

Zahra Haji Hashemi
Graduated Student of Tarbiat Modares University

Mohammad Chizari
Professor of Agricultural Extension and Education Department of Tarbiat Modares University, Peykanshahr, Tehran, Iran

Ali Mohammadi Torkashvand and Shahryar Sobhe Zahedi
Department of Horticulture, Rasht Branch, Islamic Azad University, Rasht, Iran
Research Center of Agriculture and Natural Resources, Guilan, Iran

Khadijeh Abolfathi
M.Sc. Student in Assessment and Land Use Planning, Natural Resources and Environmental Engineering Department, Faculty of Agricultural Science, Payame Noor University (PNU), Tehran, Iran

Marzieh Alikhah-Asl and Mohammad Rezvani
Assistant Professor, Natural Resources and Environmental Engineering, Faculty of Agricultural Science, Payame Noor University (PNU), Tehran, Iran

Mohammad Namdar
Forest, Rangeland and Watershed Management Organization, Tehran, Iran

Hamidreza Saberkari
Department of Electrical Engineering, Rasht Branch, Islamic Azad University, Rasht, Iran

Amir Alambaigi
Assistant Professor, Department of Agricultural Extension and Education, Faculty of Agricultural & Natural Resources, University of Tehran, Tehran, Iran

Ismael Ahangari
PhD Student, Department of Agricultural Extension and Education, Faculty of Agricultural & Natural Resources, University of Tehran, Tehran, Iran

Gholamreza Zamanian
Assistant Professor, Economics, University of Sistan and Baluchestan, Zahedan, Iran

Mostafa Khajeh Hassani
PhD Candidate, Agricultural Economics, University of Sistan and Baluchestan, Zahedan, Iran

Sepideh Rostami Dolatabadi
M.A., Agricultural Management, Department of Agricultural Extension, Education and Management, Tabriz Branch, Islamic Azad University, Tabriz, Iran

Mohammad Ghahremanzadeh
Associate Professor, Agricultural Economics, Department of Agricultural Economics, University of Tabriz

Seyyed Ali Noorhosseini
Young Researchers and Elite Club, Rasht Branch, Islamic Azad University, Rasht, Iran

Esmaeil Fallahi
Assistant Professor, Department of Agricultural Economics, University of Shiraz, Shiraz, Iran

Sepideh Gholinezhad
Young Researchers and Elite Club, Qaemshahr Branch, Islamic Azad University, Qaemshahr, Iran

Sajad Rostami, Somayeh Choobin, Bahram Hosseinzadeh Samani and Zahra Esmaeili
Department of Mechanical Engineering of Biosystem, Shahrekord University, Shahrekord, Iran

Hemad Zareiforoush
Department of Agricultural Mechanization Engineering, Faculty of Agricultural Sciences, University of Guilan, Rasht, Iran

Seyed Masoud Motamedolshariati
Former M.Sc. Student, Department of Biosystems Engineering, Ferdowsi University of Mashhad, International Campus, Mashhad, Iran

Hassan Sadrnia and Mehdi Khojastehpour
Associate Professor, Department of Biosystems Engineering, Ferdowsi University of Mashhad, Mashhad, Iran

Mohammad Hossein Aghkhani
Professor, Department of Biosystems Engineering, Ferdowsi University of Mashhad, Mashhad, Iran

Nasibeh Pourfateh
M.Sc., Razi University, Kermanshah, Iran

Nader Naderi
Assistant Professor, Razi University, Kermanshah, Iran

Farahnaz Rostami
Assistant Professor, Razi University, Kermanshah, Iran

Amir Naeimi
Assistant Professor, Faculty of Agriculture, Department of Extension, Communication and Rural Development, University of Zanjan, Iran

Mostafa Karbasioun
Assistant Professor, Faculty of Agriculture, Department of Agricultural Development, Shahrekord University, Shahrekord, Iran

Fatemeh Abbasi
Ph.D. Student of Agricultural Exatension and Education, Bu'AliSina Universtiy, Iran

Mansour Soofi
Department of Industrial Management, Rasht Branch, Islamic Azad University, Rasht, Iran

Maryam Mohseni
PhD in Industrial Management, Tehran University

Ahmadreza Ommani
Associate Professor, Department of Agricultural Management, Shoushtar Branch, Islamic Azad University, Shoushtar, Iran

Mehdi Khojastehpour
Department of Biosystems Engineering, Ferdowsi University of Mashhad, Mashhad, Iran

Amin Nikkhah
Young Researchers and Elite Club, Rasht Branch, Islamic Azad University, Rasht, Iran

Davood Hashemabadi
Department of Horticultural Science, Rasht Branch, Islamic Azad University, Rasht, Iran

Gholamreza Pezeshki Rad
Associate Professor of Agricultural Extension and Education Department of Tarbiat Modares University, Peykanshahr, Tehran, Iran

Zahra Haji Hashemi
PhD. Student of Bu Ali Sina University, Hamedan

Mohhamad Chizari
Professor of Agricultural Extension and Education Department of Tarbiat Modares University, Peykanshahr, Tehran, Iran

Vikram Swaroop Chandra, Koundinya
Postdoctoral Research Associate, Global Extension Programs & Department of Agricultural Education and Studies, Iowa State University, Ames, IA - 50011

Robert A. Martin
Department of Agricultural Education and Studies, Iowa State University, Ames, IA - 50011

Iniobong A. Akpabio, Nsikak-Abasi A. Etim and Sunday Okon
Department of Agricultural Economics and Extension, University of Uyo, PMB 1017, Uyo, Akwa-Ibom State, Nigeria

Seyyed Ali Noorhosseini-Niyaki
Department of Agronomy, Lahidjan Branch, Islamic Azad University, Lahidjan, Iran

Mohammad Sadegh Allahyari
Department of Agricultural Management, Rasht Branch, Islamic Azad University, Rasht, Iran

Shahram Saeedian
Graduate Student of Agricultural Economics, Economics Department, Sistan and Baluchestan University

Mahmoud Sabouhi Sabouni
Agricultural Economics Department, Zabol University

Javad Shahraki
Economics Department, Sistan and Baluchestan University

Ebrahim Moradi
University of Sistan and Baluchestan, Iran

Mosayeb Pahlavani
Associate Professor, University of Sistan and Baluchestan, Iran

Andiema Chesang Everlyne
Department of Agricultural Education & Extension, Egerton University, Kenya

Nkurumwa Oywaya Agnes
Department of Agricultural Education & Extension, Egerton University, Kenya

Zahra Haji Hashemi
Bu Ali Sina University

Amudavi Mulama David
Biovision Africa Trust, C/O icipe-African Insect Science for Food and Health, Nairobi, Kenya

Mohammad Chizari
Professor of Agricultural Extension and Education, Department of Tarbiat Modares University, Peykanshahr, Tehran, Iran

Index

www.ingramcontent.com/pod-product-compliance
Lightning Source LLC
LaVergne TN
LVHW060548210926
848485LV00005B/1282